D

Landmark
AS Geography
Second Edition

ert Prosser, Michael Raw, Victoria Bishop and Gill Miller

LANDMARK GEOGRAPHY

Contents

This symbol links material on this page to relevant Skills and Techniques information on other pages in this book.

HUMAN ENVIRONMENTS

GEOGRAPHICAL SKILLS AND TECHNIQUES

Every effort has been made to contact the holders of copyright material, but if any have been inadvertently overlooked, the publishers will be pleased to make the necessary arrangements at the first opportunity.

Cover photo: Tony Stone Images (Queenstown, New Zealand)

Every effort has been made to contact the holders of copyright material, but if any have been inadvertently overlooked the publishers will be pleased to make the necessary arrangements at the first opportunity.

The publishers would like to thank the following for permission to reproduce photographs (T = Top, B = Bottom, C = Centre, L= Left, R = Right):
Action Aid, fig. 7.12
Advertising Archives, fig. 8.21
Associated Press, AP, fig. 5.22
Victoria Bishop, figs 2.13, 2.25, 2.31, 2.32
Cambridge University Collection of Air Photographs, figs 3.31, 8.35
Corbis/G Rowell, fig. 1.7b, P A Souders, fig. 1.46
Ecoscene, fig. 3.3, A Cooper, fig. 3.44, C Gryniewicz, fig. 8.5
Environmental Images/J Morrison, fig. 2.11a, S Gamester, fig. 2.11b, R Brook, figs 2.52, 6.14, R Visser, fig. 2.56, P Treanor, fig. 6.19, M Bond, fig. 8.24
Field Studies Council/John Bebbington, FRPS, pages 292/293
Phil Flowers, fig. 2.24
GeoScience Features Picture Library, figs. 1.29, 2.9, 2.37, 3.6, 3.26, 3.30, 7.1, 7.4, M Hobbs, fig. 1.27c, Dr R Booth, fig. 2.41
Getty Images/Joe Cornish, fig. 6.22
Robert Harding Picture Library/W Rawlings, figs 2.51, 7.22, 8.31, P Craven, fig. 5.9, R Cundy, fig. 6.21, P Scholey, figs 6.23, 8.30, A Evrard, fig. 6.48, N Francis, fig. 7.35, L Bond, fig. 8.28, A Tovy, fig. 8.32
1997 Map and some Activities reprinted by permission of the Land-Use UK Project, based on data supplied by Ordnance Survey, fig. 9.16

London's Transport Museum, fig. 9.10
Martin Marshall, Dept of Natural Resources and Energy, Province of New Brunswick, Canada, fig. 4.30c
Gill Miller, page 292-3 (insets)
Mountain Camera/P Marrow, fig. 1.23
NHPA/S Krasemann, fig. 4.30a
Images of New Brunswick, Images du Nouveau-Brunswick/Mary Ellen Nealis, fig. 4.30b
Ordnance Survey © Crown Copyright (Licence no. PU100018599), page 5, figs 3.21, 3.34, 6.15, 7.24b, 7.26, 7.31, 7.33
PA Photos, fig. 7.40
Panos Pictures/J Horner, figs 5.13, 5.31, S Sprague, fig. 5.16, C Stowers, figs 5.18, 6.46, T Boldstad, fig. 5.38, T Beardshaw, figs 5.39, 7.11, P Wolmuth, fig. 6.37, D Tatlow, fig. 6.56c
Popperfoto/Reuters, figs 1.2, 1.21, Heinz P Bader/Reuters, fig. 4.15, M Baker/Reuters, fig. 5.32
Robert Prosser, figs 1.11c, 1.12, 1.15b, 1.15c, 1.16, 1.17, 1.18, 1.19, 1.24a, 1.24b, 1.26, 1.27a, 1.27b, 1.28a, 1.28d, 1.34, 1.40, 1.43, 1.44, 1.45, 4.1, 4.9, 4.16, 4.17a, 4.17b, 4.18a, 4.18b, 4.18c, 4.20, 4.21a, 4.21b, 4.25, 4.31, 4.32, 4.39a, 4.39b, 4.40, 4.42, 4.43, 4.45c, 4.45d, 4.45e, 4.47a, 4.47b, 4.48a, 4.48b, 4.52a, 4.52c, 4.52d, 5.14a, 5.14b, page 168, 5.23, 5.29, 5.36, 6.1, 6.2, 6.8, 6.17, 6.25, 6.27, 6.28, 6.32, 6.43, 6.44, 6.47, 6.55, 7.5, 7.6, 7.7, 7.14, page 236, 7.15b, 7.23, 7.32
Data for Land-Use Map, Oxford 1998, collected by Geography Students at Radley College, Abingdon, fig. 9.17
Michael Raw, figs 3.10, 3.12, 3.13, 3.14, 3.15, 3.16, 3.17, 3.18, 3.19, 3.20, 3.33, 3.35, 3.36, 3.42, 3.46, 3.48, 8.12, 8.15
Science Photo Library, fig. 1.10, NASA, fig. 1.8a, RESTEC Japan, 6.6
Peter Smith Aerial Photography, fig. 2.19
South American Pictures/T Morrison, fig. 8.17, C Sharp, fig. 8.19
Still Pictures/C James, fig. 8.7, M Edwards, fig. 8.8
Woodfall Wild Images/V Corbett, fig. 7.37
David Woodfall, figs 1.39, 7.35b.

Cover Photograph: Getty Images/Christopher Arnesen (Queenstown, New Zealand)

Published by Collins Educational
An imprint of HarperCollins *Publishers* Ltd
77-85 Fulham Palace Road
London W6 8JB

The Collins Educational website is:
www.CollinsEducation.com

© HarperCollinsPublishers Ltd 2003
First published 2000

ISBN 0 00 715116 0
10 9 8 7 6 5 4 3 2 1

Project management/Editing: Melanie McRae
Design: Janet McCallum, Wendi Watson
Cover design: Derek Lee
Artwork: Jerry Fowler, Richard Mann
Picture research: Caroline Thompson
Index: Joan Dearnley
Production: Kathryn Botterill
Printed and bound in Hong Kong by Printing Express Ltd.

British Library Cataloguing Data
A catalogue record for this book is available from the British Library.

You might also like to visit
www.**fire**and**water**.co.uk

NATURAL ENVIRONMENTS

The four chapters in the first half of this book introduce you to the four major realms of planet Earth. During any particular chapter, the focus may be on one of these realms, but a key understanding is that they all interact. We can illustrate this interconnectedness of the four realms through the example of water. The main store of water is the hydrosphere – oceans, lakes, streams, ice sheets – but water is stored in all other realms: as groundwater in the lithosphere; within the structure of plants and animals in the biosphere; as water vapour in the atmosphere. Water is also transferred between the four realms - for example, from the hydrosphere by evaporation to the atmosphere, by absorption to the biosphere, by percolation to the lithosphere.

Thus, the four realms work together as the global system. Today, the structure and functioning of this global system is increasingly influenced by human activities. Think, for instance, of major environmental issues – increased CO_2 emissions to the atmosphere; deforestation which reduces biodiversity and accelerates erosion; effluents from cities and agriculture which pollute the hydrosphere. Human environments are the focus of the second half of the book. However, it is important to keep in mind that the four natural realms considered first are all influenced by human activities and processes.

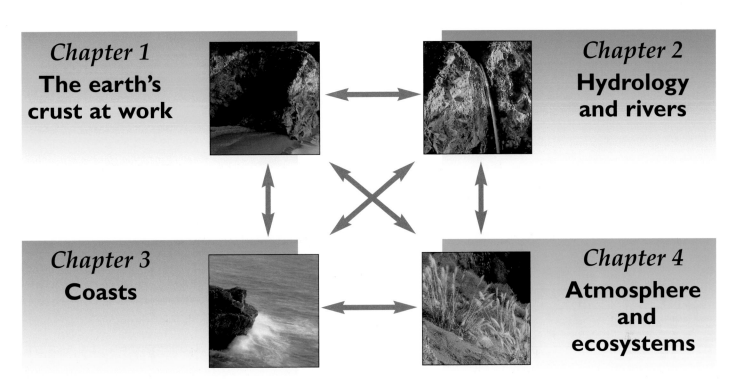

Chapter 1
The earth's crust at work

Chapter 2
Hydrology and rivers

Chapter 3
Coasts

Chapter 4
Atmosphere and ecosystems

Chapter 1 **The earth's crust at work**

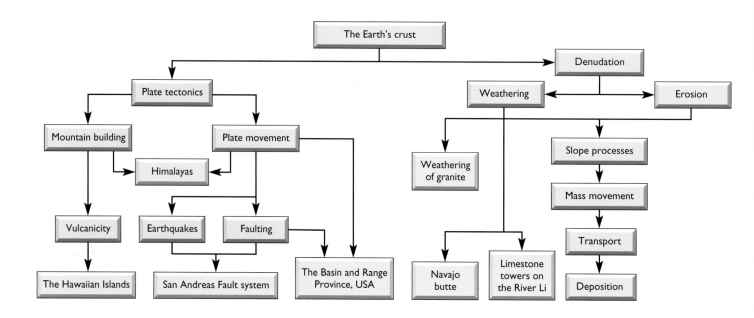

```
                            ┌─────────────────┐
                            │ The Earth's crust│
                            └─────────────────┘
```

The Earth's crust
 → Plate tectonics
    → Mountain building
       → Himalayas
       → Vulcanicity
          → The Hawaiian Islands
    → Plate movement
       → Himalayas
       → Earthquakes
          → San Andreas Fault system
       → Faulting
          → San Andreas Fault system
          → The Basin and Range Province, USA
 → Denudation
    → Weathering
       → Weathering of granite
          → Navajo butte
          → Limestone towers on the River Li
    → Erosion
       → Slope processes
          → Mass movement
             → Transport
                → Deposition

1.1 Introduction

Figure 1.2 is a vivid summary of the strong earthquake that struck north-west Turkey in August 1999. At least 40 000 people were killed. It is an extreme example of the powerful forces at work in the lithosphere – the outer layer of planet earth. Notice that the earth's crust, on which landforms evolve and the oceans lie, is the thin outer skin of the **lithosphere** (Figure 1.1).

It is fundamental to understand that the **earth's crust** is dynamic: it is constantly changing. As this chapter will show, this change occurs across a range of scales (a whole continent and an individual stone on a slope may both move 50 mm in a year) and a range of speeds, (the Himalayan mountain mass is rising at an average rate of less than 1 m in 1000 years, but a 15-metre-high sand dune in Namibia (South-West Africa) may travel 15 m in one year). Furthermore, rates of change are irregular over time, and have varying timescales. Thus, the great 1964 earthquake in Alaska caused vertical displacements of land surface of up to 15 m in less than two minutes, in comparison with an average change of 1 m in 1000 years.

The aim of this chapter is to provide an understanding of the processes at work in the lithosphere that help to explain the landforms we see around us. In general, the chapter progresses from large-scale to smaller-scale features and processes. Understanding our natural environment increases our appreciation of its beauty and our ability to conserve it, while living with the hazards it presents.

▲ *Figure 1.1* *The earth's layers*

TURKISH EARTHQUAKE

At 3.02 a.m. last Tuesday, the ground shook violently for 45 seconds under north-western Turkey, entombing tens of thousands of sleeping families. When dawn broke, the fierce August sun burned down on hundreds of square miles of the industrial heartland of the country which lay in ruin. Some 40,000 buildings were smashed by nature's power, into mountains of shattered concrete and sharp mangled steel. Ghostly voices cried out from dark holes beneath the rubble pleading for rescue.

A century of earthquakes

The magnitude of an earthquake does not determine the death toll, which depends on how densely populated the area is and on the quality of building construction.

Location	Date	Fatalities	Magnitude (Richter scale)
Izmit (Turkey)	17 Aug 1999	Exceeding 40,000	7.4
Kobe (Japan)	17 Jan 1995	5,500	6.9
California (USA)	17 Jan 1994	57	6.7
San Francisco (USA)	17 Oct 1989	63	7.1
Tangshan (China)	28 July 1976	255,000	8.0
Northern (Peru)	31 May 1970	66,000	7.8
Erzincan (Turkey)	26 Dec 1939	30,000	8.0
Yokohama Tokyo	1 Sept 1923	143,000	8.3
Messina (Italy)	28 Dec 1908	70,000-100,000	7.5
San Francisco (USA)	18 April 1906	700	8.3

▲ **Figure 1.2** *The earthquake in Turkey caused extensive destruction and killed over 40 000 people*

▼ **Figure 1.3** *North-west Turkey is situated on a convergent plate boundary; slippage between the two plates causes earthquakes.*

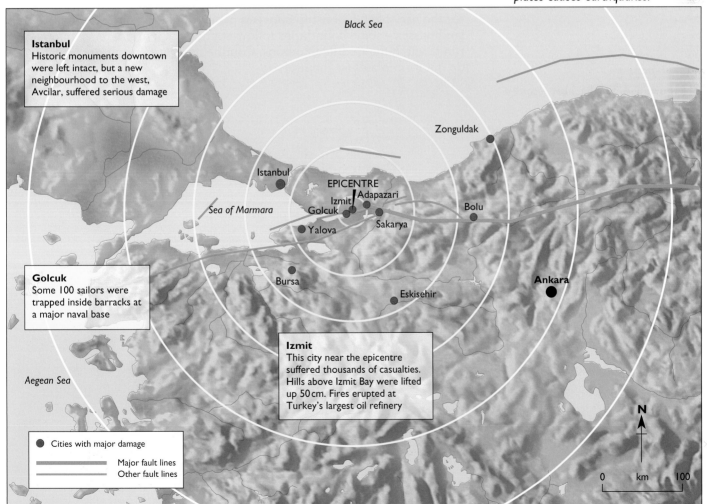

Black Sea

Istanbul
Historic monuments downtown were left intact, but a new neighbourhood to the west, Avcilar, suffered serious damage

Zonguldak

Istanbul

EPICENTRE

Izmit   Adapazari
Golcuk
Sea of Marmara
Yalova   Sakarya   Bolu

Bursa

Eskisehir

Ankara

Golcuk
Some 100 sailors were trapped inside barracks at a major naval base

Aegean Sea

Izmit
This city near the epicentre suffered thousands of casualties. Hills above Izmit Bay were lifted up 50 cm. Fires erupted at Turkey's largest oil refinery

● Cities with major damage
Major fault lines
Other fault lines

N

0   km   100

Key Terms

Fault: A fracture in the earth's crust along which the opposite sides have been displaced relative to each other.

....

Plate tectonics model: The understanding that the earth's crust is broken into a set of huge plates which move slowly in relation to each other and account for the distribution and character of land masses and oceans.

....

Focus: The site of the movement inside the earth's crust that results in an earthquake.

....

Epicentre: The point on the earth's surface that lies directly above the focus of an earthquake.

....

1.2 The moving crust

Plate tectonics

The earthquake in Istanbul in 1999 came as no surprise to earth scientists. They know that northern Turkey lies along the North Anatolian Fault (Figure 1.3) , and that earthquake distribution is closely associated with major fault zones in the earth's crust (Figure 1.4).

A **fault** is 'a fracture along which the opposite sides have been displaced relative to each other' (Skinner and Porter, 1987). A fault is said to be 'active' when displacement is still occurring, i.e. when there is movement of the rock masses on either side of the fracture. This movement may be gradual (creep), or may be sudden and abrupt. Even 'creep' movements generate minor earth tremors, but it is sudden jerks of displacement which cause major **earthquakes**. The Istanbul earthquake was triggered by an abrupt displacement along the North Anatolian Fault.

This brings us to the question of: 'why do the major fracture zones in the earth's crust occur where they do?'. The answer to this question lies in what is called the **plate tectonics** model. Tectonics refers to forces and processes that cause movement and deformation of the earth's crust on a large scale, and there are two main elements to this model.

Firstly, the earth's lithosphere is broken into seven huge plates and at least a dozen smaller plates, rather like a cracked eggshell. When we compare the pattern of lithospheric plates (Figure 1.5) with that of earthquake distribution (Figure 1.4), there is a close correlation. We can state, therefore, that earthquakes are most common along plate boundaries, i.e. along zones of major crustal fractures. If you study Figure 1.5 you will see the variable relationship between plate boundaries and continental landmasses – some boundaries follow continental margins, while others run through the middle of oceans. Continents, therefore, sit upon plates and do not form plates by themselves.

▼ **Figure 1.4** *Earthquakes recorded around the world, 1961–67*

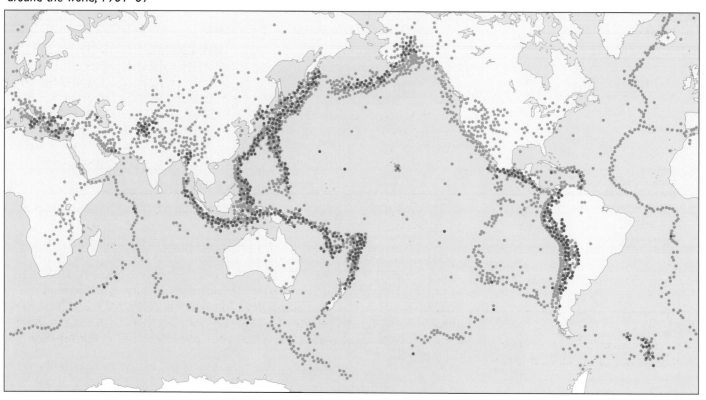

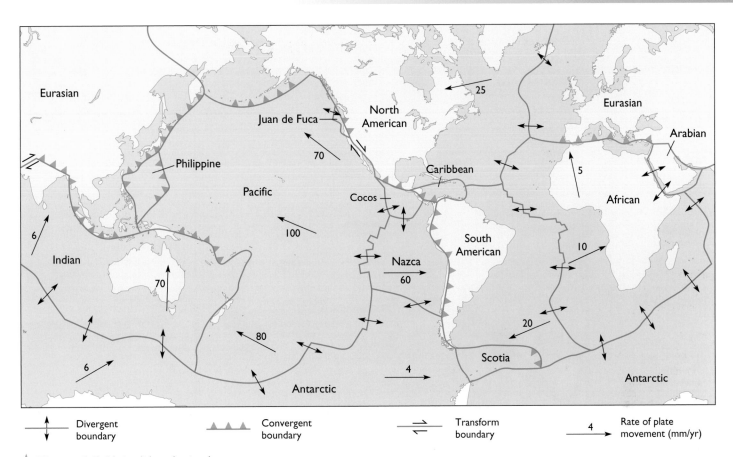

Divergent boundary

Convergent boundary

Transform boundary

4 — Rate of plate movement (mm/yr)

Figure 1.5 *Major lithospheric plates and their general direction of movement. Fault zones occur along the plate boundaries.*

1 From the text on plate tectonics.
a Define the term 'active fault'
b What do we mean by 'lithospheric plate'?
c From Figure 1.4 and Figure 1.5, and with the help of an atlas, name two countries in each continent that are at risk from earthquakes.
d Why is Turkey at risk from earthquakes?
e Use an atlas to locate the earthquakes listed in Figure 1.2. Using the world maps of Figures 1.4 and 1.5, explain why they occurred where they did.

2 Using Figure 1.5, give one example of each of the following types of plate boundary: divergent, convergent, transform (choose examples which aren't given in the text). Explain briefly why New Zealand lies in an earthquake hazard zone.

Secondly, the model shows that the plates move slowly in relation to each other. Thus, plate boundaries are active fault zones and hence, the source of earthquakes. The movements are powered by slow-moving convection currents in the **asthenosphere** which drag the plates across the surface of the earth (Figure 1.6).

Plate movements and margins

Figure 1.5 also identifies three types of plate boundary, according to the relative movement of neighbouring plates. Thus, the Antarctic and Pacific Plates are pulling apart (**divergent plate boundary**); the Nazca and South American Plates are moving towards each other (**convergent plate boundary**); and sections of the Pacific and North American Plates are moving past each other (**transform plate boundary**). It is important to remember that this a very generalised summary – along many sections of boundary, the relative movements of plates may be oblique (not directly divergent or convergent).

Along all types of boundary, there is considerable friction or drag between the plates as they grind past each other. Minor earth tremors, triggered by the release of strain energy caused by the drag, occur frequently. Major earthquakes are, fortunately, infrequent. They occur where slabs of adjoining plates 'lock' against each other for a period of years. When this happens, strain energy builds up progressively until the fault fractures once more, releasing powerful shock waves that spread outwards from the area of the dislocation, known as the **focus**. At the surface of the crust, this point is known as the **epicentre** of the earthquake, e.g. the city of Izmit was at the epicentre of the 1999 Turkish earthquake (Figure 1.3).

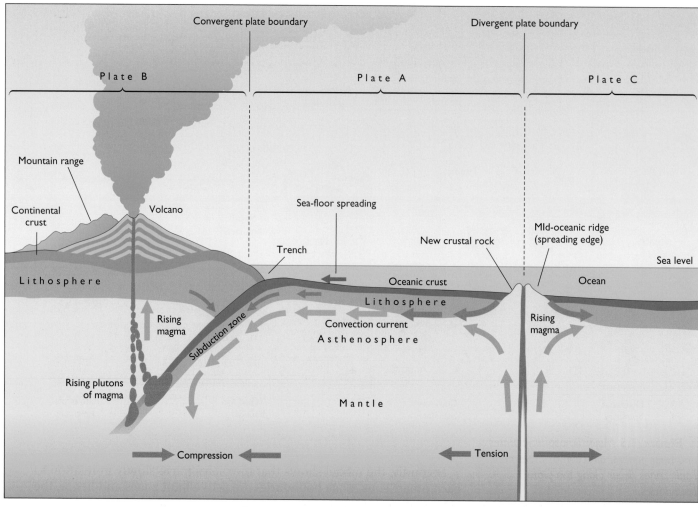

▲ **Figure 1.6** *Plate tectonics: generalised model of convergence and divergence. The diagram includes three lithospheric plates (A, B, C) and a divergent plate boundary and a convergent plate boundary.*

The rigid lithosphere of Plate A is dragged towards plate B by the slow-moving convection current in the asthensosphere. The asthenosphere is a relatively plastic zone in the upper mantle.

Plates A and C are dragged away from each other. The tensional stresses weaken the lithosphere at this divergent plate boundary, allowing magma to rise from the mantle. As the magma cools on the ocean floor it creates new crustal rock, forming a mid-oceanic ridge, e.g. the earthquake zone and plate boundary running north-south through the Atlantic ocean (Figures 1.4, 1.5).

Over time, the young crustal rocks are dragged away from the mid-oceanic ridge and are replaced by new magma. This process is called **sea-floor spreading**,

with new rock being created at the spreading edge (mid-oceanic ridge). This explains why, at a given time, crustal rocks of Plate A are progressively older as distance from the divergent boundary increases.

As Plates A and B converge at the convergent boundary, Plate A is dragged beneath the continental Plate B, creating an **oceanic trench**. *Compressional forces along the leading edge of the continental plate cause crumpling and uplift of the crustal rocks to form mountain ranges. As the lithospheric slab of the oceanic plate (A) is dragged deeper into the* **subduction zone** *beneath the continental plate (B), progressive heating destroys the crustal rocks. Some of the melted rock rises as* **plutons** *of magma, penetrates the continental crust causing further uplift and, where the magma reaches the surface, creating volcanoes. Release of compressional stresses along the subduction zone triggers earthquakes (Compare the maps of Figures 1.4 and 1.5).*

3 From Figure 1.6:

a Explain why a divergent plate boundary is also known as a 'constructive' boundary, and a convergent boundary is also called 'destructive'.

b Describe what is meant by 'sea-floor spreading'.

c Explain why the sea-floor spreading process is also known as a 'conveyor belt'.

d Define the term 'mid-oceanic ridge', and explain why it is also known as a 'spreading edge'.

e Define the term 'subduction zone' and describe briefly the processes at work in this zone.

f Give two reasons for the evolution of mountain ranges along a convergent plate boundary.

The relative plate movements play an important role in the evolution of major topographic features on the earth's crust such as mountain ranges, volcanoes, island arcs and ocean trenches.

The outcome of the type of convergence shown in Figure 1.6 is illustrated by the evolution of the Cascade Range and its line of volcanoes in the north-western USA (Figure 1.7a, b). In contrast, the example of the Himalayas shows what can happen when two continental plates collide (see page 14). We can now explain how the Istanbul earthquake, which began this chapter, happened. The African plate is pushing northwards against the Eurasian plate. Across Turkey this convergence is at an oblique angle, and strong compressional strain builds up. The earthquake, which was 7.4 on the **Richter scale**, was triggered by the release of energy along a section of the North Anatolian Fault that had been 'locked' for a number of years. This is one of a set of faults that form a fault zone along the northern edge of the Mediterranean basin. Two weeks after the Istanbul event, movement along another fault caused an earthquake which was 5.8 on the Richter scale, in Athens. Over 100 people died in this second earthquake. If you put all of this movement in a time frame of millions of years, the convergence of the African and Eurasian plates means that the Mediterranean Sea will eventually disappear.

One important characteristic of a major 'earthquake' is that the main event, which may last less than one minute, will be followed by a series of aftershocks – less powerful tremors – which may last for several weeks. Following the great earthquake in Kobe, Japan in 1995, which was 7.2 on the Richter scale, hundreds of aftershocks were recorded over a three-week period. These aftershocks indicate continuing adjustments along the fractured sections of the active faults.

▼ **Figure 1.7a** *The volcanoes of Mt Rainier and, in the distance Mt St Helens, rise above the ridges of the Cascade Range. Plutons of magma rise from the subduction zone and erupt through the Earth's crust to create the north-south line of volcanoes.*

▶ **Figure 1.7b** *The Cascade Volcanoes, USA, and the plate tectonics of the surrounding area.*

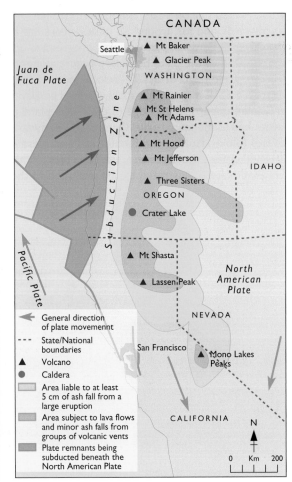

EXAMPLE: The Himalayas

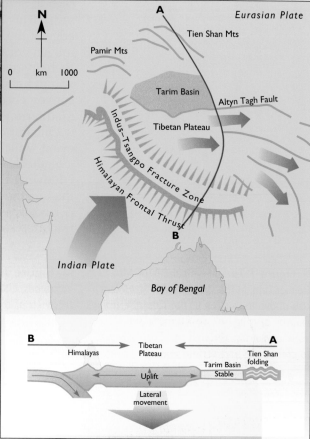

◀ **Figure 1.8a** *An oblique aerial view of the Earth's crust with the snow-capped Himalayas rising above the Indo-Gangetic Plain (left) and the Tibetan Plateau (right). As the Indian Plate moves north (left to right) it descends below the Eurasian Plate and uplifts the Himalayan Ranges.*

The Himalayan mountain ranges extend in a huge arc across the northern rim of the Indian sub-continent, 200–250 km wide, 2500 km long, with extensive areas above 8000 m in altitude (Figure 1.8a). They are geologically young, with an uplift of 3000–4000 m during the past 2 million years, and are still rising at an average rate of 70 cm per 1000 years.

This huge series of mountain ranges is generated by the collision of the Indian and Eurasian lithospheric plates (Figure 1.8b). For 80 million years, the Indian plate has been pushing northwards, and for around 35 million years has been thrusting beneath the Eurasian plate, causing progressive uplift. Notice that the location of the plate boundary occurs along the Indus–Tsangpo fracture zone (Figure1.8c). Therefore, most of the uplifted and distorted rock masses that make up the Himalayan ranges are the crumpled upper leading edge of the Indian plate, that includes layers of sedimentary rocks. Continued northward movement causes folding, faulting and further uplift as these rock masses are crushed against the Eurasian plate. The impact of the collision on the Eurasian plate is illustrated by the evolution of the Tibetan Plateau.

▲ **Figure 1.8b** *The tectonics of the Himalayan region*

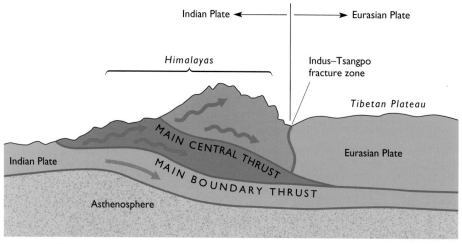

◀ **Figure 1.8c** *A generalised cross-section of the Himalayas showing how they are uplifted when the Indian Plate meets the Eurasian Plate*

4 Read the Himalayas example:
a What type of plate boundary is creating the Himalayan mountain ranges?
b Use simple diagrams to describe the processes that are forming (i) the Himalayas, (ii) the Tibetan Plateau.
c The Cascade Range (Figure 1.7b) and the Himalayas are both developed in association with lithospheric plate boundaries. In what ways are they similar and different? Explain these similarities and differences.

Transform plate boundaries

Along transform boundaries the principal characteristic is a shatter zone of slip or tear faults created by the forces that build up as plates grind past each other (Figure 1.9). These zones are identifiable in the landscape as a complex linear set of fault-defined mountain ranges, ridges, depressions and troughs (Figure 1.10). Earth tremors are frequent along active faults and these regions are at risk from major earthquakes when sections of fault become 'locked' and strain energy builds up. Sudden fracturing following periods of locking along sections of the San Andreas Fault (pages 16–17) caused disastrous earthquakes in San Francisco in 1906 (8.3 on the Richter scale) and 1989 (7.1 on the Richter scale). Volcanic activity is less common, but can occur when sections of fault pull apart, creating tensional strain, weakening the earth's crust and allowing magma to reach the surface.

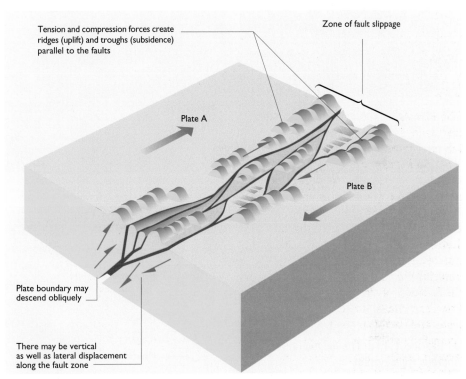

Tension and compression forces create ridges (uplift) and troughs (subsidence) parallel to the faults

Zone of fault slippage

Plate A

Plate B

Plate boundary may descend obliquely

There may be vertical as well as lateral displacement along the fault zone

► **Figure 1.9** *Features of a transform plate boundary*

► **Figure 1.10** *The San Andreas Fault, California: a conservative plate margin. The two lines of hills, showing gully erosion by ephemeral streams, are caused by crumbling rocks that flank the fault. The land to the right of the fault is slipping away from the camera; to the left, the land is moving towards the camera.*

EXAMPLE: The San Andreas Fault system (USA)

Figure 1.11a *The location of the San Andreas Fault system*

Figure 1.11b *Major fault lines through the southern part of the San Andreas system*

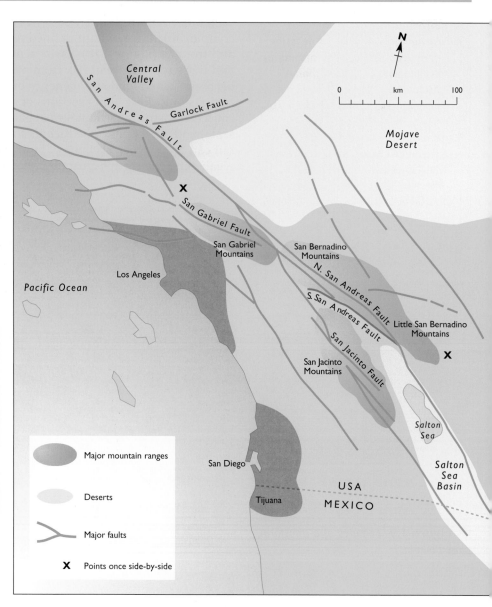

The San Andreas Fault system marks a transform section of the boundary between the Pacific and North American plates. It runs 1200 km northwards from the Mexican border to San Francisco. At present, the Pacific plate is moving north at an average of 35 mm per year, and over the past 25 million years there has been a lateral movement of 1000 km (Look at the points marked X on Figure 1.11b). It is a very complex fault zone, up to 500 km wide in places. Individual faults branch, join, bend and side-step each other, with lateral, vertical, tensional and compressional displacements causing uplift and subsidence as well as the lateral movements. For example, in Figure 1.11b, the Salton Sea Basin is a fault trough (graben), that even after thick sedimentary infilling, is still 74 m below sea level. This **graben** is the result of tension faults which cause slabs of crust to subside, and is flanked by a series of fault-defined mountain ridges, e.g. the San Jacinto Mountains (Figure 1.11c Figure 1.11d). In contrast, in the north-east of Los Angeles, the fault zone swings east–west. This causes compression against the North American Plate that has produced the San Gabriel Mountains (3000 m high). The complexity and activity of this fault zone explains why San Francisco and Los Angeles experience such serious earthquakes.

5 Read the San Andreas Fault system example.

a Why is San Francisco at risk from serious earthquakes?

b Use the San Andreas Fault system example to illustrate these main features of transform plate boundaries:

■ Movement is slow, and operates over long periods.

■ The boundary is a shatter zone rather than a single fault.

■ Major earthquakes are most likely along sections where a fault 'locks' for periods of years.

■ Along some sections of a transform boundary, the two plates may be moving obliquely towards each other or pulling apart. These oblique movements create compressional or tensional stresses, and distinctive landforms.

▲ **Figure 1.11c** *The view east from the San Jacinto Mountains. The Elsinore Fault runs through the valley in the foreground. The San Jacinto Fault, beyond the ridge, defines the western edge of the Salton Sea Basin. The San Andreas Fault runs along the base of a distant ridge and defines the eastern edge of the basin.*

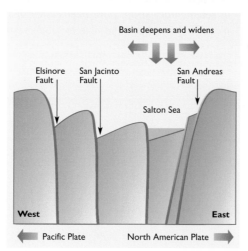

◄ **Figure 1.11d** *Cross-section of the Salton Sea Basin showing the three faults and the movement of the Pacific and North American Plates.*

1.3 Stresses within continental plates

Continental interiors, distant from plate boundaries, have often experienced long-term relative stability. Such regions are known as **cratons** – for example, the interior of North America, east of the Western Cordillera. We do feel occasional minor earth tremors in the British Isles, e.g. around Manchester in 2002. This tells us that the underlying plates do impose stresses on continental crustal masses. The main landform outcomes of these stresses are flexing and uplift (the High Plateau region of Colorado and Utah – Figure 1.12), and crustal stretching. These 'pull-apart' forces create tensional stress that produces faulting and the development of fault-block systems of **horsts** and **grabens**, accompanied in places by volcanic activity (Figure 1.13). A tectonic map of Africa shows a linear fault-block system running south from the African–Arabian plate boundary at the southern end of the Red Sea (Figure 1.14). This includes the great East African rift valley (a graben) and the volcanic mountains of Mt Kenya and Mt Kilimanjaro. This is the result of crustal stretching of the African plate. The Basin and Range Province of the western USA is a more complex example (Figure 1.15).

6 Define the terms 'horst', 'graben', 'rift valley'.

7 Use the East African rift system to illustrate the effects of tectonic forces on cratons.
Why is volcanic activity often associated with rift systems?

8 From Figure 1.12 and the Basin and Range Province example:
a Describe the landforms and landscapes shown.
b Explain the role played by tectonic forces working on the North American craton in the evolution of these landscapes.

▲ **Figure 1.12** *The 'gooseneck' meanders of the San Juan River, Utah. The thick layers of sedimentary strata have been slowly uplifted by the flexing of the North American plate. The river has been able to erode vertically to keep pace with this uplift.*

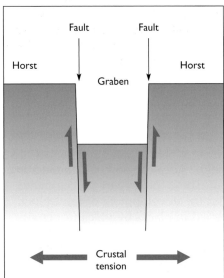

▲ **Figure 1.13** *A simple horst–graben structure*

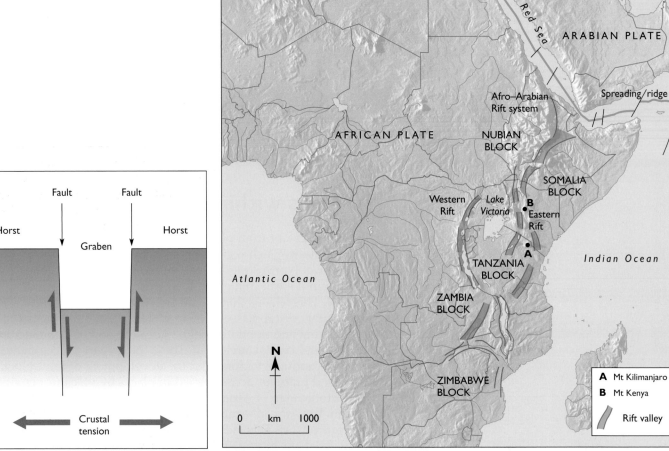

▲ **Figure 1.14** *The East African rift system is a complex set of horst–graben structures resulting from crustal tension.*

EXAMPLE: The Basin and Range Province, (USA)

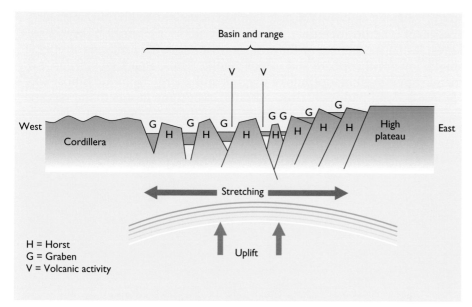

◄ **Figure 1.15a** *A generalised diagram of the Basin and Range Province, (USA)*

This huge region covers most of Nevada and extends south to Mexico and east into Utah. Structurally it is a complex sequence of north–south fault-blocks (horsts) and grabens. Each unit may be 30–40 km wide and up to 150 km long. The uplift and east–west crustal stretching began about 25 million years ago, producing north–south faulting and periodic volcanic activity. It is estimated that the crust has stretched by as much as 10 per cent. The landscape today is dominated by fault-defined north–south mountain ridges separated by broad sediment-filled valleys (graben) and extensive plateau blocks (Figure 1.5b). Into this uplifted and fractured landscape, the main rivers have carved a set of canyons like that of San Juan (Figure 1.12) and the Grand Canyon of the Colorado River (Figure 1.5c). The tensional stresses that weaken the continental crust allow magma to reach the earth's surface.

▲ **Figure 1.15c** *The Grand Canyon of the Colorado River. This is is the result of long-continued uplift and vigorous fluvial incision, and is more than 1.5 km deep, up to 12 km wide and 400 km long.*

◄ **Figure 1.15b** *Part of the Basin and Range region. Faults run between the ranges (horsts) and the basins (grabens).*

1.4 Volcanic activity and landforms

Some of the world's most spectacular and best-known landforms are the result of volcanic activity. Volcanic landforms are created by the surface extrusion of igneous materials. All are associated with forces and processes that result in a weakening of the earth's crust and with a **magma** supply from the lower lithosphere and upper mantle. Thus, an active volcano is a dynamic **open system**: outputs of magma and energy from the lithosphere store move along pathways (throughputs) to become inputs to the volcano store, from which there will be outputs by further eruptions and subsequent erosion (Figure 1.16).

► **Figure 1.16** *Mt St Helens, Washington State, USA: In May 1980 an explosive eruption destroyed 12 per cent of the mass of the volcano, reducing its height by 400 m. Much of the north-east face of the mountain disappeared and a new crater 3 km across and 700 m deep was formed. The mountain had been dormant since 1857 when the last active episode ended, but by 1978 seismologists and geologists were giving warnings that an eruption was likely. A bulge growing on the north flank, frequent earth tremors, and changes in gas type, pressure and temperatures were signals that magma was rising within the mountain. The eruption was triggered by an earthquake, 5.0 on the Richter scale, and activity continued for more than two years. This photo, taken in 1982, shows steam and gas still rising from the fresh cone being formed within the main crater.*

'If you'd been wandering near the Mexican village of Paricutin in 1943 you could have watched the birth of a cinder cone volcano. Below ground on a local farm, rapidly expanding gas sent molten lava exploding from a vent in the earth. As it fell back, the lava solidified in cinders around the vent, and if you'd had about nine years to spare you could have sat and watched as the cinders piled up to form a cone 365 m high. With a final flourish, one last explosion created a funnel-shaped crater to top off the cone.'

(Miller, Geographical, July 1999).

Eruptions and volcanic landforms occur in four locations:

■ Above the subduction zones of destructive (convergent) plate boundaries, e.g. the Cascade Range (USA), the Andes (South America). These regions are also susceptible to earthquakes, e.g. Japan, Mexico, Alaska.

■ In zones of tension and faulting in continental interiors, e.g. Mt Kilimanjaro and Mt Kenya, East Africa.

■ Along divergent plate boundaries. However, the igneous upwelling that creates mid-oceanic ridges rarely forms volcanoes.

■ Above **hot spots** in the upper mantle. A hot spot is an area of unusually high rates of magma production and volcanic activity. Hot spots can occur in a variety of locations, for instance, below a continental plate in the case of the Yellowstone Basin (USA) (Figure 1.17); below a mid-oceanic ridge (a divergent plate boundary) in the case of Iceland; below an oceanic plate in the case of the Hawaiian Islands. Hot spots also generate crustal uplift of up to 1000 m above the surrounding region, so accentuating the effect of the volcanic deposits.

► **Figure 1.17** *A geyser basin in Yellowstone National Park. The Yellowstone volcanic region lies over a hot spot beneath the North American lithospheric plate. Today, activity is limited to geysers and hot springs, but in the past there were long periods of explosive activity during which a plateau was created from thick layers of lava and ash.*

Key Terms

Igneous: Molten material and magma which extrudes through the Earth's crust and solidifies

Box 1 Volcanic eruptions

■ There are five main types of volcano:
■ **Hawaiian** (after the Hawaiian chain of volcanoes): Semi-permanent upwellings of free-flowing lavas to create shield volcanoes.
■ **Strombolian** (after Mt Stromboli in Italy): Infrequent violent eruptions, largely made up of lavas.
■ **Vulcanian**: Infrequent ejections of lava fragments and ash which settle as cinders.
■ **Pelean** (after Mt Pelée in the Caribbean): Explosive eruptions of viscous lava, ash and gas which roll downslope as rapidly moving 'nuées ardentes' (glowing clouds), like an avalanche.
■ **Plinian**: Explosive ejections of relatively viscous lavas and ash which rise high into the air before settling, as well as pyroclastic flows, forming steep composite volcanoes

Igneous materials

In studying the effects of volcanic activity, we need to take two further characteristics into account: first that the location of vulcanism has varied over geologic time; second that the activity takes varied forms. For example, much of Snowdonia and the central mass of the English Lake District are built of volcanic strata up to 4000 m thick, from an eruptive period some 450 million years ago. At a much larger scale, the Deccan Plateau of central India covering more than 500 000 sq.km is made from lava, and took 1 million years to build. Neither the British Isles nor the Indian sub-continent are active volcanic regions today.

In order to understand any particular example of a volcanic landform, we need to answer two basic questions:
■ How did the **igneous** materials emerge at the crustal surface?
■ What type of materials were ejected?
When the igneous materials move upwards through the crust via a single vent, a volcano is built from a series of eruptive episodes. When the materials rise through a linear **fissure** in the earth's crust, lava flows spill out (Figure 1.18). Over time a **plateau** may form – for example, the Deccan Plateau (India), the Columbia Plateau (USA).

► **Figure 1.18** *Vent and fissure vulcanism in New Zealand: Mt Taravera, with the vent volcano, Mt Edgecombe, in the distance.*

Figure 1.19 *Composite volcanoes are made out of layers of ash and lava like these beds in Urewera National Park (New Zealand).*

9 Read section 1.4 on volcanic activity.

a Name two main sources of magmatic material.

b Explain why volcanic activity is associated with fault zones.

c The Hawaiian Islands and the Yellowstone geyser basin are distant from lithospheric plate boundaries. Explain why they are areas of volcanic activity.

d Name two variables that influence the form of a volcanic landform.

e Define the terms shield volcano and composite volcano and suggest reasons why they are different.

▶ **Figure 1.21** *Mt Pinatubo, Philippines: In June 1991, a 600-year period of dormancy ended with one of the largest eruptions of the twentieth century. For five days, repeated explosions lifted vast volumes of tephra (debris and ash) more than 10 000 m into the atmosphere. Pyroclastic flows up to 200 m thick poured down the mountain. More than 1000 people died, and weather patterns in many parts of the world were affected for more than a year as the high-level ash cloud drifted around the globe.*

The nature of the igneous materials extruded decides how they are ejected and controls the shape of the landforms resulting from the succession of eruptive episodes. The material ejected, varies in terms of its *viscosity* (dependent on chemical composition, especially silica content) and *form* ie. the balance between plastic magmatic lava, **tephra** (**pyroclastic** material ejected in a solid form, from fine ash to large clasts), and gases. For example, basic lava, which is low in silica and has a low viscosity, flows freely and builds **shield volcanoes**. Acidic lava, which is high in silica and is viscous (slow-flowing), builds steep-sided volcanoes like Mt Pelée in the Caribbean. Most vent eruptions yield a mixture of lava, tephra and gases, and so build **composite volcanoes** (Figure 1.20). Whenever volcanic activity is explosive, the ejected materials will include rocks blasted from the existing mountain. This is what happened when the summit and north-east face of Mt St Helens was removed by a volcanic explosion in 1980. Furthermore, during its evolution, a volcano will change its morphology – dependent on the type of material ejected and the mode of eruption. At present, there are more than 500 volcanoes classified as 'active'.

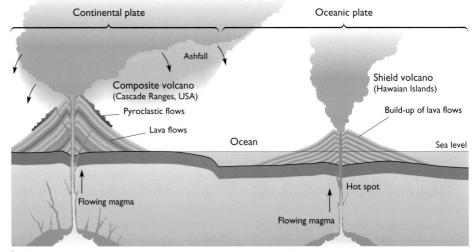

Figure 1.20 *A composite volcano and a shield volcano*

In general, shield volcanoes are built from frequent, relatively quiet eruptions such as at Mauna Loa, Hawaii (see the Hawaiian example), while **composite** and acidic volcanoes evolve through an irregular succession of violent, explosive events as in the case of Mt St Helen's. Except for the regular Hawaiian-type activity, most volcanoes experience periods of dormancy interrupted by relatively brief active episodes (Figure 1.21).

EXAMPLE: The Hawaiian Islands

► **Figure 1.22a** *The hot spot which has created the Hawaiian Islands is beneath the Big Island – therefore the age of the other islands increases the further away they are from the Big Island.*

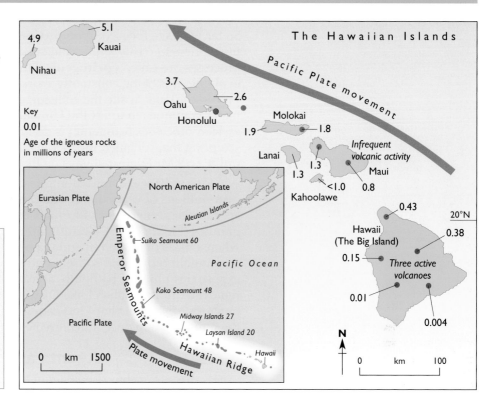

10 Use evidence from the Hawaiian Islands example to support these hypotheses:
- That lithospheric plates move slowly across the Earth's surface over millions of years.
- That plate tectonic movements both create and destroy crustal rocks.

Today a lithospheric 'hot spot' lies beneath the Big Island of Hawaii (Figure 1.22a). Magma upwellings create shield volcanoes. For more than 100 million years, the Pacific Plate has drifted slowly northwest, carrying the volcanic mountains with it and allowing new volcanoes to form in the same spot. As the distance of the volcanoes from Hawaii increases, so does the age of the volcanoes. The cross-section (Figure 1.22b) shows that coral **atolls** can develop when the materials are near the ocean surface. At greater distances, the submerged summits are known as **seamounts**. Finally, the moving plate drags the crustal materials down the subduction zone where they are destroyed, providing magma for the Aleutian volcanoes.

► **Figure 1.22b** *A cross-section showing how the Hawaiian Islands were formed by a magma plume at the hot spot as the plate moved northwest.*

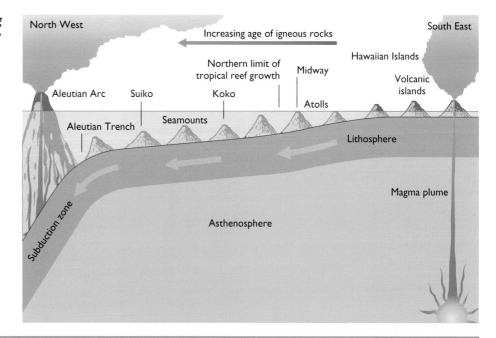

1.5 Weathering and erosion

The power of denudation

So far we have focused upon processes that work from within the lithosphere to create large-scale crustal features. These internal forces and processes are called **epeirogenic**. However, as this macro-framework is evolving, it is being constantly worked upon by a set of external (exogenic) processes. Thus, while the collision of the Indian and Eurasian lithospheric plates continues to uplift the Himalayas, a parallel set of **weathering** and erosion processes (**exogenic**) is denuding the great mountain ranges and sculpting their detailed morphology. But if denudation rates exceed rates of uplift, will Mt Everest always be the world's highest peak (Figure 1.23)?

As we shall see, denudation rates are generally slow, but when sustained over millions of years the results can be impressive. For example, the spectacular scenery of California's Sierra Nevada mountains is carved into granites (Figure 1.24b). Granites are plutonic igneous rocks that originate when a mass of molten magma (a pluton) is intruded deep into the roots of a mountain range, and subsequently cooled slowly to form a batholith. To expose this plutonic batholith, at least 4000 m of overlying rocks have been removed. In the English Lake District there are a number of local granite outcrops, (Eskdale, Shap) indicating that the mountains we see today are the remnants of a once much greater range (Figure 1.24a).

Agents and processes

Look again at the landforms and landscapes illustrated in photographs earlier in this chapter: you will see evidence of the work of one or more of the three agents of erosion and landform formation – water, ice and wind. Note how the external (exogenic) processes of these agents work in combination with the internal (epeirogenic) processes. For instance, refer back to the San Juan 'gooseneck' meanders of Figure 1.12: the depth and shape of the canyon is the result of the continued slow uplift of the sedimentary strata of the continental crust (epeirogenic processes) in combination with vigorous incision by the river and progressive weathering of the canyon sides (exogenic processes). Without the uplift, the incision would not continue; if uplift exceeds the capacity of the river to incise, then the river course becomes abandoned (see chapter 2).

We need to be clear what is meant by 'weathering' and 'erosion' and how they fit into the system of landscape formation: Figure 1.25 shows the erosion–transport–deposition system.

Weathering covers only the early stages of the erosion–transport–deposition system of landscape formation as it does not include transport of material. In general, this system of exogenic processes works at spatial and temporal scales smaller than those of epeirogenic processes, e.g. slopes and gullies on the flanks of a volcano (Figure 1.26); the growth of deltas; the infilling of a graben fault trough; the retreat of sea-cliffs (see also chapters 2 and 3).

▼ **Figure 1.23** *Mt Everest, the climbers' ultimate goal – but will it always be?*

◄ **Figure 1.24a** *Shap Granite quarry, Wasdale Head, Cumbria. This granite originated deep in the roots of a mountain range. The outcrop illustrates the power of denudation processes.*

► **Figure 1.24b** *The glacial trough of Upper Yosemite Valley, California, which is carved into the granite mountain mass of the Sierra Nevada. This is the combined effect of the works of water, ice, wind and temperature.*

► **Figure 1.25** *The erosion–transport–deposition system*

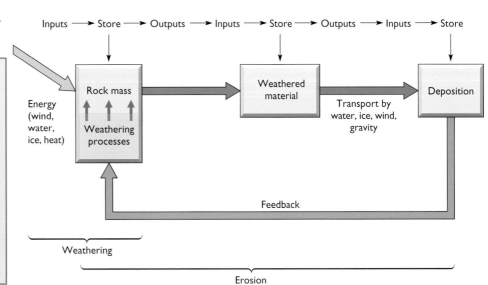

Key Terms

Weathering: The chemical alteration and mechanical breakdown of rock materials during exposure to air, moisture and organic matter.

Erosion: The complex group of related processes by which rock is broken down physically and chemically and the products removed.

25

Figure 1.26 *Vigorous gullying of a cinder cone volcano – Mt Bromo, Indonesia. The climate is tropical and humid.*

Weathering

There are two forms of weathering:

Mechanical weathering: The physical disintegration of rocks.

Chemical weathering: The decomposition or decay of rock minerals by means of processes which create chemical reactions (Box 2).

A third form, biological weathering, is really a subset of both mechanical and chemical weathering. For instance, plant roots penetrate rock fissures and so assist mechanical weathering; organic acids from plant roots assist chemical weathering.

Weathering takes place on exposed rock surfaces and at the junction between the **regolith** (the surface mantle of weathered material and organic matter) and bedrock (see the weathering of granite example).

In order to understand the role played by weathering in the evolution of landforms, we need to find answers to four interrelated questions:

- Where does weathering occur?
- What processes are involved?
- How fast do weathering processes work?
- What factors (including human activities) influence weathering?

Mechanical and chemical weathering each include several important processes, and both sets are at work in all environments, although their relative importance varies (see Weathering of granite example, page 27). For instance, as moisture availability is a crucial factor in chemical weathering, these processes are more active in humid environments (Table 1.1). We can use two examples of landforms to introduce the diversity of weathering and its role in the erosion–transport–deposition system. Both landforms are rock towers: Navajo Butte has evolved in the semi-arid environment of Arizona, USA; the towers along the River Li are found in the humid environment of south-central China (pages 28–29 examples).

Table 1.1 *Rates of chemical denudation: variations in different climatic regions (tonnes/sq.km/year)*

Region	Rate	
Very wet tropics	80	Most
Wet temperate	67	
Seasonal tropical	6	
Arid	3	Least

N.B. These four regions are the extremes of 12 global climatic regions (Maybeck, 1979)

311

11

a What is the difference between weathering and erosion?

b Draw a sketch of the volcano in Figure 1.26. Add labels to your sketch, based on the systems model of Figure 1.25.

Box 2 Weathering

Mechanical weathering processes:

- **Unloading (release of strain):** Expansion that occurs when the weight of overlying and surrounding rocks is removed. The expansion opens up cracks and fissures that weaken the rock and increase the surface area available for further weathering.

- **Frost action and hydro-fracturing:** The increase in pressure in cracks as water expands on freezing (9 per cent expansion), and from water pressure that gradually extends the cracks, leading to ultimate fracture.

- **Salt weathering:** Pressure increases caused by the crystallisation of salts in rock pores and fissures.

- **Insolation weathering:** Repeated heating and cooling of rock surfaces causes expansion and contraction, gradually weakening a thin layer. Alternate wetting and drying may produce a similar result.

Chemical weathering processes:

- Decomposition by chemical reactions with rock minerals, causing loss of internal cohesion and strength. Minerals become unstable, break down and reform to create new, more stable minerals.

- Rocks are made of an aggregate of minerals, and these minerals vary in their resistance to chemical reactions. Thus some minerals decompose more readily than others, and weaken the rock structure.

- Acidic water is a crucial agent, with its ability to attack potentially soluble minerals. For example: limestones are rich in calcium carbonate $CaCO_3$, and are not very soluble. In the presence of water containing carbon dioxide, the calcium carbonate becomes calcium bicarbonate, which is soluble in water.

- The principal processes involved in chemical decomposition and the creation of new minerals are: solution; hydrolysis; hydration; carbonation; oxidation; reduction; chelation.

EXAMPLE: The weathering of granite

Rock type is one variable influencing weathering. Yet different weathering processes produce different landforms from similar rock types as these three examples developed on granites illustrate.

◀ **Figure 1.27a** *Sheet jointing of granite, Yosemite National Park, California – mechanical weathering by stress release. As the overlying rocks have been removed, reduced pressure on the rock mass allows the development of doming and sheet jointing. The thin surface sheets then slowly disintegrate.*

▶ **Figure 1.27b** *Weathering profile of granite, Tanzania, East Africa. Thick regoliths are common on gently sloping surfaces in the humid and semi-humid tropics. Chemical weathering processes work downwards as acidic water penetrates fissures in the granite. Once the 'corestone' phase is reached, weathering attack occurs across the whole rock surface. Over perhaps hundreds of thousands of years, a deep weathering profile develops, which in places is over 20 m thick.*

▶ **Figure 1.27c** *The well-known Dartmoor tors are the result of long-term sub-surface and surface chemical and mechanical weathering. The structural and sheet joint systems increase the surface area for weathering attack.*

EXAMPLE: Weathering at work

a Navajo butte

This butte is an outlier rock tower, once part of the main plateau which has since been removed by erosion. Mechanical weathering processes are the main cause of the gradual disintegration. The weathering is the first stage of the erosion–transport– deposition system of landform evolution. The butte consists of two components set upon an extensive pediment plain. Notice that the thick debris apron, built progressively from weathered materials, protects the bedrock and slows down the rate of weathering. The thin sediment mantle provides little protection for the pediment surface.

► **Figure 1.28a** *Navajo Butte, Arizona, USA*

311

◄ **Figure 1.28b** *Formation of the Navajo Butte*

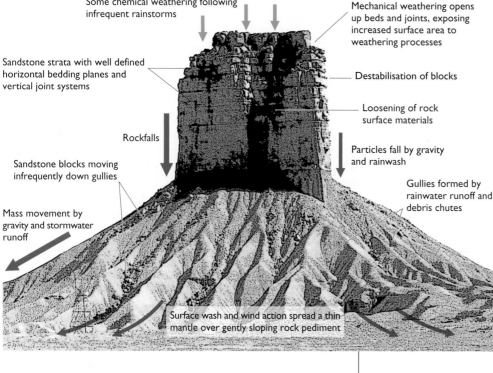

Some chemical weathering following infrequent rainstorms

Mechanical weathering opens up beds and joints, exposing increased surface area to weathering processes

Sandstone strata with well defined horizontal bedding planes and vertical joint systems

Destabilisation of blocks

Loosening of rock surface materials

Rockfalls

Particles fall by gravity and rainwash

Sandstone blocks moving infrequently down gullies

Gullies formed by rainwater runoff and debris chutes

Mass movement by gravity and stormwater runoff

Surface wash and wind action spread a thin mantle over gently sloping rock pediment

1
Rock butte

2
Debris apron

Sharp breaks of slope

3
Low angle pediment plain with thin sediment mantle

B e d r o c k

► **Figure 1.28c** *The main sections of the Navajo Butte*

► **Figure 1.28d** *Limestone towers or Mogote scenery and the River Li Gorge (China). The scale of the towers is indicated by the size of the buildings in the foreground.*

b Limestone towers along the River Li (China)

These spectacular towers are developed on well bedded and strongly jointed limestones. There is discussion among geologists over their evolution, but one popular idea is:

'In humid tropical and subtropical environments, some limestone landscapes are dominated by spectacular tower-like hills called mogotes, up to 100 m or more high, separated by broad, alluvial valley floors. This type of terrain is described as tower karst and has been thought to represent the effects of enhanced limestone dissolution (decomposition) in regions of high precipitation. Acid waters in the alluvial valleys cause rapid lateral planation and undercutting of rock faces while high temperatures promote the evaporation of water flowing across the exposed rock leading to the deposition of a protective layer of calcite (on the towers).'

(*Summerfield, 1996*)

The calcite coating reduces the rate of mechanical weathering. In contrast, the vegetation which clings to the towers assists chemical weathering through the percolation of organic acids from plant roots. The towers represent a late stage in gorge development.

12 From the Weathering at work example:
a Draw a sketch of the photo of the River Li (Figure 1.28d). (Use the Navajo Butte sketch as a guide (Figure 1.28b).
b Label your sketch to describe the features and illustrate the weathering processes.

▼ **Table 1.2** *Factors affecting weathering*

| Rock type – structure and composition |
| Availability and character of moisture |
| Temperature regimes |
| Relief and topography |
| Time |
| Location within a landscape |
| Vegetation |
| Human activities |

Weathering rates

When we try to answer the question 'How fast does weathering work?', we face some difficult problems: first, weathering rates are usually very slow, so we need long records; second, mechanical and chemical processes work alongside each other and it is not easy to separate what is going on; third, weathering occurs not only on exposed, visible surfaces, but also within and at the base of the regolith. Look again at the weathering of granite example to see how weathering of the same type of rock (in this case granite) can produce very different types of landforms. If you look at Table 1.2 you will see eight factors affecting weathering. These are also linked as in the case of location, temperature and the availability of moisture.

311

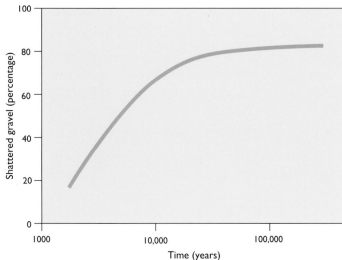

▲ **Figure 1.30** *Mechanical weathering rates in the Negev Desert, Israel*

◄ **Figure 1.29** *Chalk cliff collapse near Dover. The collapse followed a long period of slow weathering and erosion. Chemical weathering weakens the cliff front; wave action (mechanical weathering and erosion) undercuts the cliff base; the cliff becomes unstable and collapses.*

Key findings of research into weathering on a stone and gravel desert surface (Figure 1.30) show:

■ Mechanical shattering of clasts (rock fragments) is the main form of weathering.

■ Shattering rates decrease over time – for example, 70 per cent of shattering occurs within 14 000 years; it may take up to 500 000 years to achieve 80 per cent.

■ Pressure applied by salt expansion in cracks when evaporation takes place is the most important process, indicating the significance of water even in arid environments.

■ Over time, chemical weathering processes produce a protective crust which slows the mechanical processes.

The researchers conclude: 'At first the presence of a very gravelly and permeable profile allows even quite modest amounts of salts to cause shattering. However, with time, sealing of the soil surface occurs because of the development of a pavement and the accumulation of dust and gypsum. The plugging of the soil surface changes the wetting depth, temperature and moisture fluctuations of the regolith profile. At this stage, the shattered gravel horizon is no longer subject to the extreme conditions of wetting and drying that encourage shattering. The salts tend to cement and protect the gravel instead of shattering it.'

The Negev Desert example illustrates the significance of the regolith. Surface mantles develop where weathering exceeds removal by erosion. In the Navajo Butte example (p.28), weathering rates are maintained across the rock tower as fresh rock is continuously exposed. In contrast, across the debris apron weathered material accumulation exceeds removal, and the bedrock is protected from direct mechanical processes. Thus, as we study how a landscape is changing, we need to ask: 'Is there a regolith – a surface mantle of weathered material and organic matter?'

It is clear that the character and rate of weathering are controlled by the interaction of a set of variables, which helps to explain the diversity of landforms (Table 1.2 and Figure 1.29).

13 From Figure 1.30:
a Summarise in one sentence the work of mechanical weathering shown by the graph.
b What evidence is there that chemical weathering is present even in arid environments. Suggest what part it plays in controlling the rate of mechanical weathering.

14 Select any two landform examples you have studied in this chapter. Outline the possible influence of the variables listed in Table 1.2 in forming these landforms. (You may find labelled sketches or diagrams useful. A graphical software package may help.)

15 With the help of an atlas, e.g. physical, geological and climatic maps, analyse the pattern of chemical denudation in Britain (Figure 1.32).

In general we can state that:
■ Chemical weathering increases with temperature and precipitation (Figures 1.31 and 1.32).
■ Mechanical weathering rates are several times greater than those of chemical weathering (However, this may vary in parts of the tropics where chemical weathering can happen quite quickly at extreme depths). Beware however, for careful analysis of the data shows considerable variation in the relationships suggested by these generalisations. Notice, too, that weathering and denudation rates often use indirect measurements.

They are based on the solid and solute loads of rivers, i.e. estimates of the amount of material transported from the landscape and arriving in rivers in relation to the size of the drainage basin (Table 1.3 and Figure 1.33).
■ Long periods of slow mechanical weathering may trigger sudden episodes of sudden erosion (Figure 1.29).

312

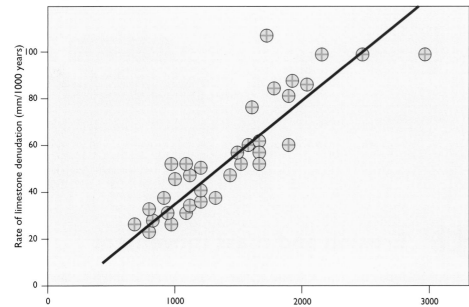

▲ **Figure 1.31** *Rates of limestone denudation in the Mediterranean Basin*

▼ **Figure 1.32** *Rates of chemical denudation in Britain. Moisture availability influences the rate of chemical weathering, hence the high readings in the uplands.*

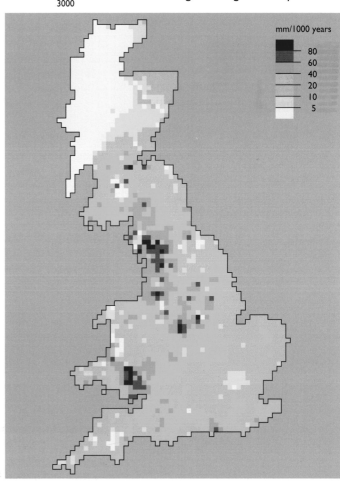

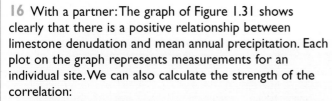

16 With a partner: The graph of Figure 1.31 shows clearly that there is a positive relationship between limestone denudation and mean annual precipitation. Each plot on the graph represents measurements for an individual site. We can also calculate the strength of the correlation:

a Number each of the sites.
b Read off scores from the graph to construct a table (Site 1 is done for you):

Site	Precipitation	Denudation rate
1	650	25

c Apply a statistical correlation technique to calculate the strength of the relationship (Use a software package if available).
d Use your findings and the graph to summarise the relationship.

323

River	Area of drainage basin (million sq.km)	Denudation (mm/1000years)		
		Total	Mechanical[1]	Chemical[2]
Amazon	6.15	70	57	13
Amur	1.85	13	10	3
Brahmaputra	0.58	677	643	34
Colorado	0.64	84	78	6
Danube	0.81	47	31	16
Mackenzie	1.81	30	20	10
Mekong	0.79	95	75	20
Mississippi	3.27	44	35	9
Murray	1.06	13	11	2
Zambezi	1.20	31	28	3

(1) estimated from sediment load
(2) estimated from solute load

▲ **Table 1.3** *Estimating denudation by measuring river loads*

▼ **Figure 1.33** *Sediment and solute loads for ten large drainage basins*

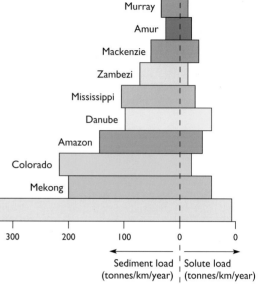

17 From Figure 1.33 and Table 1.3:

a Explain how the data tells us that weathering and denudation rates are much higher in the Brahmaputra drainage basin than in the Colorado drainage basin. (Note that the two basins are relatively similar in size.)

b By the use of a graph and a statistical correlation technique, test this hypothesis: there is no relationship between drainage basin size and denudation rate.

c On average, how much greater is (i) mechanical load than solute load, and (ii) mechanical denudation greater than chemical denudation?

d To what extent is it true to say that weathering and denudation rates are greater in tropical than in temperate and cold environments? What factors other than climate may influence the rates?

323

1.6 Erosion and mass movement

Erosion includes the removal and transportation of solid and weathered surface materials. **Mass movement**, also known as **mass wasting**, is the downslope movement of material under the influence of gravitational force and is frequently assisted by the energy of water, ice and wind. For instance, shattered debris falls by gravity from rock faces above a glacier, but further mass movement is achieved by the energy of the glacier (Figure 1.34). Erosion, therefore, involves the redistribution of material across the Earth's crust.

▲ **Figure 1.34** *Sheriden Glacier, Alaska. The brown mantle covering the snout of the glacier is morainic debris, transported upon, within and below the ice from the distant mountains. It is exposed as ablation (melting) increases.*

Rates of movement range from very slow, perhaps millimetres per year to sudden and rapid, and have been grouped into five main types of process (Table 1.4). An individual landform is the outcome of the combined work of several processes (Figure 1.35a, b). Wind energy too, can cause the transportation of large volumes of fine material.

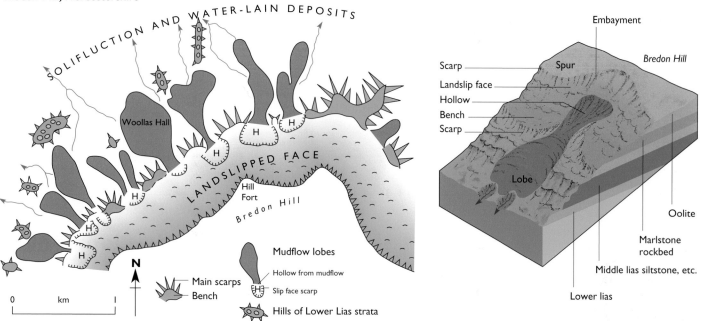

▼ **Figure 1.35** *Hillslope processes on Bredon Hill, Worcestershire*

SOLIFLUCTION AND WATER-LAIN DEPOSITS

Woollas Hall

LANDSLIPPED FACE

Hill Fort

Bredon Hill

0   km   1

N

— Main scarps
— Bench

Mudflow lobes
Hollow from mudflow
Slip face scarp
Hills of Lower Lias strata

a   *Plan view of Bredon Hill*

Embayment

Scarp
Landslip face
Hollow
Bench
Scarp

Spur

Lobe

Bredon Hill

Oolite
Marlstone rockbed
Middle lias siltstone, etc.
Lower lias

b   *Block diagram of Bredon Hill*

▼ **Table 1.4**

Main types of mass movement	
Creep and Heave:	The very slow downslope movement of individual fragments and slope aggregates. Caused by cycles of expansion and contraction in weathered debris and soil materials.
Flow:	A mass of rock and/or weathered materials moves downslope above a poorly defined shear plane. Varies in speed from slow (solifluction) to rapid (mudflow, debris flow) The flow motion means that turbulent shear stresses occur throughout the moving mass.
Slide:	A mass of rock and/or weathered material slips rapidly downslope above a well-defined shear plane, e.g. landslide. The movement may be straight or rotational, e.g. slumping.
Fall:	The free-fall of rock through the air.
Subsidence:	Either the sudden collapse of a slope segment into a subsurface cavity or the progressive settling of the ground surface, e.g. by the removal of groundwater, or following deep mining.

We can understand mass movement in terms of two sets of forces acting upon a rock fragment or a slope. There are resisting forces that work to prevent movement, and driving movements that encourage movement. Movement occurs when driving forces exceed resisting forces. A second key understanding is that movement is discontinuous and that relatively long periods of inactivity are disturbed by relatively brief episodes of movement. (Figure 1.36) Soil creep may occur for a hour or so each day; mudflows may be triggered by seasonal heavy rainstorms or annual snowmelt; cliff collapse may end centuries of stability (Figure 1.29).

18 Use Table 1.4 to analyse the landforms of Figures 1.34 and 1.35. Make labelled diagrams of the mass movement types at work on each

19 Apply the landform evolution model concept (Figure 1.36) to any two of the landforms in Figures 1.34 and 1.35 to illustrate the general principles contained in the model.

Equilibrium
Accelerated change
Recovery towards equilibrium
Equilibrium
Threshold
Amplitude of change
Time

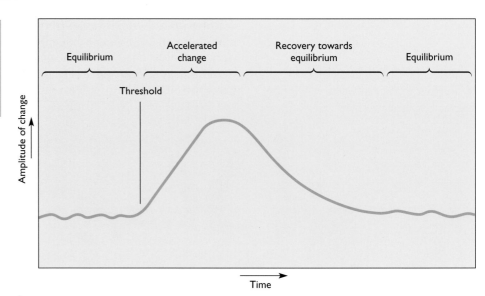

▲ **Figure 1.36** *A general model of landform evolution*

1.7 Slope processes and development

In this section we shall examine three important ideas:

- Landforms and landscapes consist of sets of slopes that change at varying rates over time.
- The slopes we see today are the result of the work of weathering and mass movement processes over a broad range of timescales.
- There are three broad categories of slope: stable, actively unstable, conditionally unstable.

A slope as a system

We know from the variety of weathering and mass movement processes at work that slopes are dynamic – there is constant movement and change, even when nothing visible is happening. We can understand this if we think of a slope as a cascading system. From Figure 1.37: weathering processes in the rock store produce an output of debris; mass movement processes transport the weathered material downslope (throughput) via a series of depositional stores.

During this journey, further erosion takes place. Output from the slope system becomes an input to the fluvial system (see chapter 2).

The sediment throughput is achieved by the interaction of the five mass movement types (Table 1.4) and the set of factors influencing erosion (Table 1.2).

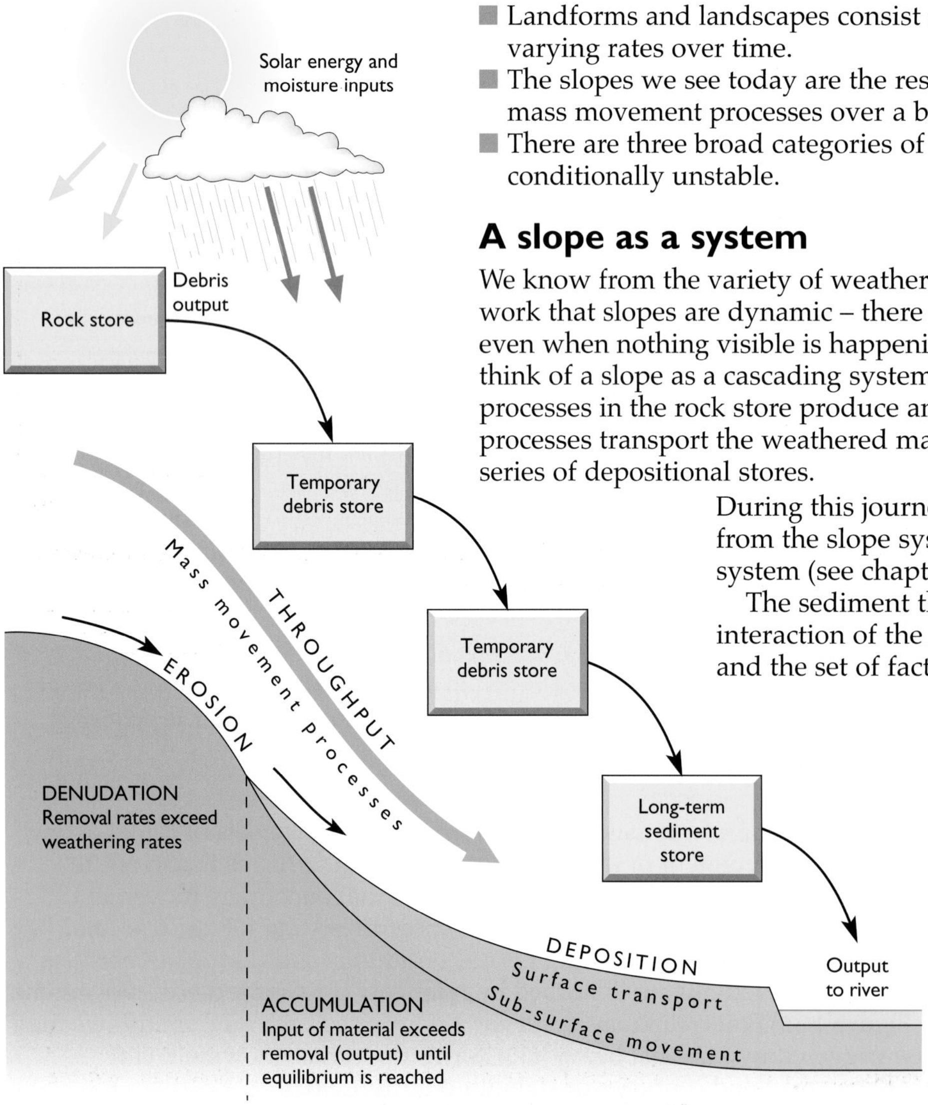

Solar energy and moisture inputs
Rock store
Debris output
Temporary debris store
Temporary debris store
Long-term sediment store
Output to river
THROUGHPUT
Mass movement processes
EROSION
DENUDATION
Removal rates exceed weathering rates
DEPOSITION
Surface transport
Sub-surface movement
ACCUMULATION
Input of material exceeds removal (output) until equilibrium is reached

◄ **Figure 1.37** *A slope as a cascading system*

20 **a** Look at Figures 1.38a–d and use the data to illustrate the statement made in each caption.
b Summarise the relationship shown on Figure 1.38e.
c Comment on the relationship between the variables in each graph, e.g. Do landslides always increase when it rains?

The graphs of Figure 1.38 illustrate the results of these complex inter-relationships. It is clear that water is a primary agent in slope development, even in arid environments. For example, the energy of rainsplash, overland flow and channel water moves surface materials downslope; sub-surface water lubricates the regolith (the weathered surface mantle), adds volume and weight, reduces the internal cohesion and strength of the materials and so energises solifluction, slumping and landslides. Movement continues while the energy of **shear stress** (driving forces) is greater than **shear strength** (resisting forces). As the energy dissipates, deposition occurs and the material is stored once more, until the next input of energy, e.g. a rainstorm, snowmelt.

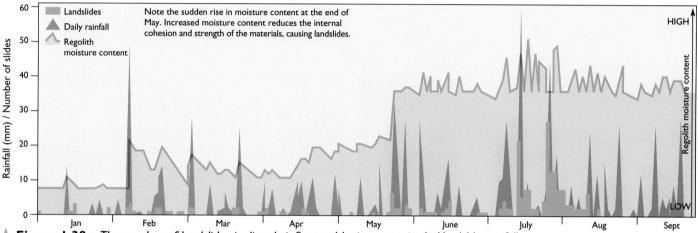

▲ **Figure 1.38a** *The number of landslides is directly influenced by increases in the monthly rainfall.*

311

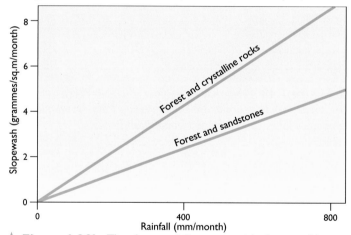

▲ **Figure 1.38b** *The slopewash increases with the monthly rainfall increases.*

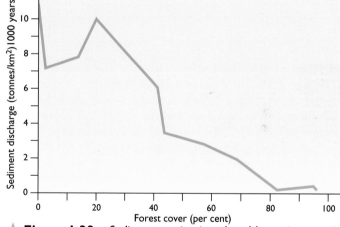

▲ **Figure 1.38c** *Sediment erosion is reduced by an increase in the percentage of forest cover.*

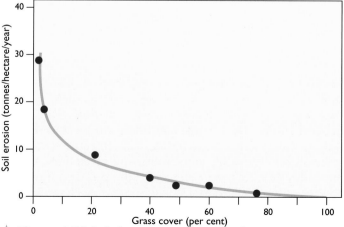

▲ **Figure 1.38d** *Soil erosion is reduced by the percentage of grass cover.*

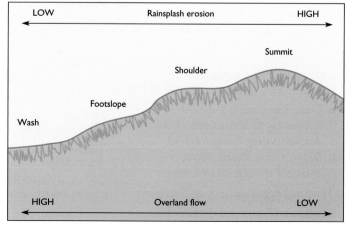

▲ **Figure 1.38e** *Soil erosion and overland flow vary with the gradient and slope of the area.*

Deposition

We have now introduced the deposition component to the **landscape** evolution model (Figures 1.36 and 1.37). Deposition takes two forms: mechanical and chemical. Mechanical deposition of solid materials occurs when there is insufficient energy from gravity, water, ice or wind for movement to continue – resisting forces exceed driving forces, e.g. the debris apron of Navajo Butte (p.28). The general term for unconsolidated surface materials is colluvium. Coarse debris that builds up along the foot of steep slope rock faces is known as talus or scree (Figure 1.39).

Chemical deposition occurs when environmental conditions change and dissolved minerals recrystallise – for example, evaporation produces salt-pans (playas) (Figure 1.40). Sub-surface chemical deposition includes the growth of stalactites and stalagmites in limestone caverns. The depositional stores on a slope may be temporary or long-term resting places.

▲ **Figure 1.39** *The Wastwater Screes in the Lake District. The gradient of scree slopes depends upon the type of debris. Each type of material has its natural angle of rest which is normally between 30° and 45°. (The screes are the paler materials. The fans spill out from gullies along the upper slope.)*

▲ **Figure 1.40** *Playa across the floor of Death Valley, California. The shining white deposits are salts resulting from evaporation of temporary lakes caused by occasional flash floods.*

How slopes evolve

Slopes evolve over time towards a shape (profile) that is in equilibrium with the erosion–transport–deposition processes at work. Once an equilibrium is achieved, the slope will continue to be reduced by erosion, but will retain its overall shape (Figure 1.41). All slope evolution involves a reduction in relief, i.e. a flatter landscape. So, if the epeirogenic uplift of the Himalayas ceased, over millions of years, exogenic slope processes would progressively reduce the relief. As a slope evolves, a regolith of soil and weathered debris can extend progressively, and rates of change slow down.

Thus, we can make this hypothesis:

Through time a slope evolves from an unstable to a stable state. On an actively **unstable slope**, shear stresses on the materials exceed shear strength, driving forces exceed resisting forces and slope evolution is relatively rapid. On a **stable slope**, the reverse conditions apply, and slope change is slow, with little evidence of active change. The unstable stage of slope development is illustrated by valley sides following glaciation. As the glacier disappears, the valley sides are over-steep and the profile out of balance with the processes at work (Figue 1.43). Aggressive mechanical weathering on the upper slopes causes denudation and slope retreat; mass movement leads to the accumulation of talus debris across the lower slopes. Through time the slope profile becomes more stable (Figure 1.44).

21 From the models shown in Figures 1.41 and 1.42:
a Describe what is currently happening to the slopes of the Alaskan valley in Figure 1.43.
b Suggest how these slopes may evolve if the ice disappears.

22 What stage of evolution has Keskadale (Figure 1.44) reached? Explain your choice.

▼ **Figure 1.41a** *As the upper slope is denuded, material accumulates on the lower slope. There is a progressive change from stage 1 to stage 3. Relief is gradually reduced.*

▼ **Figure 1.41b** *Rates of lowering vary across the slope over time, gradually modifying the slope profile. Relief is reduced.*

▼ **Figure 1.41c** *The angular profile is common in semi-arid environments. Overall relief is gradually reduced.*

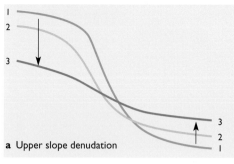

a Upper slope denudation

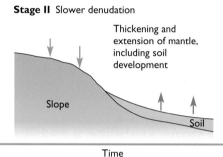

b Progressive slope lowering

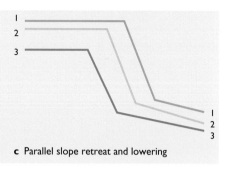

c Parallel slope retreat and lowering

Stage I Vigorous denudation

Stage II Slower denudation

Thickening and extension of mantle, including soil development

Stage III Very slow denudation. Almost complete regolith, thickening downslope and including soil

Slope

Scree

Soil

Slope

Soil

Slope

Soil

Time

▲ **Figure 1.42** *Changes in denudation rates over time*

► **Figure 1.43** *An unstable slope, Alaska (USA). The slope is being exposed by glacial movement and is oversteepened by glacial erosion. Active freeze–thaw mechanical weathering is attacking the upper slope while mass movement of the weathered debris is producing scree across the lower slope.*

◄ **Figure 1.44** *Slopes approaching stability, Keskadale, Cumbria. This valley was glaciated but the last ice disappeared 10 000 years ago. Slope processes of erosion–transport–deposition have brought this valley towards stage III of the model in Figure 1.42. Retreat of the upper slopes and infilling of the valley floor have combined to give a gently curved cross-profile.*

37

▶ **Figure 1.45** *A stable slope on the Chiltern Hills. The steady gradient and continuous vegetation cover show no visible signs of active slope development. None the less, during spells of temperature or moisture extremes, there may be some downslope movement in the regolith.*

---

23 Read the section on slope processes:

a Define the terms 'driving force' and 'resisting force'.

b Explain why a slope can be described as a cascading system. Use the diagram of Figure 1.37 to outline what might happen to the slope system if the climate becomes warmer and wetter.

c Why does mass movement cease, i.e. deposition take place? (Think of what triggers mass movement in the first place.)

d What do we mean by 'stability' and 'instability' on a slope?

e Give two examples of the relationship between 'threshold' and 'conditional instability'.

f Explain how a conditionally unstable slope fits the general landform evolution model of Figure 1.36. Give examples to illustrate your answer.

g Select examples from this chapter to support the three generalisations that end this chapter.

Is slope stability a reality?

The idea of evolution towards stability depends upon environmental conditions remaining constant over long periods. Think back to all the epeirogenic and exogenic processes that we know are at work, and it becomes clear that environmental conditions fluctuate. There are long-term trends (climatic change, tectonic plate movements), and short-term fluctuations and extreme events (hurricanes, flood surges, earthquakes). Slopes can generally adjust to the normal fluctuations of energy and material inputs, e.g. seasonal rhythms of temperature and precipitation (Figure 1.45). But beyond a certain threshold the slope system can no longer cope and sudden, perhaps fundamental, changes occur, e.g. the slope systems on and around Mt St Helens from the 1980 eruption; the slumps on Bredon Hill following heavy rains.

Many slopes are, therefore, conditionally unstable. These slopes exist where there are shifts in balance between **shear strength** and **shear stress** in the materials making up the slope. They are vulnerable to abrupt change when certain environmental thresholds are crossed. Some slopes are more sensitive than others. When you walk across a scree slope or a sand-dune, the energy applied to the slope materials may cause sudden mass movement – and may cause you some problems!

We can make three final general statements based on our model of Figure 1.36:

■ Slopes – and landforms – develop via relatively long periods of slow change, or dynamic equilibrium, interrupted by relatively brief episodes of accelerated change.

■ The disruptive episodes are generated by input surges of energy and material which take the slope beyond certain stability/instability thresholds.

■ After the disruptive episode, the slope system works to return to an equilibrium, i.e. stable state.

In this chapter we have concentrated on how natural processes work. However, as later chapters show, human activities are increasingly powerful forces in landform and landscape evolution. For instance, deforestation destabilises slopes and accelerate rates of weathering and erosion; afforestation stabilises slopes and reduces rates of change (Figure 1.46).

▲ **Figure 1.46** *Tree farming above the Mangaer, New Zealand. New trees are planted in between the mature trees. This helps to maintain slope stability after the mature trees are harvested.*

Summary

The Earth's crust consists of a set of large tectonic plates that move slowly in relation to each other. This movement is explained by the Plate Tectonics theory.

The three types of plate boundary (convergent, divergent, transform) determine the location and character of mountain building, fault systems. earthquakes, volcanoes.

Major earthquakes occur when there is a sudden stress release and movement along sections of an active fault that have been 'locked' for a period of time.

Volcanoes vary in form and character according to the type of materials ejected during eruptions.

Landforms are the result of exogenic processes of weathering, erosion, transport and deposition.

Denudation is achieved by a combination of mechanical, chemical and biological weathering processes working at different rates.

Landscape evolution tends to involve long periods of relative stability, interrupted by briefer episodes of accelerated change.

Slope development can be explained by the interaction of a set of mass movement processes.

Weathering, erosion and mass movement processes work together to produce slopes that are in equilibrium with the environmental conditions.

Chapter 2 Hydrology and rivers

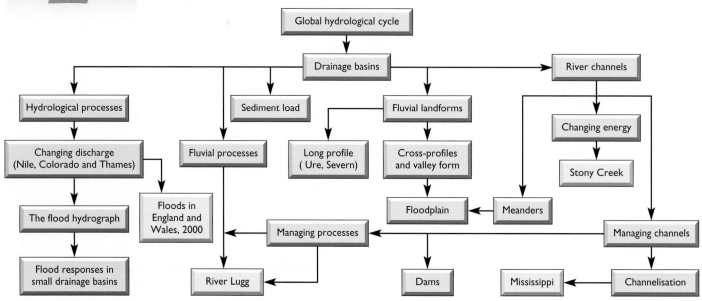

Key Terms

Hydrology: The study of water in the natural environment.

Inputs: Energy or matter that enters into a system.

Outputs: Energy or matter that leaves a system.

Open system: A system where there are inputs and/or outputs of energy and matter.

Closed system: A system where there are inputs and/or outputs of energy but not of matter.

Stores: Natural reservoirs of water in the environment, such as rivers, lakes, soil, vegetation, the atmosphere and the water table.

Flows: Paths that water follows in the land-based part of the hydrological cycle, i.e. flows through soil and flows in rivers.

2.1 The global hydrological cycle

Water is necessary for life on earth, and the study of the **hydrology** and fluvial (river) processes is vital for the management of this resource. There is a fixed amount of water circulating within our planet's atmosphere and on its surface. There are no exchanges outside the earth and its atmosphere. This global hydrological cycle is a **closed system**, which means that there are no **inputs** or **outputs** of water but only of energy (solar radiation). Water is held in **stores** such as oceans, ice caps and the soil. Water moves along pathways between these stores by processes such as precipitation, evaporation and runoff. There are huge variations in where and for how long water is stored and moved (Table 2.1).

Figure 2.1 shows the key stores and pathways or processes within the global hydrological cycle. Figure 2.2 shows these stores and **flows** as a systems diagram. A transfer occurs where water moves from one store to the next. The terrestrial (land-based) part of the cycle is linked by important processes and movements of water.

To help our understanding of how this transfer works, hydrologists divide the Earth's surface into units called drainage basins. Drainage basins are the area of land drained by a river and its tributaries (Figure 2.3). By studying Figure 2.2 you can see that if we take the drainage basin as a system on its own, it will have inputs and outputs with two other parts of the global cycle – the atmosphere and the oceans. The drainage basin is therefore an **open system** since there are movements of both energy and matter in and out of the system. These inputs and outputs will vary around the world, and in the same place at different times such as the seasons of the year.

Figure 2.4 shows how a drainage-basin system links with the global system. For any drainage basin, there is a link with the atmosphere by precipitation inputs and evaporation outputs.

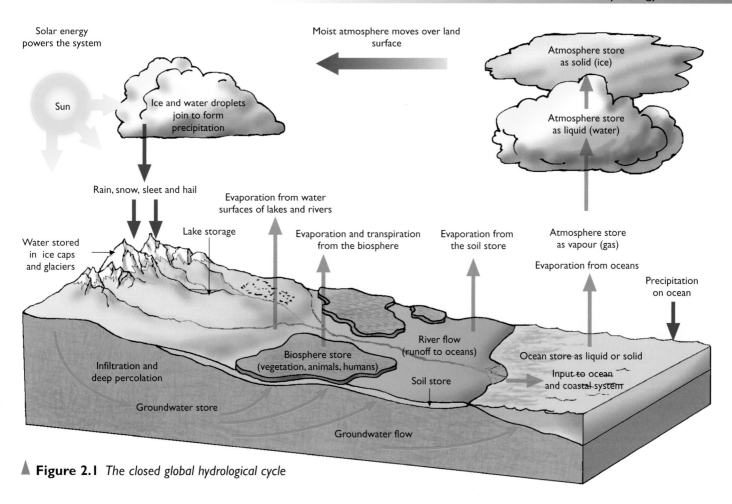

Figure 2.1 *The closed global hydrological cycle*

Table 2.1 *The amount and time of water storage in the key global stores*

Stores	per cent of water	Time stored (years)
Ice caps	77.35	9700
Groundwater	22.00	1400
Soil	0.20	1
Lakes	0.4	17.0
Rivers	0.003	16
Atmosphere	0.044	8
Biosphere	0.003	N/A
Oceans	Not freshwater	2500

Note: the figures are for fresh water, which in total makes up only 3 per cent of the water in the hydrological cycle. The salt water of the oceans makes up 97 per cent.

1 Think carefully about where and how much water will be stored in a desert environment compared with a humid environment such as the British Isles. Use Figures 2.1 and 2.2 to help you.

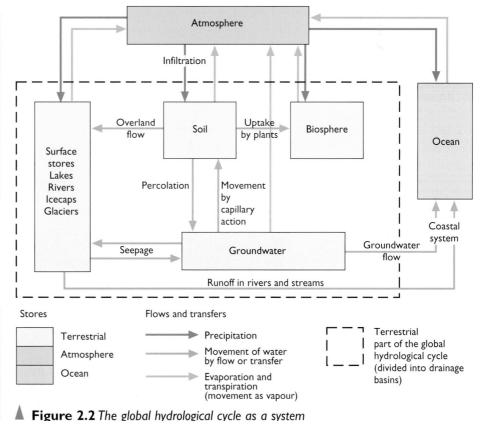

Figure 2.2 *The global hydrological cycle as a system*

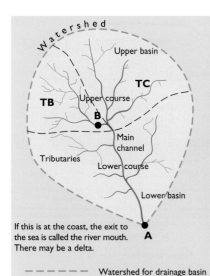

If this is at the coast, the exit to the sea is called the river mouth. There may be a delta.

– – – – – Watershed for drainage basin

– · – · – Watershed tributary basins TB, TC

● A Gauging station A will show a flow pattern which depends upon the characteristics of the whole drainage basin.

● B Gauging station B will show a flow pattern which depends upon the characteristics of the tributary TB catchment only.

Rainfall falling in the upper basin will reach the gauging station at **B** relatively quickly. It will take longer before the streamflow from this rainfall is recorded at gauging station **A**, which will also record rain falling in the lower basin.

Tributary catchment **TB** shows a relatively high number of streams per unit area. This is called a high drainage density. This is most common with impermeable rocks. Tributary catchment **TC** shows a relatively low number of streams per unit area. This is called a low density, and is most common with permeable rocks.

◄ **Figure 2.3** *The drainage basin: A simple model*

▼ **Figure 2.4** *Drainage basins are open systems and link with the global hydrological cycle*

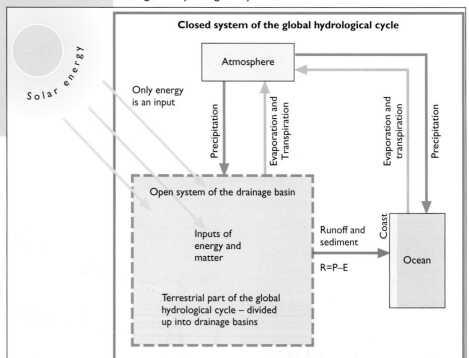

Key Terms

Regime: The seasonal rhythm of flow of a river.

Drainage basin: The area of land drained by a river.

Discharge: The volume of flow of water in a channel, measured in cubic metres per second (cumecs).

Watershed: The boundary of a drainage basin.

Evaporation: The change of state from liquid to gas from exposed water surfaces.

Transpiration: The loss of water from the leaves of plants.

Runoff: Water that moves across the surface of the land into streams rather than being absorbed by the soil.

The water balance

The balance between these inputs and outputs results in the water remaining in the **drainage basin** and becoming river runoff before going back to the oceans or to the atmosphere. This interaction can be written as a simple equation called the 'water balance':

P = E+R (Precipitation = Evaporation / Transpiration+ Runoff).

What goes in as **precipitation** equals what comes out by **evaporation** or **runoff**.

It is river runoff and surface storage which is most important to people, so another way of looking at the water balance is:

R = P–E (Runoff = Precipitation – Evaporation / Transpiration)

Since precipitation inputs and evaporation outputs vary in different places and at different times of the year, this equation gives us a basic understanding of why drainage-basin processes and river flow vary. For example, in the British Isles, precipitation inputs tend to be slightly higher in winter but evaporation outputs are low. The result is high river flow. In summer, evaporation and **transpiration** outputs are higher and the result is reduced river flow. This flow pattern over a year is called the river's 'regime'. Figure 2.5 shows the regimes of three contrasting rivers.

► **Figure 2.5** *Contrasting regimes for the River Thames, the River Nile and the Colorado River*

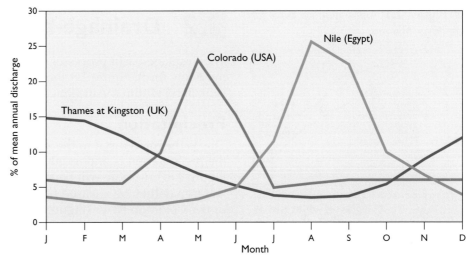

► **Figure 2.6** *Variations in the flow of the River Thames*
a in the long term
b in the medium term

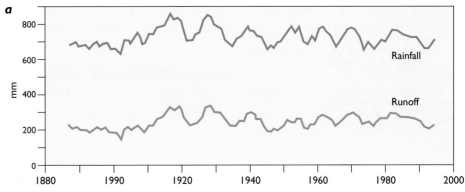

a

b

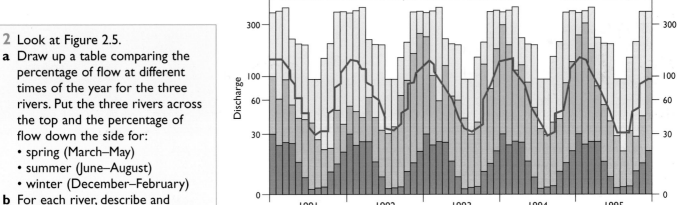

Maximum flow | Flow for the year shown | Minimum flow | — Mean flow over 5 years

2 Look at Figure 2.5.
a Draw up a table comparing the percentage of flow at different times of the year for the three rivers. Put the three rivers across the top and the percentage of flow down the side for:
  • spring (March–May)
  • summer (June–August)
  • winter (December–February)
b For each river, describe and explain the results in your table using input and output terms as fully as you can.

3 Study Figures 2.6a and 2.6b.
a Use Figure 2.6a to give the highest and lowest annual runoff and rainfall figures for the River Thames and the year in which they occurred.
b Study Figure 2.6b carefully. Which years could be described as high and low flow years? Use evidence from the graph to justify your answer.

It is important to understand that the **regime** of a river in the form shown in Figure 2.5 shows only average flow. The River Thames, for example, has a relatively simple regime with higher flow in winter and lower flow in summer. However, as Figure 2.6 shows, there are variations around these long-term averages. Another way of looking at river flow is to use a flow-duration curve. This is a graph which shows the **discharge** against the amount of time that the flow level is exceeded (Figure 2.7). Long-term and annual records can be used. Flow-duration curves are useful in flood management, since they show how often given discharges occur and therefore the size of flood defences needed to contain the flow.

Figure 2.4 also introduced another important output from the terrestrial drainage basin system – the sediment that is transferred into the coastal system. This sediment is produced by weathering and erosion in the drainage basin and transported by the energy of streamflow.

▼ **Figure 2.7** *A flow duration curve for the River Mimran*

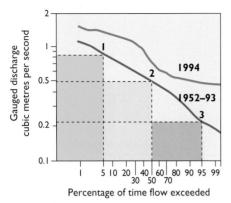

Follow the dotted lines on the graph to read off the curves, using the average flow curve for 1952–93.

1 On average a discharge of 0.85 cubic metres per second is exceeded for 5 per cent of the time. Put another way, for 5 per cent of the average year the flow would be greater than 0.85 cubic metres per second.

2 A discharge of 0.5 cubic metres per second will be exceeded for 50 per cent of the time, or for half the year will be at least 0.5 cubic metres per second.

3 A discharge of 0.2 cubic metres per second will be exceeded for 95 per cent of the year.

4 Read off the 5 per cent, 50 per cent and 95 per cent exceedance flows for the River Mimran in 1994. How do these compare with the long-term average flows?

5 One of the key features of a system is that a change in one component will have a 'knock-on' effect through the system. Apply this understanding to the drainage basin in Figure 2.8. Read Section 2.2 carefully then suggest how each of the following changes might affect how the system works.

a Precipitation remains unchanged but infiltration rates increase, e.g. as a result of ploughing

b Temperatures rise, leading to increased evaporation

c Deforestation reduces interception of incoming precipitation.

2.2 Drainage-basin processes

The drainage-basin processes shown in Figures 2.3 and 2.4 are in simple overview form. Figure 2.8 shows in more detail the stores and flows (transfers) within the drainage-basin system.

Precipitation

This is the main input to the system, along with solar energy. The nature of the precipitation is important. Rainfall will follow the routes shown in Figure 2.8, but snow will be stored on the surface for weeks or months. The water may then be released fairly rapidly by spring melting, as with the upper Colorado River (Figure 2.5). How fast the rainfall comes down – its intensity – will influence which routes the water takes, especially whether it can infiltrate. Intense downpours are more likely to cause rapid overland flow. Precipitation inputs to a drainage basin vary over time, but also over space – different parts of a drainage basin will receive different inputs, and seasonal rainfall patterns will influence river flow regimes, e.g. the Nile (Figure 2.5).

Evaporation and transpiration

All water surfaces – river channels, lakes and reservoirs – lose some water through evaporation, especially in the higher temperatures of the summer.

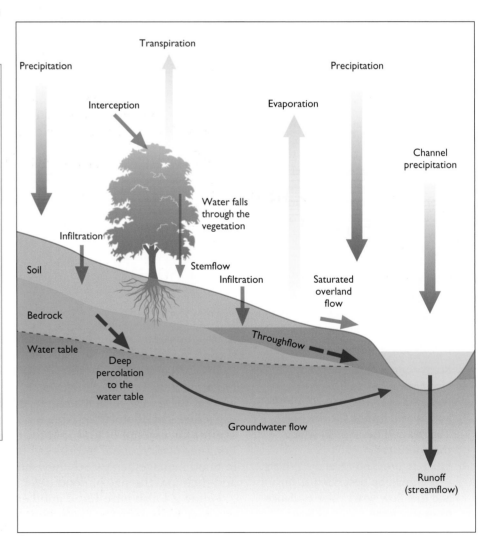

▲ **Figure 2.8** *Drainage basin hydrological processes*

Key Terms

Evapotranspiration: The output from the drainage basin resulting from evaporation and transpiration combined.

Infiltration: The process by which water percolates into the ground surface.

Percolation: The vertical movement of water in the zone of unsaturated rock or soil.

Overland flow: The movement of water over the ground surface.

Throughflow: The downslope flow of water within the soil.

Groundwater: The store of water held beneath the water table.

Interception: The process by which precipitation is trapped on the surface of vegetation.

Interception loss: The proportion of water which evaporates from the surface of vegetation and never reaches the ground.

Stemflow: Intercepted precipitation which runs down the stems of plants to reach the ground.

Throughfall: The precipitation which falls straight through the vegetation (or drips off leaves) and does reach the ground.

When plants and trees transpire (i.e. use up carbon dioxide from the atmosphere and release oxygen) some of the water that they have taken up from the soil store is lost from the stomata in their leaves. This is called **transpiration** and in well-vegetated drainage basins, is a more significant output because more water is lost this way. Like evaporation, transpiration rates are highest at high temperatures and when the humidity of the air is low. As temperatures are lower in winter, the evapotranspiration outputs will be much reduced compared with the summer.

The amount of vegetation cover and the types of plants and trees in an area influence the transpiration rates. In Britain, the natural vegetation is largely deciduous, and growth is reduced in the winter, as is the rate of transpiration. It is easy to see, therefore, how human activities such as deforestation can have a huge effect on the ways water moves through a drainage basin.

A useful concept in understanding water sources in the drainage basin is **potential evapotranspiration** (PET). This is a figure, usually in millimetres, and is the water loss which would occur by evapotranspiration assuming that the water was there to evapotranspire. PET is, therefore, the most water that could be lost from a particular environment. If this is greater than the water inputs, then there will be a water deficit and the water stores will start to empty (see Figure 4.31).

Interception, stemflow and throughfall

Vegetation surfaces, leaves, stems and branches can catch falling precipitation – this is called **interception**. Some of this will evaporate and become an output from the drainage basin. This output is called the **interception loss**. Interception can be high enough for light summer rainfall not to reach the ground at all. If the rain is heavy, or lengthy, water will run down the leaves, stems and branches, and will either infiltrate or move as **overland flow**. The balance between the amount of water which is lost by evaporation and that which moves as **stemflow** is important. **Throughfall** is the rainfall which reaches the ground, either by falling straight through the vegetation or by stemflow. Throughfall and interception loss added together make up the total rainfall on the top surface of the vegetation cover.

Human activity often changes the vegetation cover by replacing forest or natural grassland with vegetation which may not cover the ground surface fully. This will have a major impact on the amount of interception that takes place.

Overland flow

Overland flow is very important because water flowing over the ground surface reaches the river channel much more quickly than water flowing through soil and rocks. The more overland flow that takes place, the faster the river level rises and the more likely flooding is to occur (Figure 2.9). Overland flow happens for a combination of the following reasons:

■ The ground surface is relatively impermeable and the water cannot infiltrate as a result of, for example, urban development or soil compaction.
■ Vegetation cover has been removed, reducing the **infiltration** rates and the interception loss. The extra water runs over the ground surface.
■ When the soil store is already full of water, the term used is saturated overland flow. The soil store is also more likely to be full at the bottom of a slope as water infiltrates and runs down the slope within the soil.
■ The rainfall comes down faster than it can infiltrate the soil. This can occur with intense summer convectional storms.

Key Terms

Porosity: The percentage of air spaces in a certain type of soil or rock indicating how much water it could contain.

....

Permeable: A property of soils and rocks indicating their capacity for transmitting water, because of their porosity and/or perviousness.

....

Pervious: The property of rocks indicating their ability to allow water to pass through their joints and bedding planes.

....

▲ **Figure 2.9** *A flash flood in southern Iran fills a desert wadi, carving out a new route after the first heavy rainfall in 50 years.*

6 Explain briefly, the flash flood of Figure 2.9 in terms of the drainage basin system shown in Figure 2.8.

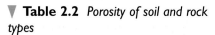

▼ **Table 2.2** *Porosity of soil and rock types*

Rock	Percentage porosity
Soils	57
Clay	50
Silt	43
Sand	35
Gravel	35
Sandstone	28
Limestone	18
Shale	10
Crystalline rock	5

Infiltration

Infiltration is where water soaks into the soil surface. This is a very important process, because water entering the soil will move slowly through the drainage basin to the river channel, whereas water which does not infiltrate will flow relatively quickly over the surface to a river as overland flow (Box 1).

Subsurface water

Subsurface water is water which infiltrates the soil and moves downwards through cracks and pore spaces. Deeper in the soil, there are fewer pore spaces due to the weight of the overlying soil. Gravity forces the water to move downhill through the soil, by **throughflow** (Figure 2.10), towards the river channels at the bottom of the slopes. Some water moves deeper through the ground into the rock layer beneath by deep **percolation**. Rocks vary greatly in the amount of water they can hold. A rock that holds water in the pore spaces between the rock grains is called porous. **Porosity** is the amount of air spaces in a rock or soil, and is a measure of the amount of water a rock could contain (Table 2.2). This water forms the **groundwater** store (Figure 2.10). Limestone, chalk and sandstones are very porous and contain a lot of groundwater. Water-bearing rocks like these are called **aquifers**.

Box 1 Factors affecting the amount of infiltration

■ **Intensity of precipitation:** The faster the rain falls, the less likely all the rain is to infiltrate.
■ **Time:** If rainfall has been over a long period of time, the soil store will have filled up. No more water will be available to infiltrate until this water has drained away. This effect is significant in Britain in winter when the pre-existing water levels are high or the store is nearly full before the next rainfall starts.
■ **Soil:** Soil which is more loosely packed, such as sandy soil, has more air spaces available to allow infiltration. Closely packed soil particles, such as clay, have fewer and smaller air spaces between them, and less room for water to infiltrate. In most soils, the speed of infiltration gets less with depth as the weight of the overlying soil squeezes out the air spaces.

■ **Vegetation cover:** The roots of plants help to create air spaces in the soil, which allows water to infiltrate.
■ **Depth of the water table:** If the water table is near to the surface, the soil store is already full so very little water can infiltrate.
■ **Human activity:** Clearing land of vegetation reduces the number of roots which penetrate the soil so that infiltration becomes less. Agricultural practices such as over-grazing by cattle, poor ploughing techniques, or bare soil tend to compact the soil and reduce infiltration. Increasing development of urban areas means that much of the soil is covered with impermeable concrete and tarmac. This completely stops the water from reaching the soil.
■ **Slope:** Water will run down a steep slope more quickly and will have less time to infiltrate.

► **Figure 2.10** *Subsurface water stores and flows. The water varies in height with seasonal variation of rainfall inputs. In Britain the groundwater store is depleted by lower inputs in the summer, and by a constant seepage into river channels. In winter, higher precipitation inputs and reduced evapotranspiration outputs allow more water to reach the groundwater store and recharge it.*

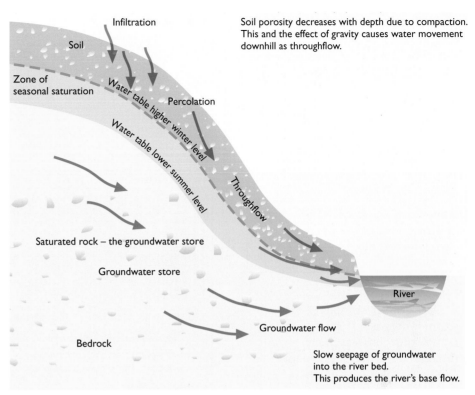

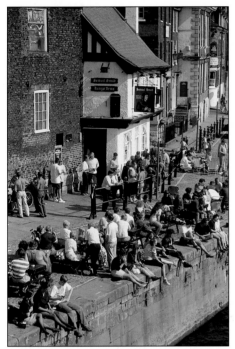

▲ **Figure 2.11a** *At discharge levels which can be contained within the channel, rivers offer many advantages to human activity – transport, water supply and recreation.*

▼ **Figure 2.11b** *When discharge levels rise and water overflows the channel as a flood, rivers become dangerous to life and property.*

Other rocks have closely packed or interlocking grains, and water can only be stored or move through the joint, cracks and bedding planes. These rocks are called **pervious** rocks. Porous and pervious rocks are called **permeable** rocks if the pores and cracks actually join up and let water pass through. Some rocks are impermeable – they do not let water through at all. They can stop the movement of water in the groundwater store, and then they are known as **aquicludes**. A few impermeable rocks, most notably clay, have pore spaces which allow water to enter, but the pores do not connect up and water cannot pass through.

2.3 The storm (flood) hydrograph

The river regime and discharge data considered in Figures 2.5 and 2.6 show average figures for each month of the year, with the record of many years averaged out as well. River flow is much more variable in reality, and river levels can change on a daily, weekly and even hourly basis (Figure 2.11a and 2.11b). The processes operating in the drainage basin – overland flow, infiltration, throughflow and groundwater flow, all transfer water to the channel. They all transfer water at different speeds. Water flowing over the ground surface as overland flow or through very porous soils as fast throughflow, transfers water to the river channel very quickly. Thus, rainfall inputs are added quickly to the river channel, the river level rises, and flooding is more likely. The flow of water through rocks is very slow and it can take weeks for groundwater to reach the river channel. This slow movement is more constant than other processes, and feeds water into the river to produce base flow. This explains why rivers can continue to flow even if it has not rained for weeks. Water flows at an intermediate speed by throughflow, or subsurface flow. These intermediate processes are grouped together as the term **interflow**. These short-term changes in the river flow result from changes in the inputs of water to the water channel. River flow is measured using a term called discharge. This figure gives the volume of water, usually in cubic metres, passing a given point (gauging station) in a set time, usually seconds, i.e. cubic metres/second.

► **Figure 2.12** *The storm hydrograph*

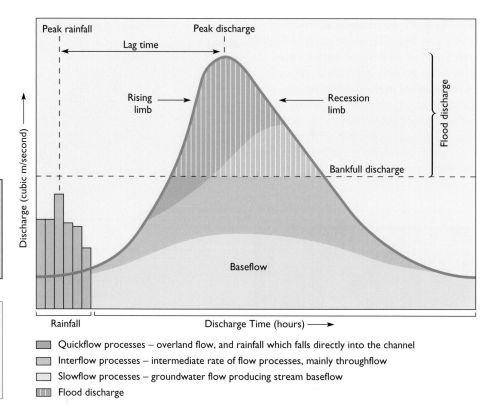

- ▭ Quickflow processes – overland flow, and rainfall which falls directly into the channel
- ▭ Interflow processes – intermediate rate of flow processes, mainly throughflow
- ▭ Slowflow processes – groundwater flow producing stream baseflow
- ▥ Flood discharge

Key Terms

Base flow: The low level of stream discharge fed largely by groundwater flow.

7 During which sections of the storm hydrograph (Figure 2.12) do (a) quickflow processes, and (b) interflow processes dominate? Explain your answers briefly.

The graph of a river which shows short-term changes in discharge is called the **storm hydrograph** (Figure 2.12). Since river flooding is an important hazard to human activity, an understanding of the nature of the storm hydrograph is vital for flood management and control (Figure 2.13).

Usually, the **rising limb** of the storm hydrograph is steeper than the falling one (**recession limb**) because during the initial stage of a storm, both the quickflow process and the interflow processes feed water into the river channel, so the level of the river rises quickly. The interflow processes are slower but continue to feed water into the river channel for longer after the rainfall stops, giving a more gentle fall on the recession limb.

► **Figure 2.13** *Flooding of rivers can have a disastrous effect on agriculture. Remember, however, that floods may leave valuable fertile sediments across floodplains.*

The amount of water that the river channel can hold is called the **bankfull discharge**. Above this amount of water, the river will overflow its banks and flood. The area above the bankfull discharge under the discharge curve represents the amount of water that will spill on to the river's floodplain. This is the **flood discharge**.

Determining the shape of storm hydrographs

The nature of the storm hydrograph mainly depends upon the relative importance of the quickflow and slowflow processes. A range of factors associated with the hydrological processes cause variations in the form of the storm hydrograph. There are summarised in Table 2.3.

► **Table 2.3** *Factors affecting the nature of the flood hydrograph*

Factor	Hydrograph dominated by quickflow processes	Hydrograph dominated by slowflow processes
	Steep rising limb / Baseflow (Discharge m³/s (Q) vs Time)	Baseflow (Discharge m³/s (Q) vs Time)
Climatic factors		
Precipitation	High-intensity rainfall	Low-intensity rainfall
	Large amounts of rainfall	Small amounts of rainfall
Snow	Fast snow melt	Slow snow melt
Evapotranspiration	Low rates of evapotranspiration outputs e.g. winter in Britain	High rates of evapotranspiration outputs eg. summer in Britain
Soil characteristics		
Soil moisture	High antecedent soil moisture conditions following prolonged rainfall	Dry soil – the soil store can hold much more water
Permeability	Impermeable soil	Permeable soil
Drainage basin characteristics		
Drainage density	High **drainage density** (large number of streams per km)	Low drainage density (small number of streams per km)
Slopes	Steep slopes	Gentle slopes
Rock type	Impermeable rocks – e.g. clay, crystalline rock	Permeable rocks – e.g. chalk and sandstone
Vegetation cover	Little vegetation cover Lack of interception and root development to open up the soil.	Forest and woodland intercept much rainfall, and root development encourages infiltration
Soil depth	Thin soil – e.g. upland areas allow little infiltration	Deeper soils provide a large soil store – e.g. slope bottoms and lowland areas.
Water stores	Lack of lakes and backwater swamps	Lakes and backwater swamps act as water stores, and slow the movement to the channel
Urban development	Urban development creates impermeable surfaces and water quickly reaches the channel via storm drains	Rural land uses intercept more precipitation and have more permeable land surfaces
Agricultural practices	Poor agricultural practices – poor soil structure, trampling by animals	Good agricultural practices which encourage soil aeration and protect the soil surface

Key Terms

Hydrograph: The level of discharge of a river plotted over time (often 1 year).

Bankfull discharge: The level of water which can be contained within the river channel before it overflows.

Flood discharge: The amount of water which overflows the banks of a river (i.e. the discharge above the bankfull discharge).

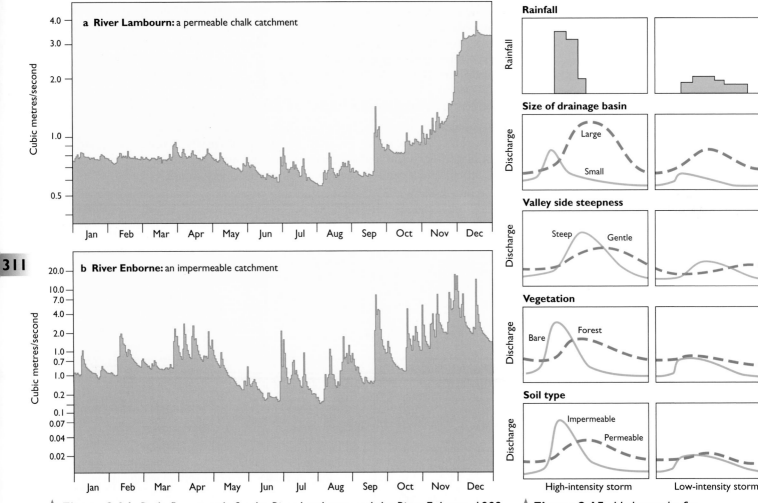

▲ **Figure 2.14** *Daily flow records for the River Lambourn and the River Enborne, 1992*

▲ **Figure 2.15** *Hydrographs for different catchments under two different storm conditions*

For example, differences in a single variable may be significant, as shown in Figure 2.14. However, most hydrographs are controlled by the interaction of several variable factors, as rock type and land uses vary within each drainage basin and slopes tend to be steeper in the upper reaches. Figure 2.15 shows how different rainfall events interact with catchment characteristics to produce a range of hydrographs.

Seasonal influences

The hydrograph can also vary at the same gauging station at different times of the year. Time of year is important in areas such as the British Isles with a well-defined seasonality of climate. A common cause of this is high **antecedent soil moisture** as a result of prolonged rainfall, particularly in the winter. Water from rainstorms occurring soon after another will not be able to infiltrate, and will reach the channel as overland flow. The result will be hydrographs which are intermediate between the typical hydrographs produced by quickflow or slowflow processes.

Characteristics of drainage basins

It is important to remember that the storm hydrograph is for one point on the river channel within a drainage basin. The data is recorded at a gauging station. The water flow will be affected by the characteristics of the drainage basin and the processes operating upstream from this point. The size and shape of the drainage basin can affect how long the water takes to reach the river channel or the gauging station from upstream. This will affect the shape of the storm hydrograph (Figure 2.16).

8 Study the hydrographs for the River Lambourn and the River Enborne. Note the difference in scale.
a How do the two rivers compare in terms of the number of peaks?
b Compare the response of the rivers to the mid-September rainfall input.
c Why do both rivers show higher than average discharges in November and December.

9 Study Figure 2.15. Describe and fully explain the variations in the storm hydrographs shown. Use hydrological terms as fully as you can.

Human activity

Human activity has a major impact on the nature of the storm hydrograph. Activities which change vegetation and land use will affect the processes operating and the routes that water takes to the channel. River water may be abstracted for urban areas and irrigation, and flow levels will be reduced. Water may also be put back into the channel from towns and factories and augment the flow downstream. Flood-control dams and reservoirs are intended to manage flood levels, and thus have a significant impact of the form of the storm hydrograph (Figure 2.17).

10 In Figure 2.16 it is assumed that precipitation is the same over the whole of each drainage basin.
a Draw the hydrographs which would result from rainfall being heavier in the northern half of the basins than in the south.
b Describe and explain the differences between your hydrographs and these in the diagram.

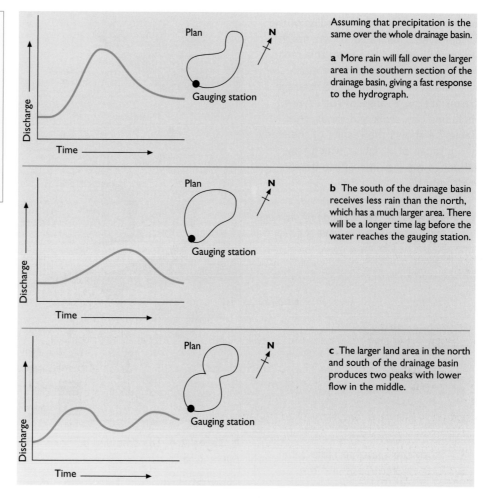

Assuming that precipitation is the same over the whole drainage basin.

a More rain will fall over the larger area in the southern section of the drainage basin, giving a fast response to the hydrograph.

b The south of the drainage basin receives less rain than the north, which has a much larger area. There will be a longer time lag before the water reaches the gauging station.

c The larger land area in the north and south of the drainage basin produces two peaks with lower flow in the middle.

► **Figure 2.16** *Storm hydrographs and drainage basin shape*

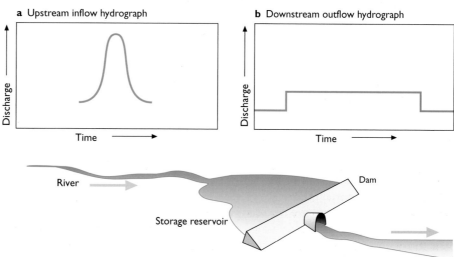

► **Figure 2.17** *Hydrographs produced by (a) unregulated and (b) regulated flow*

EXAMPLE: South-East England

Flood responses in small drainage basins

A study by the Institute of Hydrology in1994 investigated how quickly rainfall reached river channels (the lag times) from a range of small drainage basins in south-east England. All the drainage basins lie within a 75-km radius of the Chenies weather radar, to the north-west of London. Data from this radar was used to determine how much rainfall fell in the basins. The drainage basins are on a range of rock types including clay, chalk and sandstones. Table 2.4 shows the results of the study and a range of variables which may influence the lag time of the flood hydrograph.

▶ **Figure 2.18** *The location of drainage basin catchments listed in Table 2.4*

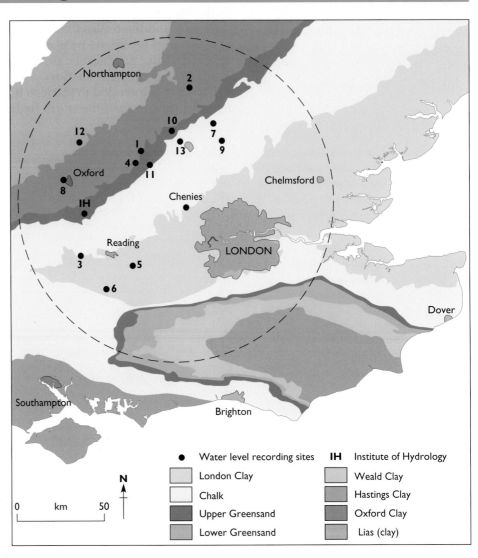

I I Study the values in Table 2.4.

a Using scattergraphs and the Spearman's Rank Correlation Coefficient, investigate the relationship between lag time and other variables in the table. Use a data package such as Excel to produce a spreadsheet of these values, if you can.

b For each of the variables compared with lag time, comment upon the degree of correlation. Suggest reasons for the strong and weaker correlations.

c What other factors need to be considered to explain the correlations that you have investigated?

▼ **Table 2.4** *Lag time and other variables for the small drainage basins located in Figure 2.18*

Catchment (drainage basin)	Lag time (hours)	Area (m^2)	Stream slope (m/km)	Percentage urban	Percentage forest
1 Aylesbury	0.5	1.74	5.6	63	2
2 Bedford	14.9	22.92	1.6	4	6
3 Beenham	2.9	3.4	20.3	2	42
4 Easton Maudit	8.8	15.76	13.8	1.7	17
5 Holme Green	3.7	9.81	13.9	15.4	39
6 Hook	6.1	2.47	14.14	8.4	19
7 Letchworth	0.8	8.52	16.1	84.5	3
8 South Hinksey	5.3	1.49	28.8	0.5	6
9 Stevenage	1.0	4.14	16.86	49.2	3
10 Toddington	1.8	0.88	38.77	38.2	0.6
11 Wingrave	7.5	5.85	12.49	0.4	5
12 Bicester	0.48	1.46	N/A	65.2	0.3
13 Luton	1.1	9.05	N/A	63	3

2.4   Changing discharge

Flooding

The discharge of a river is highly variable. Periods of high discharge can result in flooding. Flooding occurs frequently in rivers and is an important part of channel evolution under natural conditions. However, floodplains are used for settlement, agriculture and as transport corridors. This means that flooding can conflict with human activity. Some land uses are compatible with flooding, and many traditional agricultural systems have developed to gain positive benefits from flooding. For example, crops may be planted after annual flooding when soil moisture levels on the floodplain are high e.g. on the floodplain of the River Nile. Increasingly, however, flooding conflicts with human activity as traditional systems break down and population growth increases the pressure on the land.

River flooding results from two main types of events:

Flash flooding

This results from intense downpours of rainfall. The **lag time** on the flood hydrograph is very short (hours or minutes) and the effects can be devastating, but usually local in scale. Flash floods are difficult to predict, warnings are seldom possible (Look again at Figure 2.9, page 46).

Regional flooding

This usually results from high rainfall over a long time period. Discharge increases downstream as water finds it more and more difficult to infiltrate the saturated soil. Snowmelt may add to, or cause, this type of flooding. Larger areas, or even the whole drainage basin, can be affected as the floods across eastern England in 2000 illustrated (Figure 2.19, and see the Autumn floods in England and Wales example). It is much easier to predict this type of flood and so there is more time for warnings and action to be taken to reduce the loss of life or damage to property and land.

Predictions can be made about the likely occurrence of a flood, based upon statistical observations. Records of maximum discharges are made at a gauging station to produce a maximum discharge series. These are then

▼ **Figure 2.19** *Flooding in Malton, England, 2000*

▼ **Figure 2.20** *Flood recurrence intervals, plotted against discharge, for the River Bollin, Cheshire*

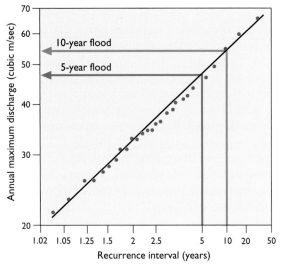

The recurrence interval shows the average interval between flood events. Thus, for a discharge level of 54 m³/sec, the average recurrence interval is 10 years from the available data. Another way of looking at this is that the discharge of 54 m³/sec would have a 10 per cent probability of occurring in any given year.

placed in rank order and the average interval between two floods of equal discharge magnitude is calculated (Figure 2.20). This is called the **recurrence interval** and is used to predict how often floods of that magnitude will happen. When flood protection measures are being planned, engineers use the recurrence interval to decide the height of flood walls, or the size of overflow channel sizes. It is important that it is understood that any flood protection will be effective up to that level, but not above it. It would be impossible, in terms of cost and environmental impacts, to stop *all* floods (see Section 2.11).

Regional droughts

Low discharge levels can also impact upon human activity. During periods of below-average rainfall, discharge will be reduced to baseflow levels and only maintained by groundwater flow. This can cause problems for irrigation and water uses, and reservoir levels will fall. Prolonged drought may result in the river drying up completely with a resulting loss of amenity value, wetland habitats and abstraction for water use. Water-use restrictions are used as a management strategy during periods of low flow – for example, during the UK droughts of 1976, 1984 and 1995. The 1995 drought resulted in 20 million people being affected by hosepipe bans. In some areas, water rotas and standpipes were introduced. By late August 1995, water storage reservoirs in the Pennines and the Lake District had declined to below 20 per cent of capacity.

EXAMPLE: The autumn floods in England and Wales, 2000

During the late autumn of 2000, extensive areas of England and Wales experienced their worst floods for many years, following unusually heavy rains (Figure 2.21). Three types of location were particularly vulnerable to flooding and property damage:

■ steep catchments with high density stream networks, giving quickflow runoff, e.g. Wales
■ river basins in eastern England with upper catchments in hills running into lower courses with broad floodplains, e.g. rivers such as the Tyne, Tees, Ouse and Trent, with headstreams in the Pennines. To the west, the lower R.Severn suffered flood surges caused by run-off from the Welsh uplands.
■ regions with high population densities, e.g. southeast England, where urban development has spread across valleys and floodplains.

In some places, several factors combined to intensify the flooding. East Sussex had prolonged rainfall onto steep slopes with extensive settlement development.

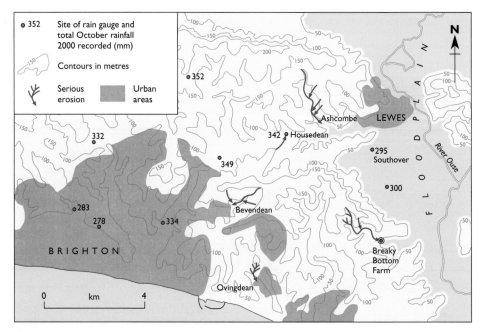

▲ **Figure 2.21** *The South Downs around Brighton and Lewes showing sites of serious erosion and rainfall totals for October 2000*

▼ **Table 2.5** *Rainfall (mm) at Southover, Lewes, 2000–01 and long-term average*

	Oct	**Nov**	**Dec**	**Jan**	**Feb**	**Mar**
Long-term average	93	98	92	92	59	69
2000–01	295	230	139	139	109	133
Daily max, 2000–01	89.8	42.7	30.0	29.0	42.0	22.6

▼ Figure 2.22 *Autumn 2000 floods in England and Wales*

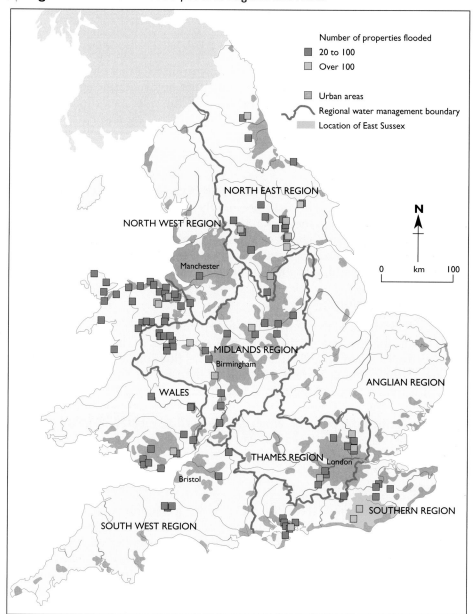

Number of properties flooded
■ 20 to 100
□ Over 100

□ Urban areas
〜 Regional water management boundary
▨ Location of East Sussex

NORTH EAST REGION

NORTH WEST REGION

Manchester

N

0 km 100

MIDLANDS REGION

Birmingham

WALES

ANGLIAN REGION

THAMES REGION London

Bristol

SOUTH WEST REGION

SOUTHERN REGION

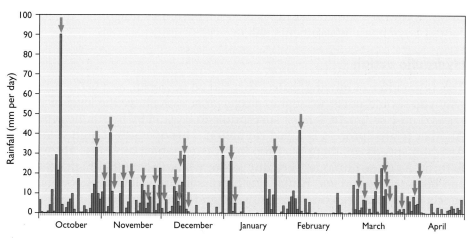

▲ Figure 2.23 *Daily rainfall at Southover, Lewes, October 2000 to April 2001. Timing of floods at Breaky Bottom farmhouse is shown by arrows.*

► Figure 2.24 *Flooding in Lewes, 2000–01*

The South Downs of East Sussex

The South Downs are a chalk upland dissected by steep-sided valleys. The Brighton area is typical of this topography (Figure 2.22). During October and November 2000 the district received over 500 mm of rain, almost three times the long-term average (Table 2.5). These unusually wet conditions continued through the winter, causing recurrent flooding (Figure 2.23). In mid-October the River Ouse flooded for the first time in 40 years, with water levels rising by up to one metre an hour, seriously affecting the town of Lewes which sits across the floodplain (Figure 2.24).

The principal causes of the floods were:

■ prolonged above-average rainfall caused by a series of vigorous depressions

■ the ground became saturated, decreasing infiltration capacity and accelerating quickflow surface run-off (or overland flow) into stream channels

■ increased run-off and sediment load combining to reduce the channel capacity of many rivers, increasing the likelihood of bankfull conditions

■ settlement development and channel management in valleys and floodplains modify streamflow and reduce the opportunities for traditional flood discharge stores across floodplains

■ changes in farming techniques have led to an increase in autumn-sown arable crops. Through the late autumn and winter, extensive areas of sloping land are largely bare of vegetation cover, leaving the land vulnerable to quickflow run-off and soil erosion.

Key Terms

Sediment load: The amount of sediment (sand, soil and rocks) carried along by a river.

2.5 Fluvial processes

Sediment erosion

A river can erode the soil and rocks which form its channel banks and bed, by the processes of corrosion, abrasion, attrition and hydaulic action. This eroded material is transported downstream by the river as sediment load, and adds to the **sediment load** derived from weathering and mass-movement processes on the valley sides.

Corrosion

This is the chemical weathering of minerals in rocks in contact with the river water. The minerals in rocks are slowly dissolved by the river water and this eventually leads the rock particles to break apart. This process will be most effective where there is a fast-flowing river which is not already saturated with minerals. Rock type is also important. Limestones, in particular, are susceptible to this kind of erosion.

Abrasion (corrasion)

This is where rocks in the sediment load which is being carried along by the river hit the rock materials on the bed and banks of the river – wearing them away. This is most effective if the river is flowing at high velocities and the particles being carried are made of hard rocks. Thus, abrasion is most effective in times of high flow and flood conditions (Figure 2.25). At these times, larger sediment particles, including boulders, are moved. Abrasion is the main process which causes vertical erosion. Rocks and boulders swirling on the river bed can produce near-circular holes called potholes by this process of abrasion.

Attrition

Attrition is a process whereby the rocks in the sediment load erode by colliding with each other as they are carried along the river. The result is that the sediment load becomes more rounded and smaller in size. Upstream sediments tend to be larger and more angular than sediments in the lower reaches since attrition hasn't been acting on them for very long. As a general rule, sediments become smaller and rounder downstream. This applies to a sediment particle on its journey from upstream to the lower reaches. Remember, sediment can also be added to the stream all along its course – if the river flows across a hard rock outcrop in the lower course, large angular sediments will be added.

Hydraulic action

This is the force of moving water. This can be powerful, and is the reason why you should never wade into a river or stream above the level of your knees. Loose sediment is most susceptible to erosion by hydraulic action. Another related process is called cavitation. This operates at high velocities. As water speeds up, there is a drop in pressure which can cause air bubbles to form. As these implode, very tiny jets of water shoot off at high speeds and hammer the bed and banks. The process is not well understood, but its effects have been seen on dam spillways and other structures.

▼ **Figure 2.25** *The North Sannox stream in Arran, at low flow. Rock outcrops and boulders are visible. At times of high flow, the boulders will be transported by the stream, adding to the abrasion of the river bed and banks.*

Sediment transportation

A river transports its sediment load in a variety of ways (Figure 2.27). The methods of transport are also used to describe the various loads of a river, i.e. the bed or traction load, suspension load, dissolved load, and wash load (for very fine particles held in suspension all the time). The balance between these processes is controlled by the relationship between the energy available in the stream and the nature of the channel material.

The sediment load varies from river to river, along the course of one river or in the same place at different times. This is because the velocity of the water and hence the energy available is crucial in determining the way that sediment is transported. The relationship between erosion, transport and deposition of sediment is complex and can be shown by the Hjulstrom diagram (Figure 2.26). This is based on experimental work rather than natural channels, but it shows the principles involved. Entrainment is the process of starting the particles of sediment moving – the opposite of settling.

- High velocities result in sediments being transported in the river flow, while low velocities result in sediment being deposited.
- Medium sand (0.25–0.5 mm in diameter) is moved at the lowest velocities.
- Larger, heavier sediments need higher velocities to start movement.
- Silt and clay need higher velocities than their size would suggest because they are cohesive (they stick together) and so, in fact, are bigger than they should be.
- Once set in motion, fine particles can be transported even if the velocity falls.
- Larger coarse particles are deposited rapidly as velocity falls. In channels with mainly boulders and gravel, transport only occurs at high flows.
- In natural channels the situation will be more complex. For example, small particles may be sheltered by larger particles and therefore they are not moved.
- Velocity of flow is variable across and vertically in a natural channel, and this will affect the processes.
- Sediment transported under lower-flow velocities, as bed load, may become suspended load under high velocities.

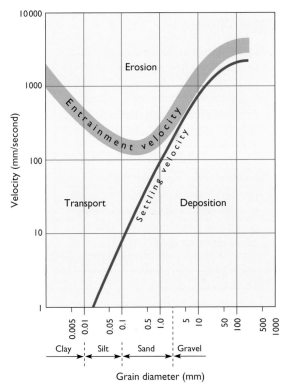

320

Reading the graph, the velocity scale and the sediment size are logarithmic. For example, a grain of sand 1 mm in diameter would be eroded at a velocity above 400 mm/second; entrained between 400 mm and 200 mm/second. It would be deposited at velocities of 90 mm/second or below.

▲ **Figure 2.26** *Velocity and particle movement (After: Hjulstrom)*

12

a At what streamflow velocity would a sand grain of 0.5 mm diameter become entrained?

b At what velocity would a gravel particle of 5.0 mm diameter settle on the channel bed?

▼ **Figure 2.27** *A channel cross-section showing the modes of transport of sediment*

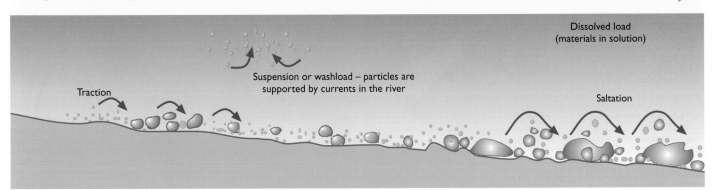

Traction – sliding and rolling of particles as they are pushed along by the drag effect of the moving river

Very fine particles will be held in suspension while there is some river flow

Saltation – the flow of water over projecting grains causes a fall in pressure above the grain. This can allow the grain to rise from the bed and move in a series of hops. As the particles return to the bed they help the process by dislodging other particles.

Direction of flow

Dissolved load (materials in solution)

Suspension or washload – particles are supported by currents in the river

Saltation

Traction

Key Terms

Equilibrium: The balance of inputs and outputs in a system.

Deposition: The process whereby sediment load is laid down as a layer of material (alluvium, soil, sand, etc.)

Graded long profile: The long profile which would be achieved if energy was balanced along a river's course and the river was able to carry all of its load.

13 Look carefully at Figure 2.25.
a The stream is in a low-energy, low-flow condition. Suggest what may be happening in terms of erosion and sediment transport.
b What might take place during a high-energy, bankfull episode?

A river must have energy available to perform the work of transporting sediment. A river's power is the energy available to overcome friction and to move sediment. This will be greatest in rivers with high discharge, high downstream gradient, and an efficient channel (see below). If the river has just enough energy in a particular section (reach) to transport the sediment available, then it is in equilibrium. If it has more power available than needed to transport its sediment load, the river will have excess energy and will erode its channel. If there is less energy available than is needed to transport its load, then deposition of the sediment will result.

Two further terms are useful in describing sediment transport. The **competence** is the maximum sediment particle size that can be carried at a particular velocity. The **capacity** is the total load of sediment that the river can carry. As we have seen, these will vary with both discharge and velocity.

Deposition

The velocity at which a sediment particle drops to the channel bed is called the settling **velocity**. This depends upon the size and shape and density of the sediment particle. **Deposition** may be temporary on the channel bed and the sediment may be moved again at a time of higher flow. In other situations there is a net deposition of sediment, and a deposition landform results, e.g. floodplains and point bars on the inside of meander bends.

2.6 The sediment load of a river

The sediment load of a river comes from two sources:
■ Erosion of the channel bed and banks.
■ Material which has been produced and moved downslope by weathering and mass-movement processes, and accumulates at the slope base. This material is added to the channel as the banks are laterally and vertically eroded, or in times of higher discharge when the river is high enough to flow over the sediment and pick it up.

The sediment load is very important because it affects the energy levels of the river. Energy is used up in transporting sediment as well as in water flow. If the sediment load increases too much, then the river does not have enough energy to carry it and deposition results and the river reaches **equilibrium** again. If the sediment levels are reduced, the river will have more energy, which it uses to erode.

▼ **Figure 2.28** Factors which affect the sediment load of a river

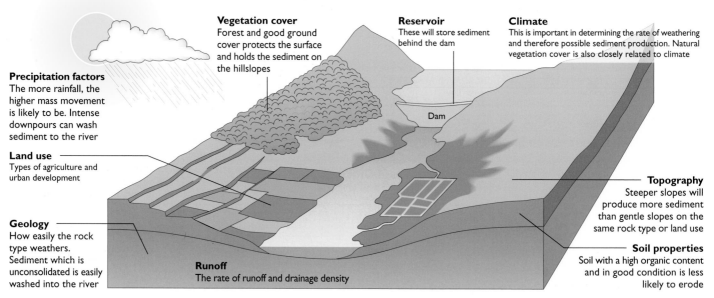

Precipitation factors
The more rainfall, the higher mass movement is likely to be. Intense downpours can wash sediment to the river

Vegetation cover
Forest and good ground cover protects the surface and holds the sediment on the hillslopes

Reservoir
These will store sediment behind the dam

Climate
This is important in determining the rate of weathering and therefore possible sediment production. Natural vegetation cover is also closely related to climate

Land use
Types of agriculture and urban development

Dam

Topography
Steeper slopes will produce more sediment than gentle slopes on the same rock type or land use

Geology
How easily the rock type weathers. Sediment which is unconsolidated is easily washed into the river

Runoff
The rate of runoff and drainage density

Soil properties
Soil with a high organic content and in good condition is less likely to erode

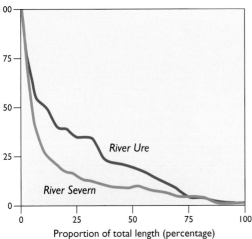

Figure 2.29 *The long profiles of the River Severn and the River Ure*

The amount and nature of the sediment load depends on several factors:

The sediment load is produced by weathering of rocks on the hillslope. Therefore, the type of rock in the drainage basin is important, as is the climate. The sediment from the hillslopes has to reach the channel. This depends on the vegetation cover, the rainfall intensity, the gradient and the land uses.

Figure 2.28 summarises the factors which interact to determine the sediment load of a river. For any one drainage basin, the range of factors can be complex. For example, a hot wet climate may produce a high amount of sediment due to fast rates of weathering, but this climate will also result in a dense forest cover which would protect the sediment from rainfall (Look again at Section 1.5, pages 24–32). Sediment is also stored in the drainage basin for varying lengths of time, with only a small proportion reaching the river.

Human activity is also an important influence on the sediment load of rivers. For example, dams reduce sediment loads downstream as sediment is stored in the reservoir. Sediment yields are increased by cultivation, removal of natural vegetation, mining activity and early stages of urban development.

The long profile of a river

A river uses energy to carry out its 'work' of erosion, transport and deposition of sediment and to move the water in its channel. This energy is produced when water flows down a slope, so the height of a river above sea level determines this. The lowest level of a river is called the **base level**, and for most rivers it is sea level. The long profile of a river is a graph drawn along the course of a river from the source to the mouth. The axes may be callibrated in conventional units of height and length or, alternatively, as percentages of the river's total drop (relief) and length (Figure 2.29). The study of river long profiles shows that they have a concave shape, with a steeper upper reach and a gentler lower reach. River processes are related to long profile because every river is trying to achieve a smooth, concave, **long profile**. This ideal profile is called the **graded long profile** and in this ideal situation, the available energy and the river processes will be in equilibrium. Where this is not the case, the river will be working to smooth out its long profile by erosion, transport or deposition, as shown in Figure 2.30.

14 Study Figure 2.29.
a Which river has the long profile that might be described as 'nearly graded'? Explain your answer.
b At what proportion of the total length along its course will the river Ure be vertically eroding? Why?

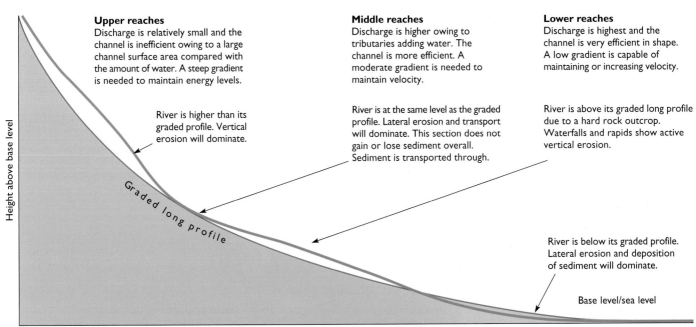

Figure 2.30 *How a river smooths out its long profile*

Upper reaches
Discharge is relatively small and the channel is inefficient owing to a large channel surface area compared with the amount of water. A steep gradient is needed to maintain energy levels.

River is higher than its graded profile. Vertical erosion will dominate.

Middle reaches
Discharge is higher owing to tributaries adding water. The channel is more efficient. A moderate gradient is needed to maintain velocity.

River is at the same level as the graded profile. Lateral erosion and transport will dominate. This section does not gain or lose sediment overall. Sediment is transported through.

Lower reaches
Discharge is highest and the channel is very efficient in shape. A low gradient is capable of maintaining or increasing velocity.

River is above its graded long profile due to a hard rock outcrop. Waterfalls and rapids show active vertical erosion.

River is below its graded profile. Lateral erosion and deposition of sediment will dominate.

Base level/sea level

Graded long profile

Height above base level

Distance downstream from river source

Figure 2.31 *Gulfoss waterfall (Iceland). There are two steps to the waterfall due to harder basalt lava outcrops and softer sedimentary layers. The overall height is 31 m.*

Figure 2.32 *The gorge cut by the Huítá River below Gulfoss waterfall is up to 70 m deep and 3 km long. This has formed since the ice sheets disappeared 10 000 years ago.*

Figure 2.33 *Cross-section of Gulfoss*

Outcrops of hard rock along the river's course will involve vertical and headward (upstream) erosion for the river to achieve its graded long profile. As the river erodes upstream to remove the resistant rock outcrop, the site is marked by a waterfall (Figure 2.31). Downstream of the waterfall, a **gorge** usually forms by the headward retreat of the rock outcrop (Figure 2.32). The water cascading over the fall is capable of much abrasion and hydraulic action. The result is a **plunge pool** which eventually undercuts the waterfall, resulting in collapse and headward retreat (Figure 2.33).

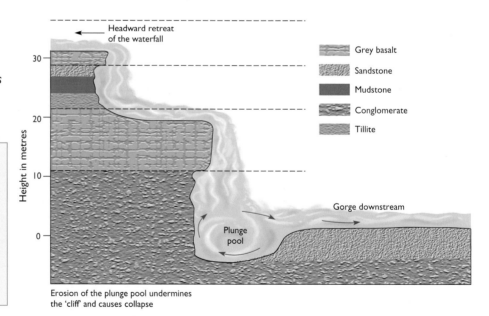

Erosion of the plunge pool undermines the 'cliff' and causes collapse

15 From Figure 2.33.
a Explain how the plunge pool was formed.
b Which rocks appear to be the most resistant and which the least resistant to erosion?
c Why is the valley that is formed by the waterfall retreat so steep-sided?

2.7 River valley form

The river valley is one of the main landforms created by erosional processes. However, the form of a river valley (**cross profile**) is not simply the result of river processes, but rather the interaction of these with weathering and mass movement on the valley slopes (see chapter 1). These processes are influenced by the climate of the area, by the geology and by time.

Key Terms

Cross profile: A cross-section of a river valley and channel.

he relationship between slope processes and river processes

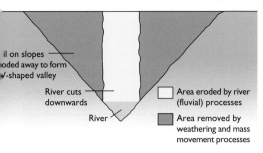

il on slopes
oded away to form
V-shaped valley

River cuts
downwards

River

☐ Area eroded by river
(fluvial) processes

▨ Area removed by
weathering and mass
movement processes

alley form produced by slow weathering and mass-movement
cesses compared with river erosion

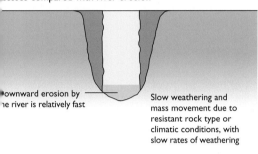

ownward erosion by
e river is relatively fast

Slow weathering and
mass movement due to
resistant rock type or
climatic conditions, with
slow rates of weathering

alley form produced by fast weathering and mass-movement
cesses compared with river erosion

Upper valley slopes have been
exposed the longest and so
have retreated furthest

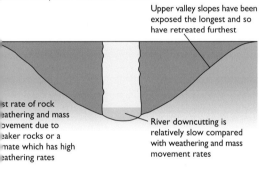

st rate of rock
eathering and mass
ovement due to
eaker rocks or a
mate which has high
eathering rates

River downcutting is
relatively slow compared
with weathering and mass
movement rates

▲ **Figure 2.34** *River valley form*

The form of a river valley is commonly described as V-shaped (Figure 2.34). In a **V-shaped river valley**, a river can only directly erode the area of its channel. The downcutting (**vertical erosion**) of the river creates a slope, allowing weathering and mass movement to occur. If these processes are slow (e.g. in arid or semi-arid climates), or if the rock type is very resistant to erosion, the result will be a valley shape dominated by fluvial processes – steep-sided and gorge-like (Figure 2.34b).

If the rock type is more easily eroded, or the climate allows faster rates of weathering, hillslope processes become more important, and the valley forms a more open V-shape (Figure 2.34c). Humid temperate and tropical climates will encourage this type of valley shape. Soft rocks will produce even more open V-shapes in these areas. Valley profiles become more complex with variations in rock type along the valley profile, or if there are geological structures such as joints and faults which can affect weathering rates The Grand Canyon of the R. Colorado, USA, is an extreme exmple of this action (Figures 2.36 and 2.37). In high latitudes, aspect may also be important: a higher rate of weathering takes place on one side of the valley and may result in asymmetrical valley profiles.

At the valley base there is an important interaction between the river, and the slope processes. The sediment may collect at the slope base where it may be removed from the slope bottom by the river. This process is called **basal removal**. If the river is actively eroding its channel, basal removal will be efficient and valley slopes will tend to be steeper. If the river is not actively eroding, basal removal is slower and the valley profiles will be gentle.

This is most common in lower **reaches**, where **lateral** (sideways) **erosion** dominates and basal removal occurs only along the hillslopes where meanders reach the valley sides (Figure 2.35). In other areas of the valley floor, material will accumulate. Lateral erosion, the movement of meanders downstream and infilling by flood deposits, create a flat valley floor. In general, meanders are best developed in the lower reaches of the river where high discharges produce larger and more meandering channels.

Therefore, the energy of a river at any point along its course will be important. This will also vary in one place at different times as discharge levels vary. Most basal removal takes place during times when a river's energy is much increased. If the river has reduced energy – for example in times of low flow – material will collect at the slope base.

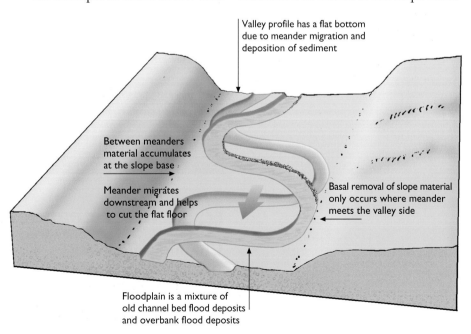

Valley profile has a flat bottom
due to meander migration and
deposition of sediment

Between meanders
material accumulates
at the slope base

Meander migrates
downstream and helps
to cut the flat floor

Basal removal of slope material
only occurs where meander
meets the valley side

Floodplain is a mixture of
old channel bed flood deposits
and overbank flood deposits

► **Figure 2.35** *A valley profile showing the formation of a floodplain*

▶ **Figure 2.36** *Profiles of the Grand Canyon at seven different points along its length, showing how the topography and the width of the chasm are controlled by the types of rocks through which the Colorado River has cut. Cliffs and steep slopes are formed by granites, schists and limestones, while gentler slopes are formed by softer rocks such as shales.*

306

Limestone/sandstone
Shale
Sandstone
Limestone
Shale
Sandstone
Granite/Schist/Gneiss

Three Mile Wash
Soap Creek Rapid
North Canyon Rapid
Vasey's Paradise
Kwagunt Rapid
Lava Canyon Rapid
Bright Angel Canyon

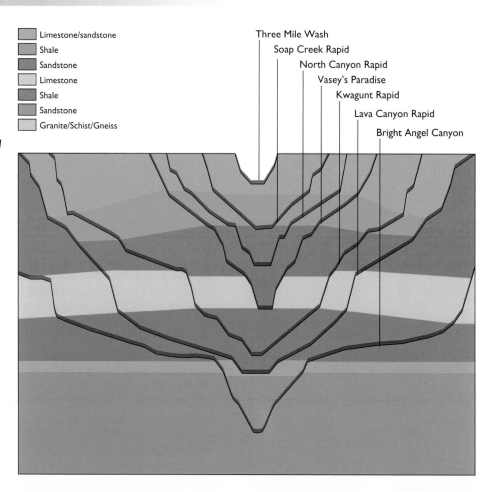

▶ **Figure 2.37** *The lower part of the Grand Canyon, Arizona, with Bright Angel Canyon in the background. Horizontal layers of sedimentary rocks can be clearly seen in the sides of the canyon.*

2.8 River channels

Channel efficiency

The cross-section of a river channel is highly variable along the long profile of a river. The shape of the channel and the depth of flow affect the relative amount of water in contact with the bed and banks, where energy is lost through friction. A channel is described as more efficient if there is a large

Key Terms

Planform: The form in plan view or as seen on a map.

....

Sinuosity: A plan view measure of the extent to which a river meanders.

....

16 Look again at Figure 2.25 on page 56.

a Is the North Sannox stream channel, at this level of discharge, efficient or inefficient?

b Would channel efficiency be different during periods of bankfull discharge?

Explain your answers by referring to Figure 2.38.

area of water in the cross-section compared with the length of the bed and banks. This can be found using the hydraulic radius (Figure 2.38). With a high hydraulic radius, the channel loses less energy through friction with its bed. There is therefore more energy available for the work of erosion and transport, or the channel can maintain its velocity over a more gentle downstream gradient. As a general rule, discharge increases downstream and the channel shape becomes more efficient. This allows the flow velocity to stay the same or increase, even though the gradient is less. This relationship is the biggest influence on the concave long profile of a river.

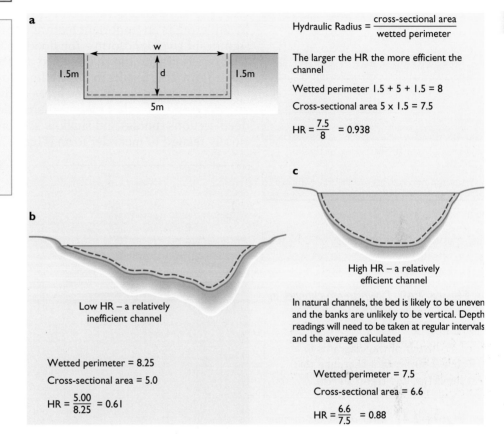

$$\text{Hydraulic Radius} = \frac{\text{cross-sectional area}}{\text{wetted perimeter}}$$

The larger the HR the more efficient the channel

Wetted perimeter $1.5 + 5 + 1.5 = 8$

Cross-sectional area $5 \times 1.5 = 7.5$

$HR = \frac{7.5}{8} = 0.938$

High HR – a relatively efficient channel

In natural channels, the bed is likely to be uneven and the banks are unlikely to be vertical. Depth readings will need to be taken at regular intervals and the average calculated

Low HR – a relatively inefficient channel

Wetted perimeter = 8.25

Cross-sectional area = 5.0

$HR = \frac{5.00}{8.25} = 0.61$

Wetted perimeter = 7.5

Cross-sectional area = 6.6

$HR = \frac{6.6}{7.5} = 0.88$

► **Figure 2.38** *Finding the hydraulic radius of a river*

▼ **Figure 2.39** *The sinuosity index*

Plan view

A

Valley floor distance

Channel length

B

$$\text{Sinuosity index} = \frac{\text{channel length (A–B)}}{\text{valley floor distance (A–B)}}$$

Planform of channels

Water under natural conditions rarely flows in a straight line. Even in rare straight sections of river channels, the line of maximum river depth and velocity (the **thalweg**) follows a curved (sinuous) path. The **sinuosity index** is a useful way of describing channels (Figure 2.39). The sinuosity index will be 1 for a straight channel. If the index is greater than 1.5 the channel is described as meandering. A very sinuous river will have values of around 3. Sinuosity can be calculated from field measurements or from OS maps. As well as straight and meandering channels, there are braided channels. Braided channels have islands or bars of sediment in the channel which divide up the flow of water. The form of a river channel is not random, but is related to the stream gradient load and to other variables, as shown in the Schumm Model (Figure 2.42).

Channels in cross-section

Channels may change their cross-sectional form as discharge levels change. For all channels, an increase in discharge results in an increase in velocity, width and depth of the channel. How the channel adjusts depends on the material making up the channel bed and banks. When these are made of cohesive materials, channels tend to deepen with increasing discharge.

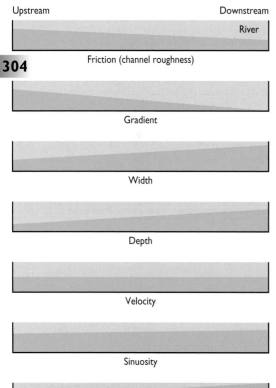

Upstream   Downstream
                                River
Friction (channel roughness)

Gradient

Width

Depth

Velocity

Sinuosity

Hydraulic radius (channel efficiency)

▲ **Figure 2.40** *Downstream changes in channel form: generalised trends*

17 For each of the diagrams in Figure 2.40, write a sentence that summarises the trend shown.

► **Figure 2.41** *River meanders forming on the chalk of the South Downs in Sussex.*

► **Figure 2.42** *Exogenic processes and landforms (the Schumm Model)*

Key Terms

Helicoidal flow: Flow where there are cross-currents forming a circular vortex running in the same direction as the main flow, rather like water spirals down a drain.

Meanders: The bends formed in a river as it winds across the landscape.

Where the channel bed and banks are made of loose material, the channel tends to adjust by increasing its width. Since, for most rivers, discharge increases downstream, there will be an increase in width, depth and velocity downstream, along with other factors (Figure 2.40).

Within a section of a river, the channel cross-section shows variations in velocity. The point of maximum velocity is found in the deepest part of the channel just below the water surface, where there is least friction with the bed.

2.9 Meanders and floodplains

Meandering is an important feature of most rivers and produces the most significant river landform – the floodplain (Figure 2.41). **Meanders** start when friction with the channel bed and banks causes turbulence in the water flow. This results in a spiralling flow of water called **helicoidal flow**.

This causes erosion in some areas of the banks where velocity is high, and deposition in other places where velocity is reduced. A sequence of deep sections (**pools**) and shallow sections (**riffles**) develops which is closely related to meander form (Figure 2.43).

The Schumm Model of channel–load relationships

CHANNEL PATTERN			CHANNEL TYPE		
			Suspended load	Mixed load	Bed load
Straight	Low (<10) Low		1	2	
Meandering	Width–depth ratio		3a	3b	4
Braided	High (>40) High		Channel boundary / Flow of thalweg / Bars		5

	HIGH	←	RELATIVE STABILITY	→	LOW
	Low (3per cent>)	←	Bed load/total load ratio	→	High (>11per cent
	Small	←	Sediment size	→	Large
	Small	←	Sediment load	→	Large
	Low	←	Flow velocity	→	High
	Low	←	Stream power	→	High

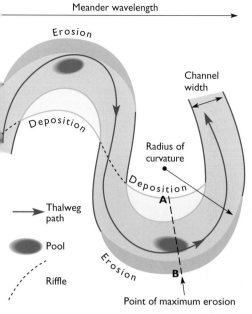

Meander wavelength

Erosion

Deposition

Channel width

Radius of curvature

Deposition

A

Erosion

B

Point of maximum erosion

Thalweg path

Pool

Riffle

▲ **Figure 2.43** *The development of a meandering channel featuring riffles and pools*

Key Terms

Pools: Hollows of relatively deep water in a river bed where fine sediment has been carried away. They occur in sequence with riffles.

Riffles: Accumulations of coarse sediment on a river bed (which is too heavy to be carried away), resulting in areas of shallow water. They occur in sequence with pools.

River cliff: A steep section of river bank formed by erosion and collapse of bank material. It is most common on the outside of meanders due to the high rate of erosion from fast-flowing water.

Floodplain: The relatively flat area forming the valley floor on either side of a river channel. During flooding (and depending on the size of the flood) the river overflows onto the floodplain.

Point bar: Sediment laid down on the inside of a meander bend due to slower-flowing water. Point bars are usually laid down sideways and then build up sideways as meander scrolls.

A terminology has been developed to describe meanders (Figure 2.42). The size and shape of meanders (wavelength, amplitude and curvature) are proportional to the discharge of the mean annual flood of the river at that location on the river's course. Meanders increase the length of the river channel for a particular downstream section. Thus, the gradient of the channel is reduced and, as a result, the available energy is reduced too. Thus, meander development is related to balancing the energy available to do the work of transporting the water and load in that section of the river. Meanders migrate laterally by eroding the outside of the beds forming a **river cliff** and depositing sediment on the inside of the beds, producing a **point bar** (Figure 2.45). This is a result of variations in flow velocity across the channel (Figure 2.44). Although meandering channels are most common, where the river gradient is low and the flow is not powerful enough to form bars and erode banks, straight channels will occur. If the gradient is very high, and helicoidal flow does not fully develop, the channel will be braided.

The **floodplain** is an area of relatively flat land on either side of the channel. It is formed by downstream lateral migration of meanders. This occurs because the point of maximum erosion is just downstream of the centre of the meander bend (Figure 2.43). Rates of erosion and meander migration are generally slow. Studies of bank erosion in British rivers show that rates vary between 0.2 and 2.58 metres per year. Work by Hooke (1980) indicates that rivers in Devon take between 6000 and 7000 years to completely traverse the floodplain.

As well as meander deposits, the floodplain consists of overbank deposits from times of flood. As the river overflows its channel, there is a reduction in velocity and any coarse sediment will be quickly deposited.

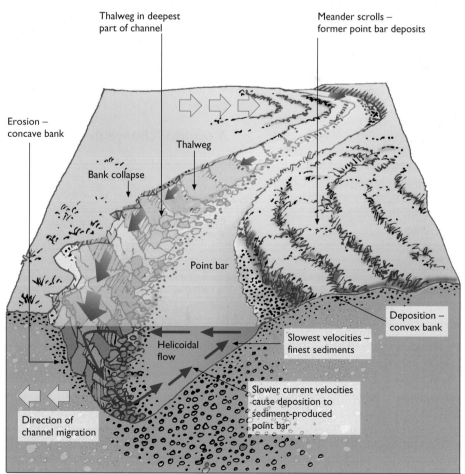

Thalweg in deepest part of channel

Meander scrolls – former point bar deposits

Erosion – concave bank

Thalweg

Bank collapse

Point bar

Deposition – convex bank

Slowest velocities – finest sediments

Helicoidal flow

Slower current velocities cause deposition to sediment-produced point bar

Direction of channel migration

▲ **Figure 2.44** *How meanders migrate laterally across the floodplain*

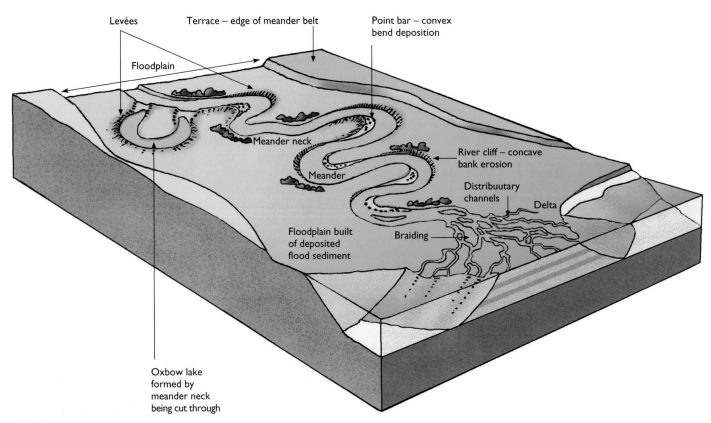

Levées

Terrace – edge of meander belt

Point bar – convex bend deposition

Floodplain

Meander neck

Meander

River cliff – concave bank erosion

Distribuutary channels

Delta

Floodplain built of deposited flood sediment

Braiding

Oxbow lake formed by meander neck being cut through

▲ **Figure 2.45** *Landforms of the floodplain. The floodplain itself consists of a complex mix of bedload meander scrolls and flood deposits.*

These deposits form distinct ridges on the floodplain, called **levées**. The finer sediment load is slowly deposited from the flood waters as they sit on the floodplain and forms a fertile deposit called **alluvium** (Figure 2.46).

Changing energy due to changes of base level

Rivers very rarely achieve a perfectly graded long profile before their work is interrupted by changing sea or land levels. These changing base levels can be divided into two categories.

A positive change in base level

This results from a net rise in sea level. It may be due to the sea level itself rising, or from the land sinking, or both. The outcome is that the lower part of the long profile is drowned and there is a loss of energy as the river reaches the new base level much sooner along its course. The response to this change by fluvial processes is to deposit material in the channel to raise the river level (Figure 2.47a). Drowned parts of the lower river valleys can

18 In a meandering stream, the thalweg (path of maximum velocity) swings from side to side of the channel. Discuss how this influences the pattern of erosion and deposition on the bed and banks of the channel.

► **Figure 2.46** *The cross-section of a floodplain showing how it was formed.*

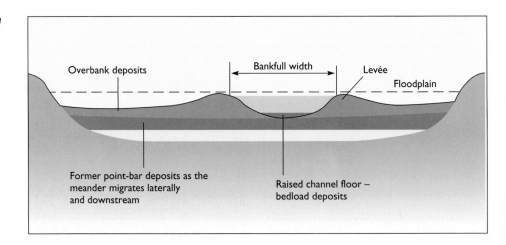

Overbank deposits

Bankfull width

Levée

Floodplain

Former point-bar deposits as the meander migrates laterally and downstream

Raised channel floor – bedload deposits

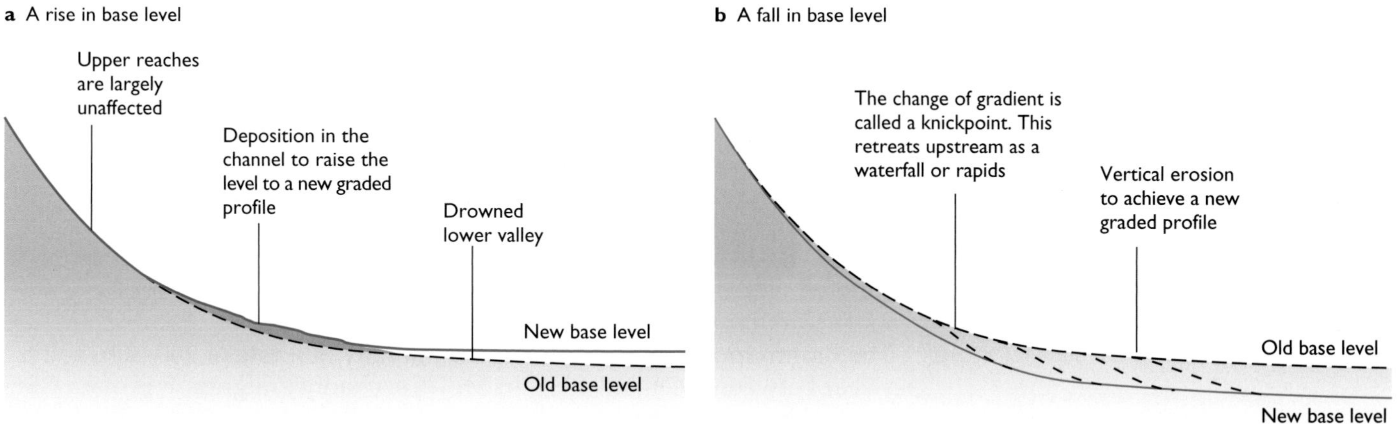

a A rise in base level

Upper reaches are largely unaffected

Deposition in the channel to raise the level to a new graded profile

Drowned lower valley

New base level

Old base level

b A fall in base level

The change of gradient is called a knickpoint. This retreats upstream as a waterfall or rapids

Vertical erosion to achieve a new graded profile

Old base level

New base level

▲ **Figure 2.47** *Changes in base level and the effects on the long profile of a river*

be identified by narrow coastal inlets called 'rias'. In Britain, sea level rose as the ice sheets from the last glaciation melted. Added to this, much of southern Britain is slowly sinking, which produces positive **base level** changes to lowland rivers.

A negative change in base level

This is produced by a net fall in sea level. The land may be rising or the sea level may be falling, or both. An important cause of negative changes in base level is **isostasy**, which followed the last glaciation. Since the new base level is lower than the old one, the river has increased energy to erode its bed and try to lower the profile to the new level. This erosion begins in the lower course and works upstream, over time. In the long profile, the downcutting can be identified by a break of slope called a knickpoint (Figure 2.47b). This may be evident in the channel as a waterfall or rapids. The incision of the channel can be seen in the valley floor as the river cuts vertically down to erode the floodplain. New landforms are produced. The remnants of the floodplain are left as higher-level river terraces on the edge of the new floodplain (Figure 2.48). Some British rivers such as the Thames show evidence of more than one fall in base level. In some cases the meanders themselves become lowered into the valley floor to form incised meanders.

> ## Key Terms
>
> **Isostasy:** The local change in sea level resulting from differences in the weight imposed upon the land by a growing or declining ice-sheet.

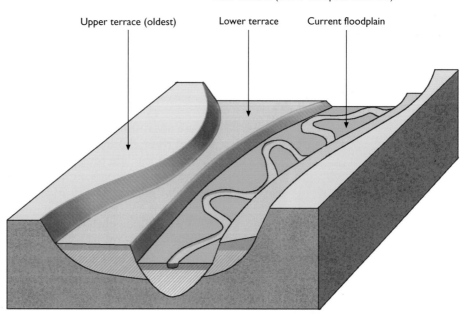

River terraces (earlier floodplain remnants)

Upper terrace (oldest)

Lower terrace

Current floodplain

▶ **Figure 2.48** *Formation of terraces by repeated negative changes of base level*

2.10 Changing energy and sediment loads in river channels

Changes to the course of a channel are a normal part of floodplain evolution. But meander migration may cause problems if floodplains have been developed and settled. Temperate rivers can show rates of bank erosion of up to 10 metres per year. However, rates of bank erosion increase rapidly when discharge increases and large rivers may experience very fast rates of bank erosion.

The Brahmaputra, in India, migrated laterally at a rate of 200–400 metres per year between 1975 and 1981. In MEDCs, channel migration is often controlled by using structures like concrete walls, gabions (stone-filled wire baskets) and stone rip-rap. Geo-textile mats and reed or grass cover can also be used. Flow can be modified near to banks to reduce flow, and therefore erosion, by using submerged vanes or hydrofoils.

Channel form is in balance, or dynamic equilibrium, with discharge and sediment supply. Changes to either of these will result in changes to the channel geometry through deposition and erosion (Table 2.6).

▼ **Table 2.6** *Human-induced changes to channels*

■ **Changes to discharge**
Increase in discharge velocity (i.e. straightening of the channel)   This will increase energy to erode and transport sediment. Erosion will increase. Deposition will be reduced. Sinuosity will increase.
Decrease in discharge (i.e. flow regulation by dams and reservoirs)   This will reduce the energy available to erode and transport sediment. Deposition will increase. Sinuosity will decrease.
■ **Changes to sediment load**
Increase in sediment load (i.e. mining, deforestation, poor agricultural practices, early stages of urbanisation)   The river will have less energy since it has more sediment to transport. Deposition within the bed will increase. Channel may change from meandering to braided as the new sediment load decreases the channel gradient and builds up bars and islands in the river bed. A new, higher floodplain may be produced.
Decrease in sediment load (i.e. due to gravel extraction, dredging for navigation, later stages of urbanisation, dam construction trapping sediment in the reservoir)   The river will have more energy to erode since it is carrying less sediment. Vertical and lateral erosion may occur. Channel gradient is reduced and bedload size increases until a new equilibrium is established.

19 For any two changes mentioned in Table 2.6:
a Describe how they create dis-equilibrium (imbalance) in a channel.
b Suggest how the stream works to re-establish discharge–load–channel form equilibrium.

There are numerous examples of how human activity has inadvertently changed river channels. Close to Alston, North Yorks, are many abandoned mines from the late nineteenth century. The mine tailings increased the bedload of the River Nent and contaminated floodplain sediments. The River Nent was originally meandering, but increasing sediment load from the mining activity resulted in gravel deposits in the channel bed. From 1861, the channel started to become braided as a response to the changed conditions. Since mining stopped in the 1930s, the extra bedload has been eroded and the channel has slowly been returning to its meandering course. A decreased bedload and flow regulation can also have significant impacts, as shown by the example of Stony Creek, California.

EXAMPLE: Stony Creek, California

Stony Creek is a tributary of the Sacramento River in California, about 200km north of San Francisco. The creek has been affected by both extraction of gravel and flow regulation (Figure 2.49b). Reducing the sediment load of a river will increase the amount of energy available for erosion. The results are channel degradation in both the long and cross-profiles (Figure 2.49c and Figure 2.49d). The flow is regulated by three dams and reservoirs, the most downstream being Black Butte Dam. The aim of the regulation was to provide water for irrigation and to reduce flood peaks. Figure 2.49a shows the impacts of these changes on the reaches of the Creek.

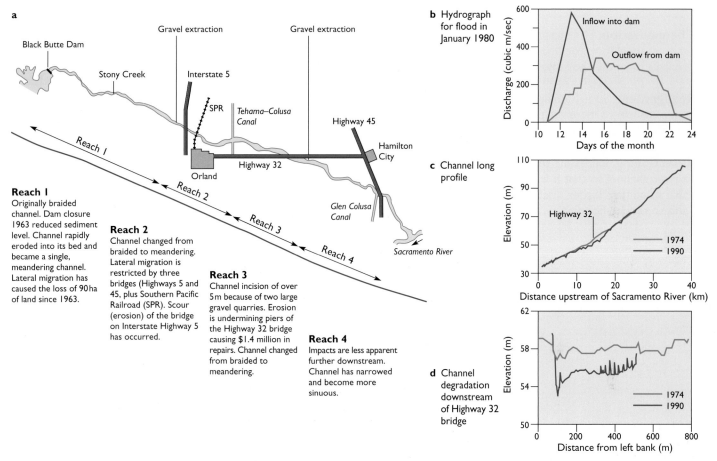

a Stony Creek

Black Butte Dam

Stony Creek

Gravel extraction

Gravel extraction

Interstate 5

SPR

Tehama–Colusa Canal

Highway 45

Hamilton City

Orland

Highway 32

Glen Colusa Canal

Sacramento River

Reach 1

Reach 2

Reach 3

Reach 4

Reach 1
Originally braided channel. Dam closure 1963 reduced sediment level. Channel rapidly eroded into its bed and became a single, meandering channel. Lateral migration has caused the loss of 90ha of land since 1963.

Reach 2
Channel changed from braided to meandering. Lateral migration is restricted by three bridges (Highways 5 and 45, plus Southern Pacific Railroad (SPR). Scour (erosion) of the bridge on Interstate Highway 5 has occurred.

Reach 3
Channel incision of over 5m because of two large gravel quarries. Erosion is undermining piers of the Highway 32 bridge causing $1.4 million in repairs. Channel changed from braided to meandering.

Reach 4
Impacts are less apparent further downstream. Channel has narrowed and become more sinuous.

b Hydrograph for flood in January 1980

Inflow into dam

Outflow from dam

Discharge (cubic m/sec)

Days of the month

c Channel long profile

Elevation (m)

Highway 32

1974

1990

Distance upstream of Sacramento River (km)

d Channel degradation downstream of Highway 32 bridge

Elevation (m)

1974

1990

Distance from left bank (m)

▲ **Figure 2.49** *Stony Creek, California*
a Stony Creek
b Hydrograph for the flood in January 1980
c Channel long profile
d Channel degradation downstream of the Highway 32 bridge

20 How has the dam changed the flow downstream (Figure 2.49b)?

a Describe and explain the impacts of the dam downstream on channel processes.

b How has the gravel extraction affected the river? Explain your answer fully, using river-process terminology and the graphs.

c How and why has the planform of the channel changed because of human activity?

Key Terms

Hard engineering: The use of engineering structures to control natural processes such as erosion or floods.

Soft engineering: Methods which work with rivers and other natural processes to control erosion or floods.

Channelisation: The changing of river channel form for flood control and other purposes.

Realignment: Straightening of a river channel to increase its ability to move water quickly.

Resectioning: Increasing the size of a river channel so that it is more able to contain higher discharges and so reduce flooding.

2.11 Managing river channels

Until now, management of river channels has focused on trying to alleviate floods, water storage, meander migration, or improving water routes for navigation. Since the early nineteenth century these schemes have employed hard-engineering structures or changes to channel shape (Figure 2.51).

Dams and reservoirs

Dams and reservoirs control flow to reduce floods and supplement low flows. Since discharge levels are evened out and sediment load is reduced, increased erosion may result downstream of the dam. This is aptly termed **clearwater erosion** (Figure 2.50). The changed sediment transport results in reduced channel slope so that the bedload is no longer transported. Fine material is removed but coarser bedload remains. Channel width is reduced and depth is also reduced, despite downcutting, because banks are unstable. Channel sinuosity increases.

Channelisation

Channelisation groups together all changes to channels in planform (realignment) or cross-section (resectioning) (Figure 2.51).

Realignment

Realignment increases the channel flood capacities by straightening the river. The straighter section is steeper than the former meandering section so the average flow velocities increase. This method can be effective for flood control but, since the channel will try to re-establish a meandering pattern, the realigned section needs to be lined with concrete walls, sheet piling or gabions. Many of these methods are visually intrusive and destroy bankside habitats. Following realignment, there is a need for constant dredging of the channel as the river tries to recreate riffle and pool sequences and meandering (see Mississippi example).

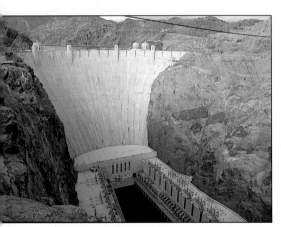

▲ **Figure 2.50** *When the Hoover Dam was built across the Colorado River (USA) clear water erosion took place on a 111km stretch of the river below the dam. The river bed was eroded by 7.1m over 14 years.*

► **Figure 2.51** *Channelised section of the River Ravensbourne at Lewisham. The river has been contained by concrete banks.*

EXAMPLE: Mississippi
Managing the Mississippi Meanders

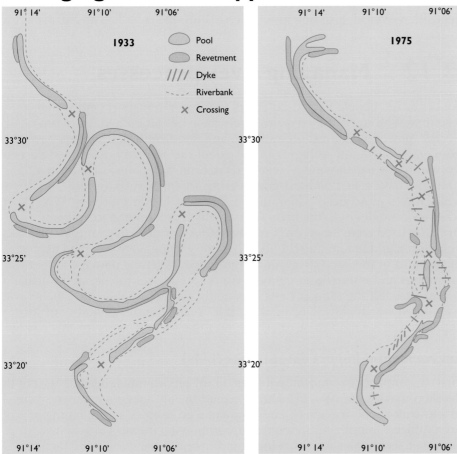

◀ **Figure 2.52** *The Greenville Reach of the Lower Mississippi in 1933 and 1975*

▼ **Table 2.7** *Dredging the Mississippi at Greenville Reach*

Period	Volume dredged (m³/km/year)
Pre-cutoff	137
1950 (cutoff)	10360
1951–64	29835
1965–73	62833
1974–77	39695

Following serious flooding of the Lower Mississippi in 1927, sections of the lower river were straightened to increase flood capacity and to reduce the distance travelled by shipping (Figure 2.52). The straightening increased the channel gradient, which caused increased erosion of the river bed. The erosion progressed upstream and affected some tributaries. The sediment load increased and was deposited in the straightened section as the river tried to re-create its natural meandering pattern. As a result, billions of dollars have been spent dredging the river to keep it open for navigation and to maintain flood capacities (Table 2.7). Spur dykes have also been built to stop erosion of the banks and sedimentation. The dykes work by diverting the river flow away from the eroding bank. The reduced flow in the area protected by the dykes allows deposition to occur and builds the bank towards the centre of the channel. Flow velocities become higher in the centre of the channel, which allow the channel to remain open for navigation.

Resectioning

This involves enlarging the channel by dredging it to widen and deepen it. The aim is to increase the cross-sectional area of the channel so that higher flows can be contained within the channel. This increases the hydraulic radius of the channel so that the channel is more efficient and reduces energy loss by friction. Most of the rivers in lowland England have been resectioned. **Resectioning** is used in urban areas where development and settlement up to the river's edge do not allow for other forms of flood defence such as floodbanks or artificial levées. It is also used in agricultural

71

areas as part of land-drainage schemes. Resectioning has a major impact on wildlife habitats. Continued dredging is needed for maintenance, which prevents habitats from re-establishing themselves.

At the upper end of the dredged section, erosion will increase because of increased gradients. However, downstream in the undredged section, deposition of eroded sediment will occur since the river has no extra energy to carry it.

2.12 Managing river processes

For centuries, people have used rivers for power, transport, and water resources. Throughout the world, rivers have been dammed for hydro-electric power, water-resource storage and flood control. Channels have been resectioned and realigned to the extent that in many developed parts of the world it is hard to find a river in its natural state. Most management strategies have relied on **hard-engineering** schemes, such as dams, flood walls, levées or artificially maintained channels, to control rivers. When such schemes were devised, the main factors which determined the type of structure used were based on 'Cost–Benefit Analysis'. For example, the costs of building and maintaining a flood-control scheme were weighed against the benefits of the flood protection it provided. Potential damage to homes, businesses and infrastructure which would occur over the life of the scheme was assessed. If the ratio of costs to benefits was favourable, the scheme was considered worthwhile. During the last 30 years, the sustainability of such an approach has been questioned. There are three main reasons for this change.

Environmental Impact Assessments (EIAs)

First, the impact of proposed schemes on the environment became part of the decision-making process. EIAs are required by law, and engineers now need to know more about ecology, environmental economics and planning. It has been difficult to put a figure on the environmental impacts of schemes. However, techniques for assessing environmental costs and benefits have improved greatly with the growth of environmental economics. Methods are now being used which can cost the longer-term environmental benefits of habitats compared with the relatively short life (about 50 years) of an engineering scheme.

Legislation

Secondly, there has been a change in the statutory duties of environmental care imposed on water-management agencies and companies. With the emergence of privatised water companies in the late 1980s in the UK, the environmental protection aspect of projects was strengthened in response to pressure from environmental groups.

Green engineering

Thirdly, better understanding of fluvial processes has allowed engineers to design schemes which work *with* river processes, rather than by just controlling them. The move towards this approach has been influenced by the costly maintenance of past schemes, which ignored fluvial processes and ended in constant dredging and bank protection for resectioned channels. Dams have also had a negative impact beyond the river channel which was not predicted by the original scheme. Experience of managing channels has shown that trying to change channel form in upland channels results in an unstable channel and long-term maintenance problems. For downstream channels, or lowland areas, the stability was less affected but they became uniform rivers, with a loss of habitat (Figure 2.53).

In the 1990s, some major flood events influenced public opinion as well.

BENDS RETURN AS RIVERS GET RID OF DIRE STRAIGHTS

This week the bends are being put back in the River Avon at Melksham in Wiltshire, after engineers conceded that attempts to improve on nature had not worked. The Avon is the latest in a series of projects across the country attempting to undo the damage done to Britain's rivers in the past 50 years.

At Melksham, the Avon looks more like a canal than a river. It was channelled into a straightjacket of steel and concrete in the 1960s when a series of 'improvements' upstream increased the risk of flooding.

Three flat shelves called berms are being built in the river channel to make the river meander and to speed up the flow of water. The berms will then be edged with coir matting studded with plants such as rush, iris and sedge which all but vanished when the river was straightened.

Shoals, riffles and pools will reappear and the faster-flowing water will keep the river and banks from silting up. As the vegetation re-establishes itself, fish and other wildlife will return, finding new habitats in shaded, shallow pools where there is now little more than mud and deep water. Ann Skinner, the Environment Agency's project manager said, 'There is an economic benefit besides the environmental one. We are restoring a self-sustaining system that looks after itself without too much intervention. Once a river has been returned to a more natural state it is amazing how quickly nature completes the restoration.

▲ **Figure 2.53** *The restoration of the River Avon*
(Source: The Times, 8 December 1999)

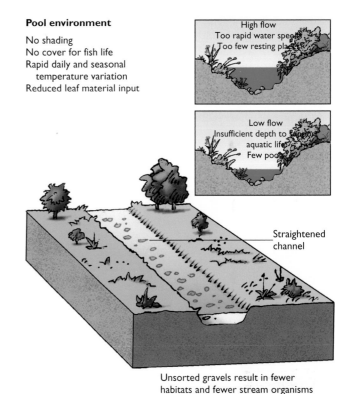

Pool environment

Adequate shading
Good cover for fish life
Minimal temperature variation
Abundant leaf material input

High flow
Diverse water speeds
Many resting places

Low flow
Sufficient water depth
in dry season

Pool with silt,
sand and gravel

Coarse gravel

Sorted gravels provide diversified
habitats for many stream organisms

Pool environment

No shading
No cover for fish life
Rapid daily and seasonal
temperature variation
Reduced leaf material input

High flow
Too rapid water speeds
Too few resting places

Low flow
Insufficient depth to support
aquatic life
Few pools

Straightened
channel

Unsorted gravels result in fewer
habitats and fewer stream organisms

▲ **Figure 2.54** *Characteristics of a natural river and a channelised river*

21 Give three reasons why there has been a trend towards 'soft' engineering approaches to management. Support your answer with examples.

22 Illustrate briefly evidence to support the claim that modern river management techniques work with, rather than against, natural processes.

▼ **Figure 2.55** *The Mississippi flood of 1993*

The 1993 flood of the Mississippi (Figure 2.55) caused widespread public outrage at the failure of defence levées and dams to control the flood. Even though not all the defences were designed for a flood of such magnitude, there was concern that water levels had been raised by water being concentrated down the river with confining levées. If flooding had occurred upstream, water levels in the river would have been reduced by the storage of excess water on the floodplain. In Britain, widespread flooding in early 1994, in southern Britain and Northern Ireland, caused much public concern – particularly the flooding at Chichester, which received a great deal of media attention.

In addition, environmental concerns have increased in importance over recent years. Channelisation for example, results in loss of diversity of habitats (Figure 2.54). One example of such a change in approach is the Kissimmee River in Florida. In 1962, the river was channelised for flood control. At a cost of $24 million, and taking nine years to complete, the river was straightened and lined with concrete. By the 1990s, the river was degraded environmentally, water pollution had increased, and there were drainage problems. About 40 000 acres of wetland habitat had been lost. In addition, the goal of flood control had not been completely achieved. Work has now started on restoring parts of the river to its original condition.

Soft engineering

More and more, schemes are being designed to work with river processes and to preserve habitats. This approach is called **soft engineering**. Since the visual amenity, flora, fauna and fisheries are more likely to be preserved, and maintenance is minimal, these approaches are more sustainable in the longer term. Soft-engineering schemes include setting flood banks further back so that limited flooding can occur, or constructing flood relief channels to take high flows and to prevent floods, while keeping the original channel in its natural form. Increasingly, development of floodplain land is restricted to uses which are more compatible with flooding such as parks, sports fields and nature reserves (see Upper River Lugg example).

EXAMPLE: The Upper River Lugg, Leominster

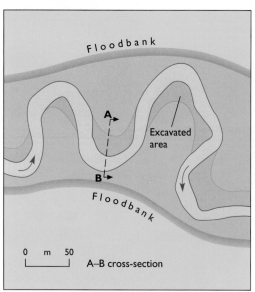

▲ **Figure 2.56a** *Excavating meanders at the Upper River Lugg*

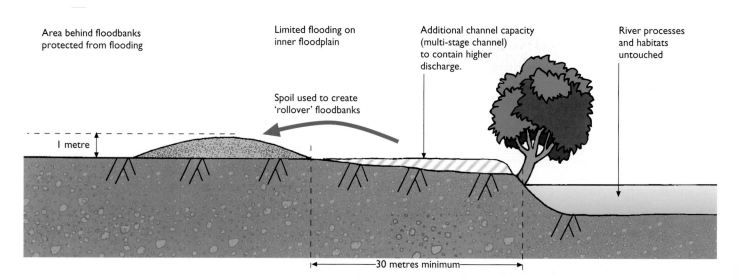

▲ **Figure 2.56b** *Typical cross-section of meanders at A–B*

A				B
Level (m)	12	11	10	12
Chainage (m)	18+	12	6	0

▲ **Figure 2.56c** *Idealised cross-section of the improvement*

Improvement of the River Lugg

The scheme was designed to reduce flooding and meander erosion. The features of the scheme included

- The inside of meander bends, which were excavated and the spoil used to build the floodbanks. The excavated meanders created a multi-stage channel so that higher discharges could be carried with the channel (Figure 2.56a).
- Outer banks of the meanders were protected from erosion with willow (Figure 2.56b).
- Floodbanks set back from the channel to allow flooding of the inner floodplain only. The floodbanks connect to the weir.
- Downstream of Leominster, several miles of multi-stage channel and flood embankments have been used to retain the main channel untouched (Figure 2.56c).

■ Constructing four stone weirs with the aim of creating deep-water pools to reduce flood-water velocities (Figure 2.56d).

Although there was a lot of disturbance to habitats during the works, the channel design simulates a natural meander cross-section and will require little maintenance. A variety of habitats have been retained in the meanders. The stone weirs provide a new stable rocky habitat and more turbulent oxygenated water for a variety of plants, invertebrates and birds. In long profile, there is a varied riffle and pool sequence with still and fast water habitats. The scheme allows the channel to follow its original course and retains most of the riverside vegetation and trees.

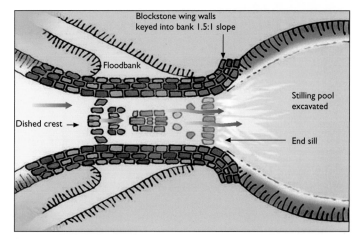

▲ **Figure 2.56d** *Plan view of a typical weir built during the River Lugg improvement*

Summary

The global hydrological cycle is a closed system which can be divided into drainage basins on the land area of the Earth.

The water balance is the result of the relationship between inputs of precipitation and outputs of evaporation and runoff, variations occur in time and space. Rivers have distinctive flow patterns as a result of this balance, as shown by the regime and the flow duration curve.

The processes operating in the drainage basin control the patterns of river flow and storm hydrograph events and flooding.

Rivers carry out geomorphological work by the processes of erosion, transport and deposition.

Sediment is an important output from the drainage basin and is important in fluvial processes and landform development. The factors affecting the sediment load relate mainly to climate, geology, vegetation, mass-movement processes and human activity.

Rivers produce distinctive long and cross-profiles which relate to energy availability and the characteristics of the drainage basin.

River channel characteristics depend on the efficiency of the channel.

River channels vary in planform due to energy characteristics. Meanders and floodplains are important features of these processes.

Changes in energy result in changes in the balance of channel processes, because of geomorphological changes in base level or human activity.

Humans attempt to manage channels to reduce flooding and channel change. These have impacts on channel form and processes.

There has been increasing concern in recent years about the negative impact of channel management. Increasingly, management schemes aim to work with fluvial processes and conserve habitats.

Chapter 3 Coasts

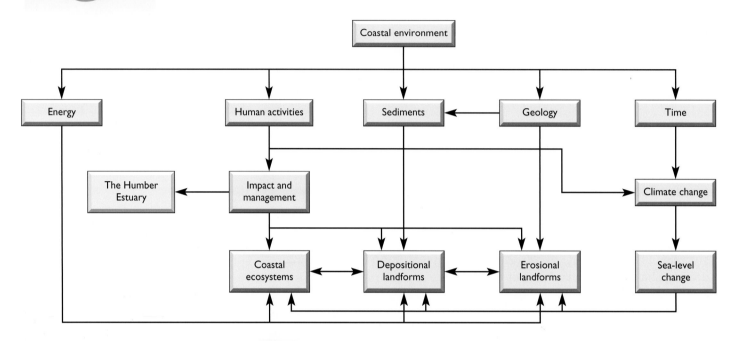

3.1 The coastal system

The coast is the interface between land and sea. One way to study the coast is to view it as a system (Figure 3.1). Energy inputs from waves, tides and winds drive the system. They interact with the geology, sediments, plants and human activities along the coastline. The evidence for this interaction are the processes of erosion, transport and deposition. These processes give rise to the system's main output: coastal landforms such as cliffs, beaches and salt marshes.

▼ **Figure 3.1** *The coastal system*

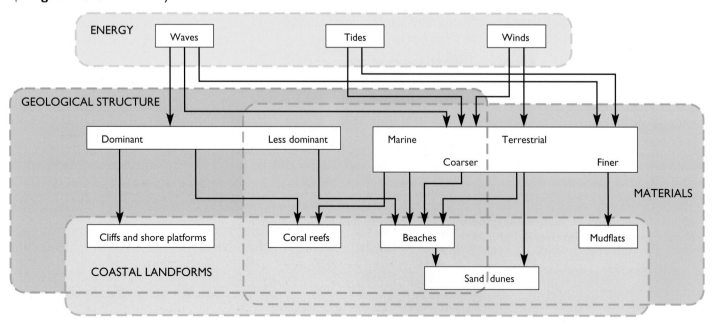

Key Terms

Equilibrium: The long-term condition of balance or stability in a system, with inputs of energy and matter equal to outputs of energy and matter.

....

Negative feedback: The process of self-regulation which restores equilibrium to a system.

....

Positive feedback: The process by which an initial change in a system is amplified and causes further change: e.g. damage to vegetation on a sand dune by trampling causes erosion of sand, resulting in further loss of vegetation, more erosion, etc.

....

I Cyclonic weather conditions generate powerful waves which result in a net transport of sand and shingle offshore. Draw a diagram similar to Figure 3.2 to show how a beach might adjust its profile to an increase in wave energy caused by cyclonic conditions.

The concept of dynamic **equilibrium** is central to our understanding of natural systems. A system is in dynamic equilibrium when its inputs and outputs of energy and matter balance. In these circumstances, a system remains in a steady state for long periods of time. Of course, short-term changes will still occur. Systems adjust to these changes by a process of **negative feedback**. For exmple, short-term events, such as storms, greatly increase energy inputs to the coastal system. This starts the movement of sand and shingle, and creates new beach forms. When energy input equals output, sediment transport comes to an end (Figure 3.2). The beach has now reached equilibrium.

Landforms such as beaches can adjust to changing energy inputs in just a few hours. In contrast, hard rock landforms, such as cliffs, may take thousands of years to achieve equilibrium. As we shall see in section 3.2, today's sea-level, and therefore the position of the coastline, is only 6000 years old. This means that large parts of the coastline have not had sufficient time to achieve equilibrium. We notice this by events such as rockfalls and landslides which can dramatically alter the coastline (Figure 3.3).

► **Figure 3.2**
The impact of changing wave energy on beach profiles

▼ **Figure 3.3**
Rockfall at Beachy Head 11.1.99

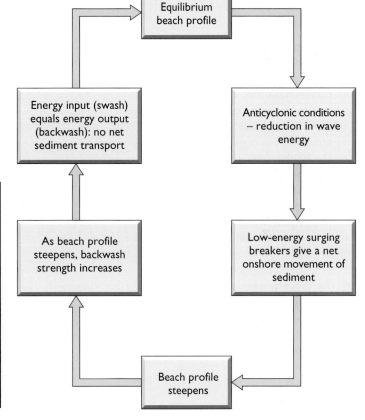

3.2 Sea-level change

Over the last two million years, huge fluctuations have occurred in global sea-levels. Sea-level changes at a global scale are the direct result of glaciation and reflect changes in the absolute volume of water in the oceans. We refer to these changes as **glacio-eustacy**. During glacial periods, water is shifted from the oceans to ice sheets and glaciers, causing a eustatic fall in sea-level. In warmer inter-glacial periods, melting ice sheets and glaciers lead to rises in sea-level. Thus, in the most recent Devensian glacial, sea-level was between 100 and 125 metres lower than today, and large areas of the continental shelf around the British Isles were dry land.

▼ Figure 3.4 *Rising sea-level in the last 20 000 years*

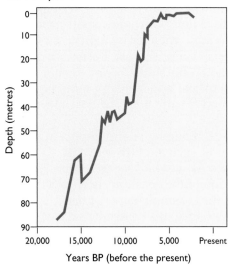

Twenty thousand years ago, the world's climate started to warm. Ice sheets and glaciers receded polewards and sea-level began to rise steadily. Rising sea-level caused coastlines to retreat – a process known as **transgression**. In the British Isles, the last transgression was called the Flandrian. It flooded the North Sea basin and the English Channel and ended just 6000 years ago (Figure 3.4). Our present-day coastline, therefore, is geologically fairly recent. Regression is the opposite to transgression. It leads to advancing coastlines and occurs during glacials when huge volumes of water are locked up in ice sheets and glaciers. At the height of the Devensian, ice sheets and glaciers in the northern hemisphere occupied an area three times greater than they do today.

At a local scale, sea-level change is associated with the land rising or sinking relative to sea level. This type of sea-level change results from either isostatic or tectonic movement (Box 1).

Box 1 Isostatic and tectonic sea-level change

■ During a glacial period, great masses of ice load the continental crust, causing it to sink by several hundred metres. When the ice melts, unloading occurs, and the crust very slowly rises by a similar amount. This is **glacio-isostacy**. In the British Isles, the ice was thickest in Scotland, and it is there that **isostatic recovery** has been greatest (Figure 3.5). This movement, which continues still, has raised shorelines around Scotland's west coast and has created **raised beaches**.

■ Tectonic movements such as faulting in active earthquake and volcanic zones also cause vertical displacements of the crust. Depending on the direction of movement, this process may produce a local relative rise or fall of sea-level.

2 From Figure 3.5:

a Describe the pattern of recent sea-level change.

b Explain the zone of falling sea-level in Scotland.

c Describe and explain the possible effects of rising sea levels on coastal landforms in the British Isles.

▼ Figure 3.5 *Estimated current rates of crustal movement in the British Isles*

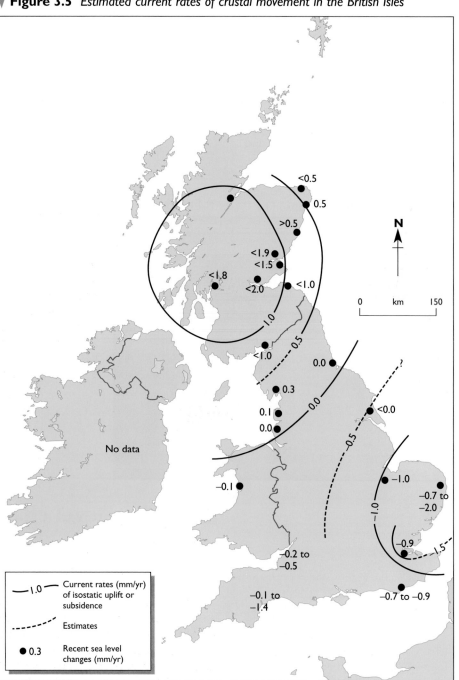

▼ **Table 3.1** *Sea-level changes and landforms*

Glacio-eustacy	
Rising sea-level	
Shingle beaches (spits, bars, etc.)	River sediments were deposited on the dry continental shelf during the last glacial period when the shelf was above sea-level. Over the last 20 000 years, rising sea-level swept up the sediment and deposited it on present-day coasts (e.g. Chesil Beach).
Estuaries	These are drowned, shallow, lowland river valleys (e.g. the Severn, Humber and Thames estuaries). Transgression flooded low-lying areas around the Wash and Somerset Levels.
Rias	These are drowned, incised river valleys on upland coasts (e.g. River Dart, Devon; River Fal, Cornwall; Dingle Bay, south-west Ireland).
Fjords	These are drowned glacial troughs (e.g. Sognfjord, Norway).
Falling sea-levels	
Ancient shore platforms	Cut by wave action in the last interglacial period when sea-level was 8–10 metres higher than it is today (e.g. Start Bay, South Devon).
Glacio-isostacy	
Falling sea-levels	
Raised beaches	Ancient beaches and cliff lines which were elevated above sea-level when the glacial ice sheets in northern Britain melted (e.g. Applecross Peninsula, north-west Scotland).
Tectonic movements	
Earthquakes/faulting	Localised tectonic movements which lead either to submergence or emergence.

Key Terms

Surfing breakers: Powerful storm waves which are high and steep and have short wavelengths.

Surging breaker: Low-energy waves which are shallow and have long wavelengths.

Fetch: The area of open ocean over which waves develop because of wind.

Dominant waves: Waves from the direction of longest fetch. These waves cause the bulk of erosion and sediment transport along a coastline.

3
a Log on to the National Data Buoy Centre website: http//seaboard.ndbc.noaa.gov/index.shtml
b Investigate the relationship between wave height, fetch and wind speed for buoys moored around the coastline of the British Isles. Use the Spearman's Rank correlation coefficient in your investigation.

3.3 Energy inputs

Waves

Waves are undulations of the water surface caused by winds blowing across the sea. They consist of orbital movements of water molecules which diminish with depth. In fact, water in a wave only shows significant forward movement when it approaches the shore and breaks. We use wave energy to recognise two types of waves: **surfing breakers** and **surging breakers**. Surfing breakers are high-energy waves (Figure 3.6). They are steep and have short wavelengths (see Box 2). Surging breakers are low-energy waves. They are shallow and have long wavelengths. Waves are the main source of energy which drives the coastal system.

Spatial and temporal variations in wave energy

Wave energy varies in both time and space. Four factors influence wave energy: **fetch**, water depth, wind strength and wind duration. Fetch is the open expanse of water where **dominant waves** develop before reaching a coastline. If the the fetch is long, the waves are more powerful because they have more time to gather energy before reaching the coast. The south coast of Hawaii's Big Island (19 degrees North) has a fetch which extends for several thousand kilometres to the south. Not surprisingly, the wave energy environment on this coast is one of the most powerful in the world.

Along shallow-water coastlines (e.g. the coastline of Morecambe Bay) waves break some distance offshore and lose much of their energy.

The wind is the main source of wave energy. Both the wind strength and its duration (i.e. the length of time it blows) are important determinants of wave power. A gale force wind blowing over 1500 kilometres of ocean for 60 hours, can generate waves 35 metres high.

Whereas wind strength and duration vary from day to day, the direction in which a coast faces, and the depth of water offshore, are both fixed. Open, deep-water coastlines with a long fetch (e.g. north Cornwall) are usually high-energy environments dominated by erosional landforms (Figure 3.8). On the other hand, sheltered coastlines protected from

323

▲ **Figure 3.6** *High-energy waves breaking against the sea wall at Hythe, Kent*

powerful waves (e.g. estuaries, coves) and shallow-water coasts, are low-energy environments. Here, tidal processes are often more important than wave action in shaping the coast.

Waves in shallow water

Waves in shallow water behave differently from waves in open ocean. As waves approach the shoreline they begin to 'feel' the sea bed. Friction between the wave and the sea bed causes a decrease in wave velocity and a shortening of wavelength. However, the rate of energy transported by waves is the same in deep water and in shallow water. As a result, waves in shallow water compensate for the reduction in velocity by increasing their height and steepness (Box 2). Eventually, the wave becomes unstable and breaks. Water runs up the shoreline as **swash** and returns to the sea as **backwash**.

Key Terms

Tidal range: The vertical difference in height between the mean high tide and mean low tide marks.

Box 2 Wave characteristics

■ Wavelength (Figure 3.7) is the average distance between successive wave crests.
■ Wave height (Figure 3.7) is the vertical distance between a wave trough and a wave crest.
■ Wave steepness is the ratio of wave height to wavelength. Powerful waves are steep because they are high and have short wavelengths.
■ The energy in a wave is equal to the square of its height. Thus a wave which is two metres high contains four times as much energy as a one-metre-high wave. Wave power takes account of velocity as well as wave height. Thus:

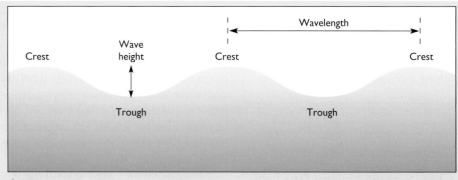

▲ **Figure 3.7** *Wave characteristics*

Wave power = $H^2 \times V$ where H is wave height and V is velocity.
■ Phase difference is another measure of wave energy. It is the ratio of swash time to wave period. Swash time is the interval (in seconds) between a wave breaking and the swash reaching its highest point on a beach. Wave period is the average interval (in seconds) between waves. Surging breakers (low energy) have phase differences of less than 1. Surfing breakers (high energy) have phase differences of 1 and above.

Tides

Tides are caused by the combined gravitational pull of the moon and the sun and **centrifugal force**. The gravitational attraction of the moon and the sun piles the ocean water nearest the two bodies into a tidal wave. On the opposite side of the globe, there is a second tidal wave. Here, centrifugal force exceeds the gravitational force of the moon and sun. The moon takes 29 days to orbit the Earth. After one revolution of the Earth, the moon has moved on from its position 24 hours earlier. It takes the Earth another 50.47 minutes to catch up with the moon each day. As a result the interval between high tides is 12 hours 25 minutes, not 12 hours.

When the moon, sun and Earth are in a straight line (i.e. twice a month, at full moon and at new moon), the tide-raising force is the strongest. This produces the highest monthly **tidal range** (Figure 3.9) (spring tide). Also, twice a month, the moon and sun are positioned in relation to the earth at 90 degrees to each other. This alignment gives the lowest monthly tidal range or neap tide. The highest spring tides of the year occur around the equinoxes, on the 21 March and 21 September.

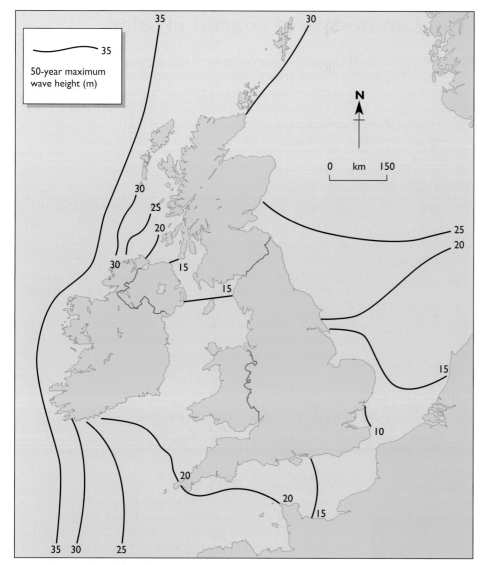

Figure 3.8 *Wave characteristics around the British Isles*

4 Study Figure 3.8.

a Describe the distribution of 50-year maximum wave heights around the British Isles.
b State and explain two possible reasons for this pattern.

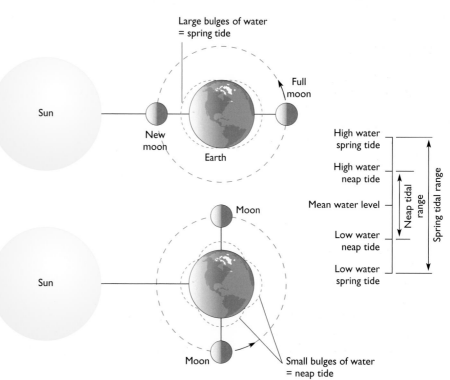

Figure 3.9 *The generation of tides*

Key Terms

Cliff profile: The cross-section of a cliff from its top to its base.

Coherent rocks: Rocks with minerals which are strongly bonded (e.g. crystalline rocks, well-cemented sedimentary rocks, etc.) and which resist erosion, weathering and mass movement.

Incoherent rocks: Rocks such as clay and shale, with weakly bonded structures, and which have little resistance to erosion, weathering and mass movement.

Lithology: The chemical composition (e.g. minerals) and physical characteristics (e.g. joints, bedding planes, faults, etc.) of rocks.

3.4 Landforms of coastal erosion

Wave action on coastlines causes three types of erosional process: abrasion (also known as corrasion), hydraulic action, and corrosion (Table 3.2). These processes are most effective when high-energy waves, associated with storm conditions, strike coasts made of less resistant rocks such as clay and shale. Concentrated wave action on cliffs, around the high water mark, leads to undercutting, the development of a wave-cut notch and, eventually, cliff collapse.

▼ **Table 3.2** *Marine erosion processes*

Abrasion	High-energy waves pick up shingle and abrade the base of cliffs. The result is a wave-cut notch. The cliff is undermined and retreats through rockfall.
Hydraulic action	Air and water are forced under pressure into joints and bedding planes in the rocks. The rocks weaken and collapse. The effectiveness of hydraulic action depends on the density of joints.
Corrosion	Some rock minerals are susceptible to solution. For example, calcareous cements which bind sandstone particles may be dissolved, leading to disintegration of rocks.

Cliff profiles

The shape of a cliff in cross-section is known as the **cliff profile**. Cliff profiles owe their form to the geology (**lithology** and **structure**), **sub-aerial processes** and wave energy along a given stretch of coastline.

▼ **Figure 3.10** *Till cliffs at Atwick, Holderness. Wave erosion undercuts the base of the cliffs leading to slope failure by rotational sliding. The resulting cliff profiles are very steep.*

Lithology

Lithology describes the mechanical and chemical properties of rocks. Coherent rocks which have interlocking crystals, strongly cemented particles and a few lines of weakness (e.g. joints and faults), resist erosion and support steep-angled slopes. Most igneous and metamorphic rocks are highly resistant. So, too, are some sedimentary rocks such as chalk, Carboniferous Limestone and Old Red Sandstone. In contrast, **incoherent rocks**, such as clay and sands, erode more easily. Sometimes this results in low-angled cliff profiles. However, weak rocks undercut by wave action often fail by slumping. The result may be almost vertical cliff profiles (Figure 3.10).

Structure

The angle of dip of sedimentary rocks is one element of rock structure which has an important effect on cliff profiles (Figure 3.11).

- Vertical cliffs develop in horizontally bedded sedimentary rocks and in volcanic rocks formed from horizontal layers of lava and ash. These cliffs, undercut by wave action, retreat parallel to themselves, and maintain their steep slope angle by rockfall (Figure 3.11).
- Seaward-dipping strata have profiles which correspond to the angle of dip of the bedding planes. Blocks weakened by erosion and weathering

a Horizontally bedded strata

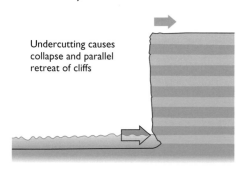

Undercutting causes collapse and parallel retreat of cliffs

b Seaward-dipping strata

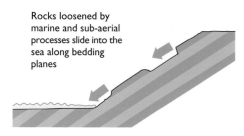

Rocks loosened by marine and sub-aerial processes slide into the sea along bedding planes

c Landward-dipping strata

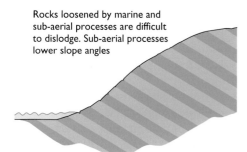

Rocks loosened by marine and sub-aerial processes are difficult to dislodge. Sub-aerial processes lower slope angles

▲ **Figure 3.11** *The effect of rock structure on cliff profiles*

▶ **Figure 3.12** *The cliffs at Blackhall, County Durham formed by rocks of contrasting lithology. The vertical profile at the base of the cliffs comprises resistant horizontally bedded Magnesian Limestone. A thick capping of till is responsible for the low-angled slope above the limestone.*

fail along these planes and slide into the sea. Many cliffs comprise more than one rock type of different lithology and structure. On England's north-east coast, till is widespread and covers more resistant rocks. The resultant cliff profiles have two distinctive slope elements. For example, at Blackhall in Co. Durham (Figure 3.12), the lower 20 metres of hard, horizontally bedded Magnesian Limestone forms a vertical face. Above it, the 30 metres of till form gentler slope angles of 25–30 degrees.

■ Landward-dipping strata form less steep cliffs. This is because eroded and weathered rock particles are not easily dislodged from the cliff face (Figure 3.11).

Sub-aerial processes

The influence of sub-aerial processes such as mass movement and weathering is particularly evident on incoherent rocks. At Filey Brigg in Yorkshire, gulleying, surface wash, slumping, and mudflows on a thick deposit of till, dominate the cliff profile, the base of which is a resistant platform of Corallian Limestone (Figure 3.13).

Wave energy

Coastlines with long fetches experience high wave energy. Wave energy is also higher on windward coasts than on sheltered leeward coasts. Many coastlines on tropical islands, which are exposed to the prevailing trade winds receive higher wave energy. On higher-energy coasts, the effect of erosion on cliff profiles tends to be greater than the effect of sub-aerial processes. Along such coastlines, rockfalls are easily removed by wave action, allowing erosion to begin anew.

Low-energy coastlines have one or more of the following characteristics: a short fetch, a sheltered situation, and shallow water offshore which absorbs wave energy. In low-energy situations, sub-aerial processes may begin to dominate. Rock debris may accumulate at the cliff base, and the cliff slope angles may be lowered by weathering and mass movement.

▲ **Figure 3.13** *Cliffs made from till, resting on a low platform of Corallian Limestone at Filey, North Yorkshire. Active sub-aerial processes such as surface wash, rain splash and gulleying explain the absence of vegetation on the cliffs.*

▼ **Figure 3.14** *Slope-over-wall cliffs near Dartmouth, South Devon. The long convex coastal slope has been formed by sub-aerial and mass movement processes. This slope is covered by a thick layer of solifluction deposits. Marine action is only responsible for the low vertical section at the base of the cliffs.*

Other influences

Relief determines the height of cliffs and the area over which sub-aerial processes operate. In Co. Clare, Ireland, the coast rises more than 300 metres to form the Cliffs of Moher – the highest in the British Isles. In contrast, at Lytham in Lancashire, the till cliffs are part of a lowland till plain, and reach only modest heights of between 10 and 15 metres.

Many cliff profiles owe their shape to processes no longer operating today. Slope-over-wall or bevelled cliffs have a long-convex upper slope, and a short, vertical lower section (Figure 3.14). They are common on many coastlines in southern Britain. The long, convex slope is the result of freeze–thaw weathering and solifluction during the last glacial. At that time, with the sea-level lower than today, these slopes were well inland. Only the base part of this convex slope has been trimmed back by wave action in the last 6000 years.

Human activities can also influence the appearance of cliff profiles. Coastal protection works such as sea walls and groynes, the lowering of cliff-slope angles to reduce the risk of mass movement, and the dumping of sediment such as mining spoil on the coast, may slow marine erosion to the point where sub-aerial processes dominate. On the other hand, human interference in the coastal system may accelerate erosion. This can happen when sand and shingle which normally protect cliffs from wave attack is deliberately removed by mining. The construction of groynes and piers, which obstruct movement of sand and shingle by longshore drift, may starve beaches downstream of sediment. Again, the result will be accelerated erosion on downdrift coasts.

5 Study the cliff profiles in Figures 3.15, 3.16 and 3.17.
  a Find the location of each cliff in an atlas and comment on its likely energy environment.
  b Draw a simplified sketch of each profile which includes information on lithology, structure and possible sub-aerial processes. (Look again at Figure 3.11.)
  c Attempt an explanation of the form of each cliff profile, and the possible influence of marine erosion, lithology, structure, and sub-aerial processes.

▼ **Figure 3.16** *Chalk cliffs and shore platform at Flamborough Head, North Yorkshire. The shore platform has probably formed by cliff recession in the past 6000 years.*

▲ **Figure 3.15** *Cliffs at Stair Hole, near Lulworth, Dorset, cut in landward-dipping Purbeck Limestone*

▼ **Figure 3.17** *Cliffs at Cap de Caballeria, Menorca, formed from seaward-dipping limestone*

▲ **Figure 3.18**
*Stages of cliff
recession: basalt
cliff, arch and stack
at Dyrholaey,
southern Iceland.*

▼ **Figure 3.20**
*Shore platform cut
into Jurassic shales
at Staithes, North
Yorkshire*

▲ **Figure 3.19** *Blow hole on a basalt
cliff at South Point, Hawaii. The blow hole
has formed by collapse of a cave roof
along a master joint.*

6 With reference to the OS map
extract (Figure 3.21) and the
geological map (Figure 3.22) describe
and explain the main erosional
features on the Gower coast.

7 Study the OS map extract (Figure
3.21) and the geological maps
(Figure 3.22) of the Gower peninsula
and explain how geological structure
has affected the planform of the
Gower coast.

Landforms resulting from cliff recession

The landforms of cliff erosion and recession associated
with upland, resistant rock coasts include **caves**, **arches**,
stacks, **blow holes**, **geos** and **shore platforms**. Each
landform merely represents a stage in the retreat of
cliffs (Figure 3.18).

Caves develop below the mean high water mark.
They occur along lines of weakness such as joints and
bedding planes. Here hydraulic action loosens blocks
along the joints, causing rockfall, and abrasion scours
rock surfaces to form hollows.

When part of the roof of a tunnel-like cave collapses
along a master joint, it may form a vertical shaft which
reaches the cliff top. This is a blow hole (Figure 3.19). If
the entire roof of a cave running at right angles to the
cliff line collapses, it forms a narrow inlet called a geo.

You can see in Box 3 (page 00) that wave refraction
concentrates wave energy on headlands. Caves which
form on opposite sides of headlands may form arches
such as Durdle Door in Dorset and the Green Bridge of
Wales in Pembrokeshire. A combination of marine and
sub-aerial processes eventually leads to arch collapse,
leaving isolated rock pinnacles or stacks.

The final stage of cliff recession is a wide rock
platform. This is known as a wave-cut or shore platform
(Figure 3.20). Shore platforms are the remains of the
cliffs, and slope gently seaward. They are abraded by
wave action, and weathered by biological and chemical
processes. Some shore platforms are too wide to have
been cut in the last 6000 years, and may have formed in
previous inter-glacial periods when the sea-level was
roughly at the same height as today.

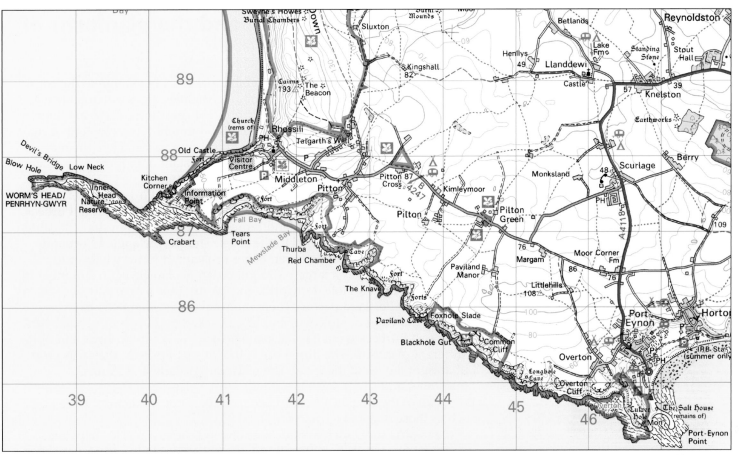

▲ **Figure 3.21** OS 1:25000 map extract of the Gower peninsula, South Wales

▼ **Figure 3.22** Geological map of the Gower peninsula

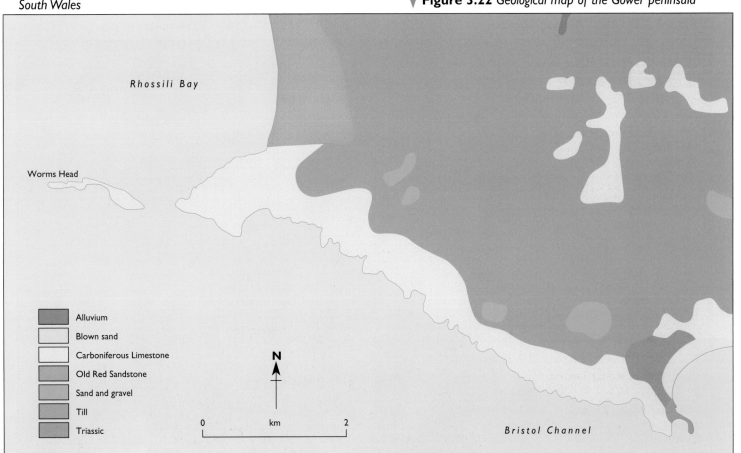

Alluvium

Blown sand

Carboniferous Limestone

Old Red Sandstone

Sand and gravel

Till

Triassic

N

0   km   2

3.5 Rock structure and the planform of coastlines

At a regional scale, the direction in which rocks run in relation to the coast has a strong influence on the planform of coastlines. Rock outcrops which run parallel to the coast often produce straight coastlines. These are known as **accordant (Pacific) coasts**. When rocks of different types crop out at right angles to the coast, the resultant planform is more varied. Here the more resistant rocks form headlands and the less resistant ones form bays. Such coastlines are described as **discordant (Atlantic)** types.

The area of south-east Dorset known as the Isle of Purbeck has examples of both accordant and discordant coastlines (Figure 3.23). In south Purbeck, the principal rock types – Portland Limestone, Purbeck Limestone, Wealden Beds and Chalk – run parallel to the coast. The steeply angled Portland Limestone is exposed at the coast. Being resistant, it forms a barrier to erosion, and produces a largely straight coastline. However, in a number of places, such as Stair Hole, Lulworth, and Worbarrow, waves and rivers have breached the Portland barrier. There the weaker Wealden Beds have been exposed and carved into coves and bays, backed by impressive chalk cliffs.

The east Purbeck coast has the same geology as south Purbeck, but quite a different planform. The difference is that at east Purbeck, the rocks crop out at right angles to the coast. The resistant Portland and Purbeck Limestones and Chalk form headlands such as Durlston Head and Peveril Point, while the weaker Wealden Beds and Bagshot Sands form wide bays such as those of Swanage and Studland.

Folded structures such as anticlines and synclines can also shape the planform of coasts. Where they trend at right angles to the coast, anticlinal or upfolded ridges may form headlands. Corresponding synclinal or downfolded areas form valleys. Rising sea-levels may flood the lower part of these valleys to form broad coastal bays.

▼ **Figure 3.23** *Isle of Purbeck, Dorset*

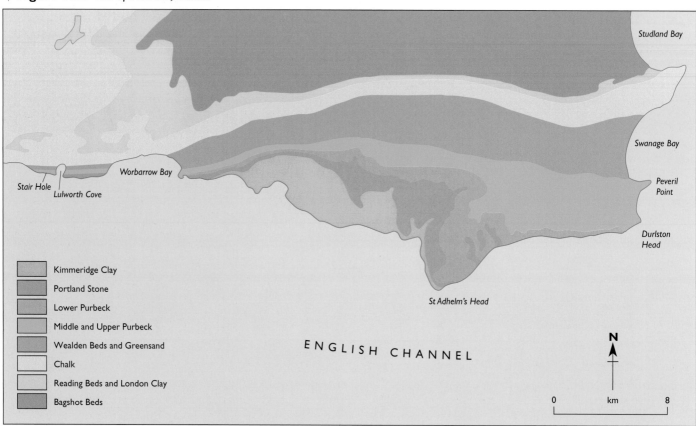

88

3.6 Beaches

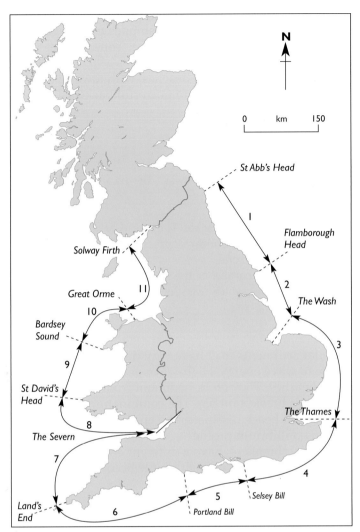

Figure 3.24 *Sediment cells around the coastline of England and Wales*

Sources of beach sediments

Beaches are accumulations of sand and shingle deposited by waves in the shore zone. The sediments which form beaches enter the coastal system from rivers, cliff erosion and wave transport.

Rivers account for around 90 per cent of beach sediments. Sand and shingle transported as bedload enter the coastal system through river mouths.

Although cliff erosion is very active on some coastlines, it provides less than 5 per cent of all beach sediments.

Some beach sediments have been combed from the sea bed. Eighteen thousand years ago northern Europe was in the grip of an ice age and the sea level was around 100 metres lower than today. As a result, a large part of Britain's continental shelf was dry land. Rivers swollen by summer meltwater deposited vast spreads of coarse alluvium on shelf areas such as the English Channel and the North Sea basin. Then, as the climate warmed and the sea levels rose, waves gradually pushed these sediments on shore to form modern beaches.

Coastal sediments remain within well-defined stretches of coastline known as **sediment cells**. Sediment cells are self-contained – there is very little movement of sand and shingle into adjacent cells. Eleven major sediment cells have been defined in England and Wales (Figure 3.24). Each cell contains several sub-cells. Sediment cells have very important implications for geomorphological processes and coastal landforms. For this reason, they are the basic unit of shoreline management in England and Wales (section 3.11).

Beach profiles

A beach profile is the cross-section of a beach from the mean (average) high water mark to the mean low water mark. The main features of beach morphology (Figure 3.25) are **berms**, **beach faces**, **breakpoint bars** and **ridges** and **runnels** (Table 3.3).

Two factors influence beach profiles: sediment size and wave type.

Figure 3.25 *A typical beach profile*

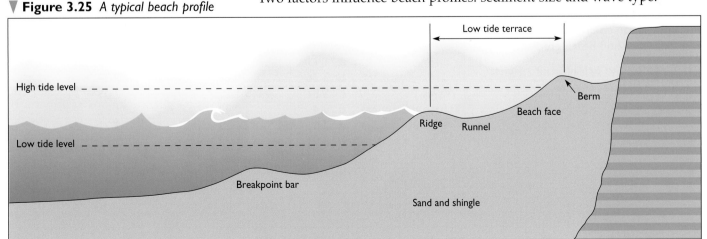

Sediment size

Shingle beaches are usually steeper and narrower than sand beaches. The main reason for this is the higher **percolation** rate of shingle compared with sand. On shingle beaches, percolation is so rapid that the swash is short, and the backwash is insignificant. So, while the swash piles up shingle, there is no backwash to drag the shingle back to the sea. Thus, the net sediment movement is in one direction, and this produces a steep beach angle (up to 10 degrees). In comparison, sand beaches, with lower percolation rates, have a longer swash and more powerful backwash. The result is a beach with a lower slope angle.

▼ **Table 3.3** *Features of beach profiles*

Feature	Description
Berm	A flat-topped ridge which develops at the limit of swash. Berms are found on steep beaches and are backshore features.
Beach face	The sloping part of a beach, below the berm.
Ridges and runnels	Linear features of the foreshore of shallow beaches. They form parallel to the shoreline. Ridges are sandy bars; runnels are linear depressions. They form as the tide migrates across the surf and swash zone.
Breakpoint bars	Ridges of sand and shingle which form parallel to the shore in the breaker zone. Associated with powerful surfing waves which transport sediment from beaches and from offshore to form breakpoint bars.

8 Table 3.4 shows data for eight samples (a–h) from each of three beaches – Crimdon, Easington and Marsden, on the Durham coast in north-east England.

a State and explain the types of waves affecting Crimdon, Easington and Marsden.
b Describe and explain the differences in beach gradients between Crimdon and Marsden.
c Test the hypothesis that wave type influences beach gradient by:
– plotting a scattergraph for phase difference (x) and beach gradient (y) for the Marsden and Easington data (N.B. show the two sites on your graph with different symbols).
– calculating the Spearman's Rank correlation coefficient for phase difference and beach gradient.
d Describe and explain your results.

Wave type

Beaches made up of similar sediments may have very different profiles. These differences are due to wave energy. High-wave-energy waves (surfing breakers) flatten beaches. They erode sediment from beaches and transport it offshore to form a breakpoint bar. The combination of a breakpoint bar and a wide, flat beach dissipates the energy of powerful surfing waves to create an equilibrium profile.

Low-energy waves (surging breakers) have a different effect. They cause a net transfer of sediment onshore, so the beach profile becomes steep, with a prominent beach face and berms.

▼ **Table 3.4** *Beach gradients and phase differences for beach profiles in Co. Durham*

			Crimdon (sand)	Easington (sand and shingle)	Marsden (sand and shingle)
Beach gradient		a	0	2.3	2.5
		b	1.5	6.5	3.66
		c	0.16	6.5	4.5
		d	2.5	5.16	5.5
		e	0.08	4.66	4.33
		f	1.33	3.83	4.66
		g	0.33	3.33	3.5
		h	1	5.6	5.5
Waves (phase difference)		a	1.11	0.68	1.55
		b	3.26	0.66	9.32
		c	5.87	1.22	2.05
		d	9.4	0.82	1.52
		e	7.81	1.22	3.21
		f	8.91	1.28	2.97
		g	10.23	1.8	2.39
		h	7.65	1.52	1.83

23

Key Terms

Percolation rate: The rate (cm^3/sec) at which water soaks into a beach. Normally, coarse beach sediment such as shingle has a higher percolation rate than sand.

....

Accretion: The accumulation of sediments (mud, sand, shingle) as a result of deposition.

....

Drift-aligned beaches: Beaches which owe their form to waves which are not refracted and whose crests are angled as they approach the shore. As a result, a lateral movement of sediment known as longshore drift occurs along the beach.

....

Beach plans

Beaches in planform may be either swash-aligned or drift-aligned. The names given to the types of beaches (e.g. spits, tombolos, barrier beaches, etc.) are not generic. This means they describe the planform of beaches but give no indication of how they have formed. For instance, tombolos may be either drift-aligned or swash-aligned features.

Swash-aligned beaches

Swash-aligned beaches form when waves approach the coast parallel to the shoreline. The swash and backwash push sediments up and down the beach along the same path. Under these conditions, there is no net movement of sediment along the beach. Waves responsible for swash-aligned beaches are fully refracted (Box 3).

Crescent-shaped bay-head beaches develop on indented coasts where waves are fully refracted. Swash-aligned beaches are usually straight, and lack the **recurved laterals** which indicate longshore drift.

Chesil Beach, Dorset

Chesil Beach (Figure 3.26) is the longest shingle ridge in the British Isles (30 kilometres). It joins mainland Dorset with the Isle of Portland. For much of its length, Chesil Beach encloses a narrow lagoon known as the Fleet. Beaches linking the mainland to an island are known as **tombolos**. Because of its straightness and the absence of any recurves, Chesil is thought to be a swash-aligned feature. It probably originated around 20 000 years ago as a bar of flint and chert in the English Channel. At that time, sea-level was 100 to 120 metres lower than today, and most of the English Channel was dry land. With the end of the Ice Age came rising sea-level; slowly, waves rolled the bar to its present position, which it reached around 6000 years ago. It seems no more than chance that Chesil finally came to rest between an island and the mainland. If it had formed across a bay (like Slapton Sands in Devon) it would be classified as a barrier beach instead of a tombolo.

► **Figure 3.26** *Chesil Beach, Dorset*

▼ **Figure 3.27** *Slapton Sands, south Devon*

Slapton Sands, Devon

Slapton Sands extends from Beesands to Strete in south Devon (Figure 3.27). It is part of a barrier beach: a shingle ridge which stretches for 9 kilometres across Start Bay. In planform, Slapton Sands is approximately 100 metres wide, up to 8 metres high and slightly concave. On its landward side, the beach has impounded a number of streams to form small lakes. Slapton Ley is the largest of these lakes.

Eighty-six per cent of the barrier consists of flint. There are no local sources of this rock, though deposits containing flint crop out on the floor of the English Channel. The origin of the barrier is thought to be similar to Chesil Beach. Wave processes and rising sea-levels in the post-glacial period combined to push the shingle towards the modern coastline, where it came to rest 6000 years ago.

Box 3 Wave refraction

Wave refraction describes the bending of oblique waves in the foreshore zone until they break parallel to the shore (Figure 3.28). Refraction depends on the depth of water offshore. Where the sea is shallow, waves are slowed through frictional drag. In deeper water, the wave moves faster. This causes the wave fronts to bend and take on the shape of the coastline. When the waves are fully refracted, and parallel to the coastline, the swash and backwash follow the same path on the beach. The result is swash-aligned beaches.

However, waves are not always fully refracted. Some waves break obliquely and their swash follows a diagonal path across the beach. This produces a net lateral movement of beach sediment along the beach. This process is known as 'longshore drift', and is responsible for drift-aligned beaches such as spits.

Wave refraction also leads to an uneven distribution of energy on coastlines. Refraction concentrates wave energy on headlands but disperses energy in bays (Figure 3.28). Thus, headlands are areas of concentrated erosion, while bays are areas of **accretion**.

▼ **Figure 3.28** *Wave refraction and the distribution of energy on a coast*

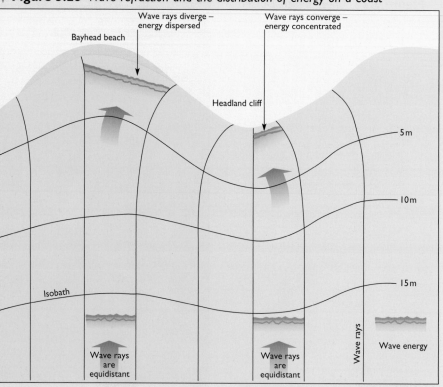

Key Terms

Swash-aligned beaches: Beaches that owe their form to waves which are fully refracted and whose crests approach parallel to the shore. Many swash-aligned beaches have a crescent-like outline in planform.

Longshore drift: The lateral transport of sediment along a beach and offshore caused by waves which are not fully refracted. Longshore drift is responsible for the recurves on spits.

► **Figure 3.29** *Longshore drift: the flow of coastal sediment produced by wave and current action when the angle of wave approach is oblique to the coastline.*

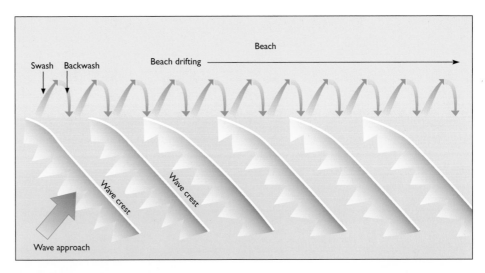

Drift-aligned beaches

Drift-aligned beaches such as spits and **barrier beach islands** are features of open coastlines. Here, waves are rarely fully refracted and longshore drift (Figure 3.29) takes place. Spits often form across estuaries (e.g. Spurn Head, Dawlish Warren) or where there is an abrupt change of direction in the coastline (e.g. Hurst Castle). Evidence for growth of spits by longshore drift is provided by recurved shingle ridges or laterals. Each ridge represents the former seaward (distal) end of the spit. Its recurve shape is due to the refraction of waves and currents around the end of the spit. The distribution of spits in the British Isles correlates closely with coasts of low tidal range. A low tidal range concentrates wave action in a narrow vertical band of coast. This appears to be important in shaping sand and shingle into spits and other drift-aligned beach.

Blakeney Point, Norfolk

Blakeney Point is a large spit on the north Norfolk coast (Figure 3.30) which has formed across the mouth of the River Glaven. It consists of a shingle bank (mainly flint pebbles) which is 15.5 kilometres long, 200 metres wide and 9 metres high. A series of recurved shingle ridges mark the former distal end of the spit. They suggest that Blakeney Point has gradually extended westwards by longshore drift. At the same time, Blakeney Point appears to have been moving onshore at a rate of about one metre a year. The source of shingle at Blakeney is uncertain. One possibility is that it was originally glacial outwash deposited north of the present-day coastline during the last glacial period when the sea level was lower.

◄ **Figure 3.30** *Blakeney Point*

Scolt Head Island, Norfolk

Scolt Head Island (Figure 3.31) is situated about 15 kilometres west of Blakeney Point on the north Norfolk coast. In planform, it is similar to Blakeney Point, except that it is detached from the mainland at its eastern end by a deep, narrow channel known as Burnham Harbour. This makes Scolt Head a barrier beach island. Scolt Head's shingle recurves also show that, like Blakeney, it has extended in a westerly direction by longshore drift. However, its further growth is restricted by the deep water of Brancaster Harbour. There is evidence to show that Scolt Head is moving landwards at a rate of one metre per year.

◀ **Figure 3.31** *Scolt Head Island*

Key terms

Pioneer communities: The plant communities that are first to colonise new areas of land such as mudflats and sand dunes.

....

Glacial outwash: Sand and shingle deposits laid down by meltwater streams draining icefields and glaciers.

....

3.7 Mudflats and salt marshes

Mudflats and **salt marshes** are landforms of sheltered coastlines where wave action is weak. In low-energy environments such as estuaries, and on the landward side of spits, tidal currents deposit fine sediment. This accretion of sediment leads to the growth of mudflats and salt marshes. Mudflats and salt marshes are most typical of coastlines which have large tidal ranges. This is because large tidal ranges generate powerful currents which can transport huge amounts of sediment. The sediments which form mudflats and salt marshes come from rivers, the sea bed and cliff erosion.

Morphology of mudflats and salt marshes

In cross-section, mudflats and salt marshes have a distinctive morphology (Figure 3.33). They have a shallow gradient which slopes down to the low water mark. This is interrupted in two places. A low cliff, up to a metre high, separates the vegetated low marsh from the unvegetated mudflats. There is also an abrupt break of slope between the high marsh and the low marsh. A dense network of creeks and tributaries drain the marsh and bring water in on the flood tide. The very high drainage density of the creeks reflects the huge volumes of water which flood into, and drain away from, the marsh on the tide each day.

The morphology of mudflats and salt marshes results from the interaction between tidal currents, sediments and marsh vegetation. Strong currents are found in the creeks and on flood and ebb tides (i.e. in the middle of the tidal cycle). Thus, rates of accretion are low in the creeks and in the middle marsh. On the other hand, slack water occurs at high tide and low tide. Hence, accretion rates are highest around the mean high tide mark (i.e. in the high marsh) and mean low tide mark (i.e. on the mudflats).

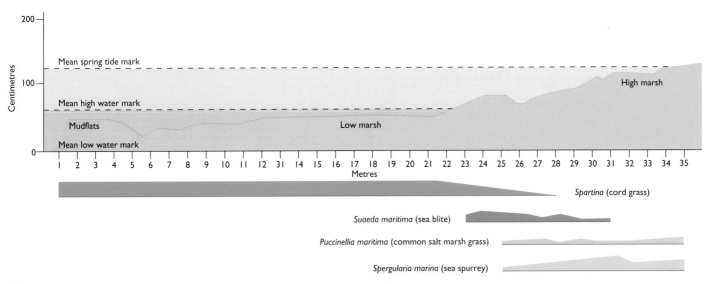

▲ **Figure 3.32** *Cross-section of a salt marsh at Alnmouth, Northumberland*

▲ **Figure 3.33** *Pioneer vegetation (cord grass and sea blite) on a salt marsh at Alnmouth.*

Ecological succession on salt marshes

Vegetation plays a vital role in the development of salt marshes (Figure 3.33). Many marshes show a clear zonation of species which is closely related to height above mean sea level. Such a plant succession in a saline, waterlogged environment is known as a **halosere**.

▼ **Table 3.5** *Plant zonation on salt marshes*

	Environmental conditions	Plants
Mudflats	High salinity levels; low oxygen levels in mud; high turbidity; long periods of flooding on each tidal cycle.	No plants, only algae
Low marsh	Less hostile conditions compared with mudflats. Salinity and turbidity still high and oxygen levels low.	Cord grass (*Spartina*) and glasswort (*Salicornia*) are two common pioneer species. Sea blite and sea purslane grow on better-drained areas.
High marsh	Flooding only occurs on spring tides. Salinity levels are relatively low and soil develops.	A wide variety of species grow, including salt marsh grass, sea rush, sea lavender, sea aster, sea blite and sea purslane.

Ecological succession on salt marshes (Table 3.5) follows a number of stages:

■ Colonisation by pioneer species such as cord grass and glasswort. These plants slow the movement of water and encourage rapid accretion (1–2 cm per year). Their roots help to stabilise the mud.

■ Accretion of sediment causes the marsh to increase in height. Environmental conditions improve; salinity falls and the period of flooding on each tidal cycle shortens. All of this allows less-tolerant species to invade. Biodiversity and plant cover increase, and species such as sea rush, sea aster, sea plantain, sea lavender, salt marsh grass and common scurvy grass dominate.

■ The marsh height stabilises a metre or so above the mean water mark. With only occasional flooding on the highest spring tides, vertical accretion ends. Salinity levels are low and soil develops.

► **Figure 3.34** *Alnmouth, Northumberland 1:25 000 map extract*

9 Using the evidence of Figures 3.34 and 3.35:

a Suggest two possible sources of sediment for the formation of salt marshes in the Aln estuary.

b Name and explain two possible reasons for the growth of salt marshes at the mouth of the River Aln.

c Explain how the growth of the salt marsh in Figure 3.34 is limited by the shape of the Aln estuary.

Key Terms

Plant succession: The sequence of vegetation changes which occurs as plants establish themselves in an environment.

....

Zonation: The geographical expression of plant succession, with older and more diverse plant communities located at increasing distance from the shore.

....

▼ **Figure 3.35** *The Aln estuary*

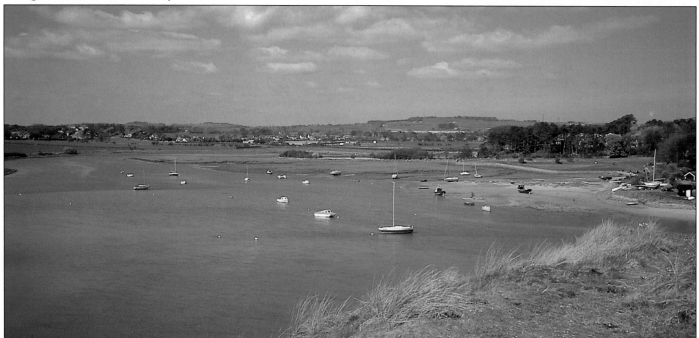

3.8 Sand dunes

Sand dunes are the only important coastal landforms produced by the wind. The wind blowing across a dry sand surface is able to transport sand grains. Larger particles roll intermittently across the surface, in a process known as **creep**. Smaller particles move by a skipping process up to a metre or so above the ground. When they hit the ground they set other sand grains in motion. This is called **saltation**.

Blown sand accumulates around inanimate objects such as driftwood, which reduce wind speeds near the ground. However, once buried, there is no further resistance to wind flow and sand deposition ends. Sand dunes only form when vegetation provides the obstacle to blown sand. Marram grass thrives in dry sand dune environments. When submerged by sand it grows rapidly to provide further resistance to the wind and encourages more sand deposition.

Conditions for dune formation

Extensive areas of sand dune can only develop where conditions are favourable. These favourable conditions include:
- A plentiful supply of sand.
- A shallow offshore zone with gentle gradients, where large exposures of sand dry out at low tide.
- An extensive backshore area where sand can accumulate.
- Prevailing onshore winds.

▼ **Figure 3.36** *The Ainsdale dunes from the first dune ridge*

All these conditions are found on the Merseyside coast between Liverpool and Southport, where they give rise to the largest dune complex in England. The Ribble and Mersey estuaries are the main sources of sand, and the lowland coast allows ample space for dune building. Meanwhile, the large tidal range (over 4 metres), shallow seas and prevailing onshore westerlies all favour dune development (Figure 3.37).

The primary **plant succession** on the sand dunes often shows a clear **zonation**. Nearest the shore, where the youngest (yellow) dunes are found, vegetation cover is patchy and few species survive. With increasing distance inland, the dunes get older (grey dunes) and the environment changes (Table 3.6). As a result, plant cover, productivity and biodiversity all increase.

▼ **Figure 3.37** *Cross-section through a sand dune complex*

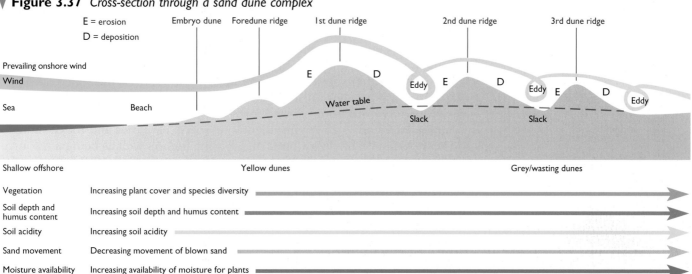

Dune systems often form a series of ridges parallel to the coastline (Figure 3.38). The dunes decrease in height inland as sand supply diminishes. Troughs or **slacks**, formed by localised eddying of the wind, separate the dune ridges. Fresh water seeps to the surface in the slacks and supports aquatic plants such as reeds, rushes and flag iris. The dunes usually have steep windward slopes and a less steep leeward slope. Sand eroded from the exposed windward slopes is deposited on the sheltered leeward slopes, so that gradually the dunes migrate inland. For this reason, the dunes increase in age with distance from the shoreline.

Sand dunes are a fragile environment. Destruction of the vegetation cover (e.g. grazing by rabbits or human activities causing trampling or firing) can lead to massive wind erosion.

▼ **Table 3.6** *Vegetation succession on sand dunes*

Dunes	Environment	Plant species
Embryo dunes	There is little fresh water for plants and the sand is extremely porous. Plants are blasted by blown sand. Further problems for plants are caused by salt spray and shifting sand.	Sea couch grass, which can extract fresh water from salt water, is one of the few species found here.
Foredune	This is usually 2–3 metres above the beach where conditions are less saline. The ridge is exposed and unstable.	Marram grass begins to colonise and competes with sea couch grass. Marram is well adapted to sand dunes: it has deep roots; sand deposition stimulates its growth; it has thick, shiny cuticles on its leaves to reduce moisture loss.
First dune ridge	This is the most prominent relief feature, and can be 10–30 metres high. The sand here is more stable and contains some humus.	Almost all marram grass. Marram slows down the wind speed and reduces sand movement.
Older dune ridges	Dunes become progressively lower inland as the sand supply diminishes. Dunes become grey as humus builds up and moisture retention increases. Soil acidity increases.	Environment is less harsh. There is shelter, fresh water and soil. Many new species appear (creeping willow, sea buckthorn, ragwort, fescue grass). Marram grass dies out.
Dune slacks	These are depressions between the dune ridges. They are sheltered and often marshy where the water table reaches the surface. They are waterlogged after heavy rain.	A wide range of plants grow here. Some, such as flag iris and bog myrtle, are typical wetland species.

3.9   The coast as a resource and a hazard

The coast is both a resource and a hazard. As a resource, the coast has great value for many economic activities. It provides sites for ports and harbours; tidewater locations for heavy manufacturing industries; reclaimed land for farming; and beaches for tourism. Unfortunately, the coast has also been exploited as a dumping ground for waste products. The coast has enormous value as an environmental resource. Estuaries and rocky shores are important habitats for birds and marine life. As well as being a resource, the coast is also a hazard. Along many stretches of coastline, flooding and erosion pose serious threats to settlements, industrial sites and farmland.

Human intervention in the coastal system takes two forms. First, intervention may aim to exploit and develop economic resources. Land reclamation in estuaries, and the construction of harbours and resorts fall into this category. Second, intervention may be necessary to protect resources. The most obvious example is coastal defence, designed to control flooding and erosion.

In the past, human intervention has tended to disrupt the natural processes of erosion, transport and accretion of sediment, causing new problems. Future intervention must be sustainable in the long term and not damage natural resources. It must recognise that the coastal system is **holistic**: that changing one part inevitably has an impact somewhere else. And finally, it must take account of environmental as well as economic costs and benefits.

3.10 The threat of global warming

During the twentieth century, average global temperatures increased by 0.6°C. A further increase of between 1.6°C and 4°C is expected by 2100. Global warming has two important consequences for the coast. First, it has caused sea-level to rise by 10–25 cm in the last 100 years. The Intergovernmental Panel on Climate Change (IPCC) forecasts a further rise of 50 cm by 2100. And second, as a result of global warming, the frequency and power of storms which hit the coast are likely to increase in the future.

These changes will increase the hazards of coastal flooding and erosion. Worldwide, up to 100 million people living in low-lying coastal areas could be at risk. Densely populated deltas in poorer countries, such as the Nile Delta in Egypt, and the Ganges–Brahmaputra Delta in Bangladesh, will be hardest hit. Some tropical island groups, which are only a few metres above sea-level (including the Maldives in the Indian Ocean and the Cook Islands in the Pacific Ocean) could disappear altogether (Figure 3.39). Even rich countries like the Netherlands, which has large tracts of land close to sea-level, will have to invest heavily to keep the sea out.

The UK is not immune to these problems. Rising sea-level and more frequent storms are likely to increase damage to coastal settlements, harbours and other infrastructural features. In some places, the coastline might retreat by up to a kilometre or more. South-east Britain is most vulnerable to flooding. Rising sea-levels in the south-east would be made worse because the region is also sinking as the process of isostatic adjustment continues. Beaches, sand dunes, mudflats and salt marshes – the coast's natural defences – are threatened by rising sea-level. Indeed, almost one-third of EU beaches are already thought to be eroding.

▼ **Figure 3.38** *Some places could completely disappear beneath the sea if sea levels continue to rise because of global warming*

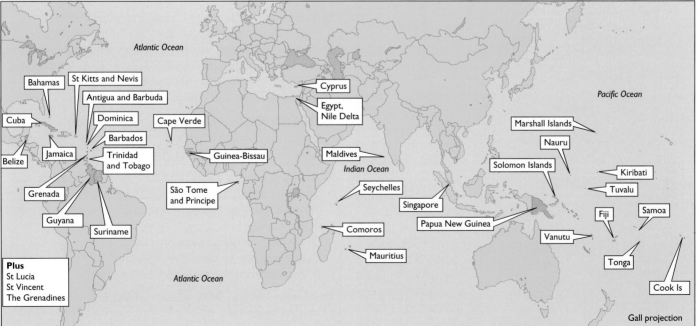

3.11   Shoreline management plans

In England and Wales, local government authorities and the Environmental Agency are developing Shoreline management Plans (SMPs) to co-ordinate planning of the coastline. The basic spatial units for these plans are sediment cells. A sediment cell is defined as 'a length of coastline which is relatively self-contained as far as the movement of sand and shingle are concerned and where interruption of such movement should not have a significant effect on adjacent cells'. There are 11 sediment cells in England and Wales. Each cell is divided into sub-cells. The east coast of England from Flamborough in Yorkshire to Gibraltar Point in Lincolnshire is one such sediment cell (Figure 3.39). The Humber estuary is a sub-cell within this larger unit.

Viewing this stretch of coast as a single unit makes sense in terms of the physical processes operating there. At Holderness, in East Yorkshire, the till cliffs are eroding at an average rate of 2 metres per year. Hard-engineering structures, including sea walls, groynes and armour blocks protect important settlements (Hornsea, Mappleton and Withernsea) and infrastructure threatened by erosion. Elsewhere on the Holderness coast erosion is unchecked, with loss of farmland.

Current policies make no attempt to halt the coastal erosion of farmland. Erosion may be undesirable, but stopping it could have severe knock-on effects elsewhere in the sediment cell. Sediments eroded from the Holderness cliffs are transported south to the Humber estuary and to the coast of Lincolnshire. Here they build up mudflats and salt marshes. This is vital. The accretion of sediment helps to protect the densely populated and industrialised areas along these coasts from flooding. The **intertidal areas** also reduce wave and tidal energy and prevent erosion.

Therefore, it is essential to understand how the coastal system within each sediment cell operates. Stopping localised erosion at Holderness would simply create problems elsewhere. The cost of losing farmland at Holderness must be set against the cost of flooding and the loss of estuarine habitats in the Humber estuary and on the Lincolnshire coast.

▼ Figure 3.39 *Sediment cell 2: Flamborough Head to Gibraltar Point*

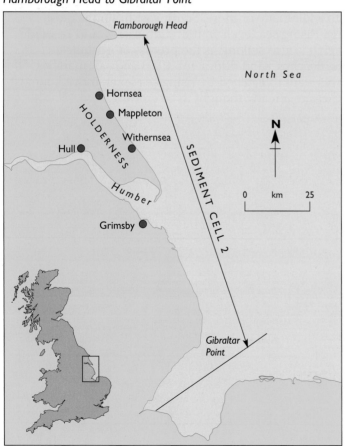

Box 4 Sediment cells

■ **Sediment cells** are the basic unit of shoreline management (as drainage basins are in river management). Because sediment cells have 'closed' sediment budgets, it is recognised that removing sediments or interrupting sediment movements can lead to increased risks of coastal erosion and coastal flooding.

■ **DEFRA**, the Environmental Agency and local authorities produce SMPs for each of the 11 sediment cells in England and Wales (Figure 3.23).

■ **SMPs** provide a sustainable, long-term strategy for the coastline of England and wales. There are four options for managaing any stretch of coastline:
1 do nothing
2 hold the existing defensive line
3 advance the existing defensive line
4 retreat from the existing defensive line.

■ Where possible, SMPs favour options that work with natural processes.

Example: The Humber estuary

Managing the Humber estuary

Background

The Humber is one of Europe's largest estuaries, 120 km long and 14 km wide at its broadest. It drains about one-fifth of England's river waters (Figure 3.40). Humberside has a population of nearly 500 000, three important ports (Hull, Immingham and Grimsby) and major tidewater industries such as oil refining, metal smelting and electricity generation. Most of the region's population and economic activity is situated between 0 and 2 metres above sea-level. The rising sea-levels forecast in the next 50 years will greatly increase the risk of flooding in the region.

The estuarine system

Salt marsh represents only 2 per cent of the Humber estuary's total area. This is well below the average for all estuaries in the UK. Because Humberside is a highly urbanised and industrial region, large areas of salt marsh and mudflats have been reclaimed for industry, agriculture and housing. Today, the remaining intertidal areas face a double threat: rising demand for housing and economic activities; and rising sea-level caused by global warming.

Protecting the Humber's intertidal areas is crucial to the economy of the region and the ecology of the estuary. The intertidal areas absorb pollutants from organic and inorganic waste. Erosion can lead to the release of these toxic substances which can then enter marine food chains.

Intertidal areas also absorb the energy of waves and tides, and reduce erosion.

Salt marshes and mudflats are essential for coastal defences. Natural accretion of sediment builds up the intertidal areas and counters the rise in sea level. Recent studies have shown that the rates of net accretion in the estuary have slowed to almost zero in the last few years.

The estuary has international importance for migrating birds, especially waders and wildfowl. Each winter, the estuary is home to 35 000 knot and 27 000 golden plovers.

The mudflats and salt marshes are rich in invertebrates (such as worms) which form the basis of the estuary's food chain.

Many fish spawn in the shallow water of the estuary.

Planning responses

Strategies for managing the Humber estuary must take account of conflicting interests. Industry needs land adjacent to tidewater for new investments in ports and manufacturing activities. Farmers want action to protect their land from flooding. Conservationists want to see an expansion of the intertidal areas (Figure 3.41). Their continued loss would be calamitous for wildlife in the estuary. Meanwhile, local authorities have a duty to provide protection against flooding for the 500 000 people who live on Humberside.

▼ **Figure 3.40** *The Humber estuary*

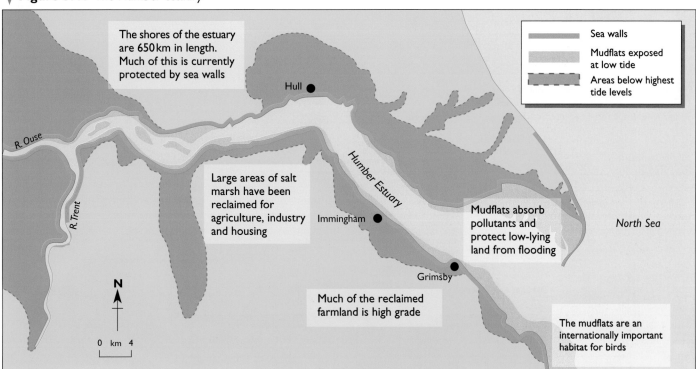

Sea walls

Mudflats exposed at low tide

Areas below highest tide levels

The shores of the estuary are 650 km in length. Much of this is currently protected by sea walls

Hull

R. Ouse

R. Trent

Humber Estuary

Large areas of salt marsh have been reclaimed for agriculture, industry and housing

Immingham

Mudflats absorb pollutants and protect low-lying land from flooding

North Sea

Grimsby

N

Much of the reclaimed farmland is high grade

The mudflats are an internationally important habitat for birds

0   km   4

The local authorities around the estuary and the Environmental Agency are currently developing a shoreline management plan to tackle these problems. A balance will be difficult to achieve. Planning is influenced both by government and EU policies. For instance, under the EU's European Habitats Directive, there is an obligation to protect the estuarine environment, which is of international importance. UK government policies insist on sustainable management and that they must also work with natural processes rather than fight against the natural forces of the sea (Box 4). However, the need to protect people, key properties and land from flooding is acknowledged.

Management options

There are four available strategies for coastal management (Box 4). These strategies can be summed up as:

1 defend the existing coastline
2 allow natural rpocesses to occur unhindered, leading to further erosion, flooding and loss of land.

The two main approaches to coastal defence are 'hard' engineering and 'soft' engineering. In the past coastal defence concentrated exclusively on building hard structures, such as sea walls, groynes and flood embankments. However, hard engineering is costly, non-sustainable and often has adverse environmental effects (Box 5).

The alternative approach, 'soft' engineering, seeks to work with nature, allowing natural systems to adjust to the energy of waves, tides and wind (Box 5). Examples of 'soft' engineering include managed retreat and beach nourishment. Managed retreat involves abandoning existing sea defences which encourage erosion of mudflats and salt marshes. Instead, a new line of sea defences is set back inland (Figure 3.41). This creates an extensive intertidal area between old and new defences. Managed retreat is controversial because it allows some land (and settlements) to flood. Its advantages are: extensive intertidal areas absorb wave energy and provide habitats for wildlife; it is sustainable (continually raising flood embankments is not); and there is less spending on maintenance of existing coastal defences.

306

Key Terms

Sustainable management: Management strategies which do not damage coastal ecosystems and which in the long term are both affordable and practicable.

Intertidal areas: The shore exposed at low tide. It may comprise shingle, sand and mud. The extent of intertidal areas depends on the tidal range and the gradient of inshore areas.

10 Write a summary report of the problems caused by rising sea-levels in the Humber estuary.

a Set out what you consider should be the main guidelines for managing the estuary (e.g. sea defences should be sustainable, etc.).

b Devise a stategy to realise these policies (e.g your choice of coastal defences, areas to be protected, etc.). Explain the costs and benefits of your proposals.

11 In the context of coastal management, what is meant by 'sustainability'? Why should sustainability be a key objective of many coastal management schemes?

12 Use the internet as a data source and, with the aid of a sketch map, describe and explain the management trategies used along a given stretch of coastline (sediment sub-cell scale).

▼ **Figure 3.41** *Managed retreat*

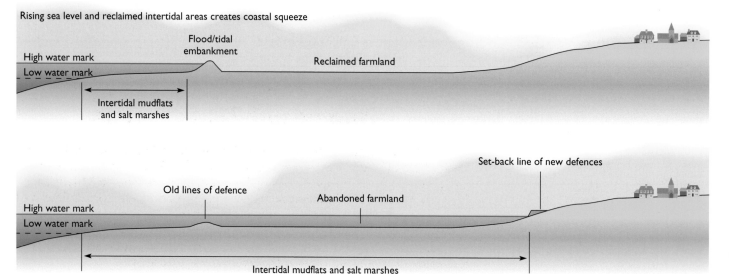

Box 5   Hard and soft engineering approaches to coastal defence

■ Hard engineering:

Hard-engineering structures include sea walls, revetments, groynes, armour blocks, flood embankments and flood barriers (Figure 3.42). Their environmental impact is generally greater than soft engineering. In particular, they disrupt the movement of sediment in the coastal system. Hard-engineering structures often cause a reduction in the width of the shoreline as low-lying backshore areas are reclaimed behind sea defences. This leads to a decrease in the size of shore habitats, a phenomenon known as **coastal squeeze**.

Defensive structures designed to absorb wave energy (e.g. revetments) lead to the build-up of sediment in the lee of the structure. Sea walls which stop cliff erosion may disrupt longshore sediment transport. The result may be beach starvation and accelerated erosion on adjacent coastlines. Sea walls also cause wave reflection – waves bouncing off sea walls hit oncoming waves which sets up strong vertical currents. These currents may erode the foreshore and undermine sea walls themselves. Groynes, designed to stop the longshore movement of sediment and encourage beach accretion, also starve coasts of sediment and cause accelerated erosion there. Some structures, such as revetments, groynes and armour blocks, are visually intrusive. Others may seriously limit public access and recreation on beaches. Flood barriers in estuaries can severely disrupt natural processes and destroy saltwater ecosystems in estuaries and other coastal inlets.

▲ **Figure 3.42** *Hard engineering sea defences (sea walls, groynes, armour blocks) at Lyme Regis, Dorset.*

■ Soft engineering:

Soft engineering is more environmentally friendly than hard engineering and works *with* natural systems. It requires less space and therefore reduces coastal squeeze. Soft-engineering approaches include beach nourishment (adding sand and shingle to beaches from elsewhere); the set-back of structures (managed retreat); and plantations of osier hedges and marram grass. Current policies favour soft-engineering approaches. Governments encourage planners to work with the coastal environment rather than fighting against the forces of the sea. Soft-engineering has less impact on the environment, is more sustainable, and requires less maintenance.

3.12   Flood protection: the Netherlands Delta Project

The Rhine, Maas and Schelde rivers enter the North Sea in the south-west Netherlands (Figure 3.43). Over the centuries, the Dutch have reclaimed large areas of land in this low-lying delta region. However, on 31 January 1953, an unusual combination of events – gale-force winds, spring tides and low pressure in the North Sea – overwhelmed the region's coastal defences and caused disastrous flooding. The floods inundated 2000 km² of farmland; 1800 people lost their lives; thousands of livestock were killed; and bridges, roads and other infrastructure were destroyed.

Following strong political pressure, in 1957 the Dutch government, proposed an ambitious hard-engineering solution to prevent future flooding of the delta. With the exception of the New Waterway (which links the busy ports of Rotterdam and Europoort with the North Sea), all other inlets in the delta were to be sealed off with dams (Figure 3.44). The plan, which involved the construction of 10 dams and 2 bridges, reduced the length of the coastline from 800 kilometres to just 80. A new road network using the dams and shortened coastline opened up what hitherto had been an isolated region. Freshwater lakes created behind dams offered opportunities for recreation.

Because flood control had priority, the delta plan had both environmental and economic costs. The conversion of salt-water inlets to freshwater lakes destroyed important habitats for birds and invertebrates, and the spawning grounds for many fish. It also destroyed important fisheries (including shellfish) and left some fishing communities and harbours cut off from the sea.

The final stage of the project – enclosing the eastern side of the Schelde – was completed in the mid-1980s (Figure 3.44). Pressure from environmentalists forced the government to change its original plan. A storm surge barrier, with sluice gates which could be lowered to prevent flooding, replaced the original plan for a permanent barrier. With the sluice gates left open for most of the time, the eastern Schelde remained tidal and salt water. Valuable areas of salt marsh and mudflats were preserved for wildlife, and the important local oyster fishery continued.

► **Figure 3.43** *To prevent future flooding in the Netherlands delta area, new dams were built across inlets, shortening the coastline.*

13 With reference to Figure 3.43, suggest one possible reason why managed retreat would not have been a practical option for coastal managers in the south-west Netherlands.

▼ **Figure 3.44** *East Schelde (Oosterschelde) storm barrier has sluice gates which can be closed during times of flooding.*

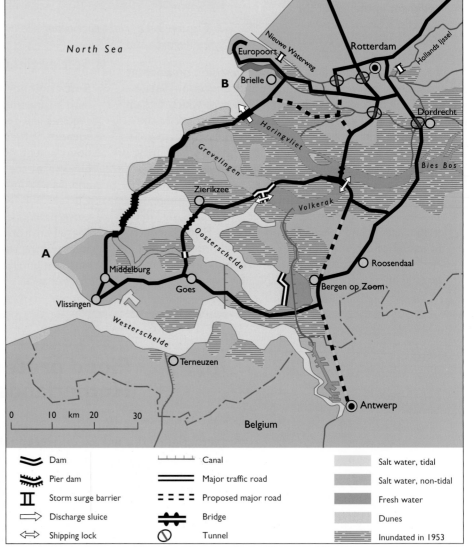

3.13 Coastal pollution in Easington, Co. Durham

The coal mining industry dumped colliery waste on Durham's beaches from 1906 to 1993. In the 1980s, this dumping amounted to 2 million tonnes per year. Half of this was tipped into the sea at Easington, along with liquid waste from tailings and washeries. Tipping colliery waste into the sea was the cheapest method of disposal. The coal industry argued that alternative methods, although less damaging to the environment, would raise production costs and put mining jobs at risk.

The environmental impact of coastal tipping was disastrous. For most of the twentieth century, the Durham beaches were the most polluted in Europe. A Royal Commission in 1984 called them 'a national disgrace'. Only since the end of deep mining has attention turned to the reclamation and rehabilitation of the coastline and beaches between Seaham and Blackhall (Figure 3.45).

Natural response of the coastline

After 1945, rates of tipping far exceeded erosion. Six collieries tipped 2.5 million tonnes of waste per year into the sea along the Durham coast. With the rate of input exceeding the rate of transport through wave action, a huge artificial beach made of colliery spoil developed at Easington (Figure 3.46). Its effect was to protect the Magnesian Limestone cliffs from erosion. For nearly a century the cliffs were affected by sub-aerial processes only. Rates of erosion became minimal and vegetation colonised the cliffs. In some places, cliff profiles were further modified by tipping over the cliffs themselves.

With the closure of Easington colliery in 1993, tipping came to an end. This produced major changes in the **sediment budget**. Now, output of sediment exceeded input. The coastal system adjusted to the new conditions by eroding the beach. Between 1993 and 1998, nearly 80 per cent of colliery spoil was removed by the natural processes of erosion. In planform, the beach underwent a dramatic narrowing. A line of low cliffs made from colliery shale formed at the back of the beach. In 1995, these cliffs were around 2 metres high. Three years later, as the wave action cut the beach even further back, these cliffs reached a height of 4 metres (Figure 3.47). By 1998, waves had again reached the base of the limestone cliffs, causing renewed marine erosion (Figure 3.48).

▼ **Figure 3.45** *Pollution on the Durham coast*

Area of pollution

● Former coastal collieries tipping waste into the North Sea

0   km   5

► **Figure 3.46** *The impacts of colliery waste tipping on the coast at Easington, County Durham. The photograph shows the coast in 1987, after 80 years of tipping over the cliffs and by conveyor belt into the sea.*

► **Figure 3.47** *The coast at Easington in 1997, four years after tipping ended. The changes are due almost entirely to natural processes. Wave action has removed large amounts of colliery spoil from the beach and has exposed the base of the Magnesian Limestone cliffs to erosion.*

▼ **Figure 3.48** *The response of Easington Beach to the cessation of tipping and the onset of marine erosion*

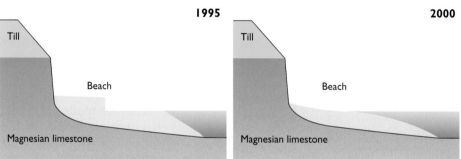

1993

Till

Beach

Magnesian limestone

1995

Till

Beach

Magnesian limestone

2000

Till

Beach

Magnesian limestone

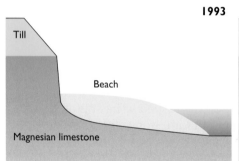

14 From the section on Easington:
a Explain, in terms of systems, how the beach at Easington changed in the period 1993 to 1998. Use the following terms: input/output of sediment; input/output of wave energy; sediment store; feedback; equilibrium.
b Describe the likely circumstances (i.e. inputs, outputs, etc.) when the beach at Easington reaches equilibrium.

15 Using examples in this chapter, comment on the view that human intervention in the coastal system has often created more problems than benefits.

Reclamation: turning the tide

The Durham coastline, with its Magnesian Limestone cliffs, is unique in the British Isles. In the past, the cliffs and the immediate coastal fringe supported flower-rich grasslands with notable plants, including bloody cranesbill and several species of orchid. Despite its status as a Site of Special Scientific Interest (SSSI), years of pollution and industrialisation mean that this natural environment survives in just a few isolated pockets.

The decline of the coal industry provided the opportunity for environmental improvement. A £10 million project called *Turning the Tide* began in 1997. It aimed to clean up the environment along a 20km stretch of coastline between Crimdon and Sunderland and restore the limestone grasslands. The project is part of the East Durham Task Force, a regeneration initiative. Its participants include Durham County Council, Easington District Council, the Countryside Agency, English Nature and the National Trust (which owns nearly 8 km of the coastline). The project, awarded a grant by the Millennium Commission, was largely complete by 2001.

In 1997, work began to remove two shoreline spoil heaps of 1.5 million tonnes at Easington and Horden. Unless these were removed they would continue to pollute the beaches. Some of the spoil went back to the Easington Colliery site which was landscaped and where a visitor centre is planned. Removing all the spoil is impracticable, however, and the sea will be left to complete the task over the next 15–20 years.

Turning the Tide is not just concerned with the impact of coal mining on the coast. Much of the coastal strip is intensively farmed arable land. The use of chemical fertilisers has damaged the natural flora and fauna of the limestone grasslands along the cliff tops. The project aims to acquire as much of the coastal strip as possible and return it to natural grassland. A cycle route through the area has also been completed, which links into the national cycle network.

Summary

The coastline is an open system with inputs and outputs of energy (from waves, winds and tides) and materials (geology, sediments and living organisms). Interaction between these components and the physical environment (i.e. processes) results in coastal landforms.

Sea-level change in the last 20 000 years has had a major impact on coastlines. Sea-level rose rapidly between 20 000 and 6000 years ago. Rias and fjords formed on submerged upland coasts, estuaries formed on submerged lowland coasts. Rising sea-level swept up sand and shingle from continental shelf areas to form beaches along the present-day coastline. Stable sea-level in the last 6000 years has allowed sand dunes, salt marshes and beaches to form.

Waves are the main energy input to the coastal system and this energy is distributed unevenly on coasts. Three factors influence wave energy: fetch, wind speed and wind duration. Waves are responsible for erosion, transport and deposition. Erosional landforms include cliffs and shore platforms; depositional landforms comprise various beach forms. Wave refraction concentrates wave energy on headlands where erosional features such as cliffs, caves and arches develop. Wave refraction disperses wave energy in bays where depositional features such as beaches dominate.

Cliff profiles are the outcome of marine erosion, sub-aerial processes and coastal geology over time. Along some coasts, sub-aerial processes such as mudslides and rotational slides dominate cliff forms. Some cliffs owe their form to past processes (e.g. slope-over-wall cliffs).

Rock structure controls the planform of some stretches of coast. Rocks which trend parallel to the coast give rise to accordant (Pacific) coastlines. Such coastlines are uniform and straight in planform. Where rocks crop out at right angles to the coast, indented coastlines form, with headlands (made of more resistant rocks) and bays (made of less resistant rocks).

Beach profiles result from the interaction of wave energy and sediments (sand and shingle). High-energy waves (surfing breakers) produce wide, low-angled beaches, and well developed break-point bars offshore. Low-energy waves (surging breakers) form steep beaches, with prominent beach faces and berms. Rapid percolation rates on shingle beaches are responsible for steep beach profiles and narrow planforms. Sand beaches are both less steep and wider than shingle beaches.

Wave refraction is a major influence on the planform of beaches. Waves which are not fully refracted set up a longshore drift of sediment and form drift-aligned beaches such as spits, with distinctive recurves. Fully refracted waves create straight or concave beaches. Barrier beaches and bay-head beaches are associated with refracted waves.

Vegetation influences the development of some coastal landforms. Salt-tolerant plants such as cord grass and glasswort promote the accretion of silt and clay to form mudflats and salt marshes. Plants such as marram and sea couch grass encourage sand deposition and the formation of dunes.

Natural hazards such as erosion and flooding threaten human activities and ecosystems along some stretches of coastline. People respond to coastal hazards by managing coastlines. Hard-engineering structures such as sea walls, groynes, revetments and flood barriers provide protection against erosion and flooding. Soft engineering includes beach replenishment and managed retreat. It has less impact on the environment, is cheaper and is more sustainable.

Coastal management increasingly recognises the coast as a dynamic interactive system. Shoreline management plans in England and Wales are based on natural units known as sediment cells. Each sediment cell is a discrete unit and is best managed as a whole.

Chapter 4 Atmosphere and ecosystems

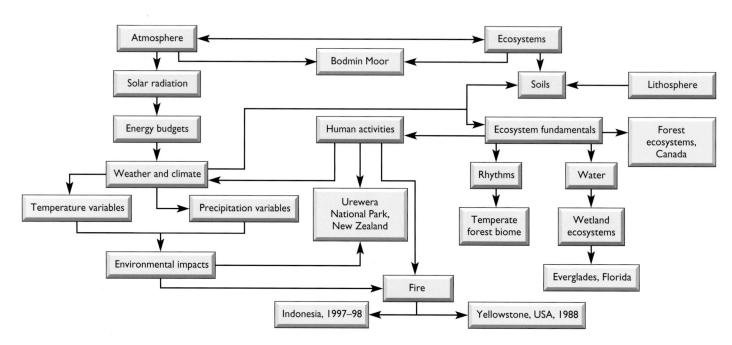

4.1 Introduction

Box 1
The following definitions of an ecosystem emphasise the centrality of relationships:
- 'A self-regulating biological community in which living things interact with their physical environment' (Raw and Shaw, 1998).
- 'Any group of interdependent, interacting species, populations and their environment' (Tivy, 1995).
- 'All of the organisms in a given area interacting with the physical environment so that a flow of energy leads to exchange of materials between living and non-living parts of the ecosystem' (Bradshaw and Weaver, 1993).

The deer in Figure 4.1 are enjoying the early morning sunlight as they graze and move slowly to a stream where they will drink. As the heat of the day increases they will drift back upslope into the shade of the forest. In the evening they will repeat their trip. Through the night they will sleep and perhaps move on to another grazing area. During the winter they will migrate to valleys for food and shelter, before moving back to the mountains as temperatures rise and vegetation grows in the warm spring and summer months. This example of animals at home in their **habitat** illustrates the interconnectedness of climate and **ecosystems**, and introduces three themes central to this chapter – *regimes, rhythms, relationships*. From this example and the definitions in Box 1 we can identify the common attributes of ecosystems: (i) an ecosystem includes the living world (**biosphere**) and parts of the non-living, physical world (**atmosphere, lithosphere**); (ii) the words 'interdependent', 'interacting', 'self-regulating', 'flow', 'exchange'. An ecosystem is, therefore, an open system, with inputs from and outputs to the atmosphere and the soil (the outer skin of the lithosphere).

The atmosphere–ecosystem relationship is demonstrated by the two basic variables which are used to classify the world's major **ecozones**: climate and vegetation (Figure 4.2), e.g. the grass steppes sub-class of the arid mid-latitudes ecozone. These two variables also have a strong influence upon soil type. The direct climate–ecosystem relationship is illustrated by the graph which shows the pattern of **biomes** in relation to mean annual temperature and precipitation (Figure 4.3). A biome is a large-scale ecosystem equivalent to an ecozone. Notice that each biome produces an individual shape on the graph. This tells us that biomes vary in their sensitivity to temperature and moisture inputs.

108

▶ **Figure 4.1** *Mule deer, Manti-La-Sal National Forest, Utah. Their life-rhythms are controlled by the seasonal rhythms of climate and vegetation. In this scene, it is autumn. The deer are grazing intensively in order to build up reserves to survive the cold, snowy winters of Utah.*

Key Terms

Biome: A world-scale community of plants and animals, characterised by a particular vegetation type, and covering a large geographical area.

••••••••••••••••••••••••••••••••••••••

| For each ecozone shown on the world map (Figure 4.2), name two countries in which the ecozone is found (an atlas will help you).

All life forms depend upon the availability of solar energy, moisture and **nutrients**. In consequence, the first part of this chapter examines how the atmosphere works to produce regional climate and weather patterns. The second part focuses upon the living organisms–soil relationships, including the stores, flows and exchanges of energy, moisture and nutrients. It is important to keep in mind throughout that, while we shall be discussing 'natural' vegetation, animal populations, soil types, etc., there is no ecosystem in the world that is not influenced in some way by human activities (see Bodmin Moor example, page 110).

▼ **Figure 4.2** *World ecozones (After: Huggett, 1998)*

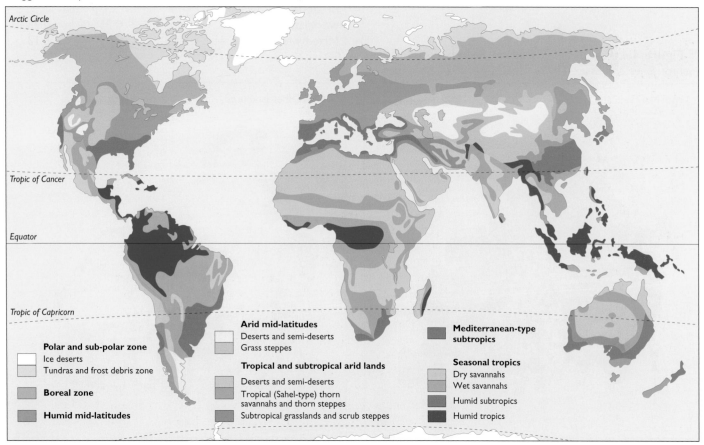

Arctic Circle

Tropic of Cancer

Equator

Tropic of Capricorn

Polar and sub-polar zone
Ice deserts
Tundras and frost debris zone

Boreal zone

Humid mid-latitudes

Arid mid-latitudes
Deserts and semi-deserts
Grass steppes

Tropical and subtropical arid lands
Deserts and semi-deserts
Tropical (Sahel-type) thorn savannahs and thorn steppes
Subtropical grasslands and scrub steppes

Mediterranean-type subtropics

Seasonal tropics
Dry savannahs
Wet savannahs
Humid subtropics
Humid tropics

► **Figure 4.3** *World biomes delimited by temperature and precipitation characteristics*

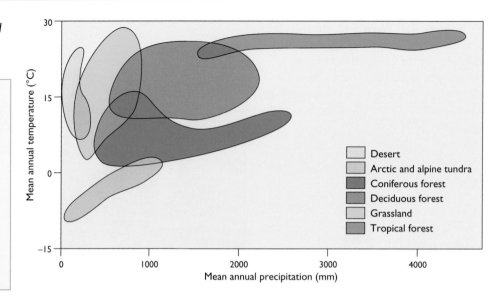

2 From the graph of Figure 4.3:
a Describe the environmental range of the tropical rainforest biome in terms of its temperature and precipitation requirements.
b By comparing the tropical rainforest and grassland biomes, illustrate this statement: 'Biomes vary in their environmental requirements'.
c Explain why the biomes produce different shapes on the graph.

EXAMPLE: Bodmin Moor

Vegetation changes on Bodmin Moor, Cornwall

Under natural conditions, much of Bodmin Moor would be under some form of woodland cover. Many centuries of felling and grazing removed almost all of this woodland, which was replaced by moorland. In the second half of the twentieth century, land-management policies caused further change, as the maps and table show (Figures 4.4a–d, Table 4.1).

▼ **Table 4.1** *Main land-use and land-cover changes, 1946–92 (Percentage of total area)*

Type	1946	1992
Moorland/wetland	74	53
Land improved for farming	24	35
Coniferous forest (plantation)	–	5
Broad-leaved deciduous woodland	0.25	0.75
Mixed woodland	0.03	0.13

▼ **Figure 4.4** *Vegetation and land use, Bodmin Moor, 1946–92 (Source: Jones and Essex, 1999)*

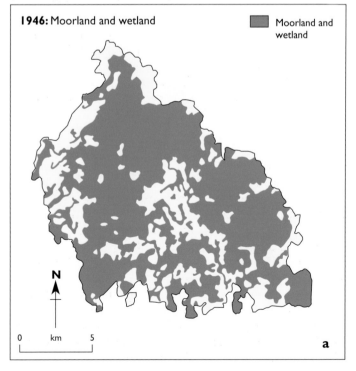

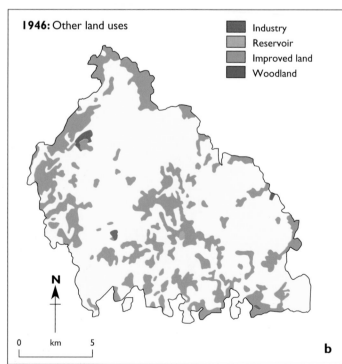

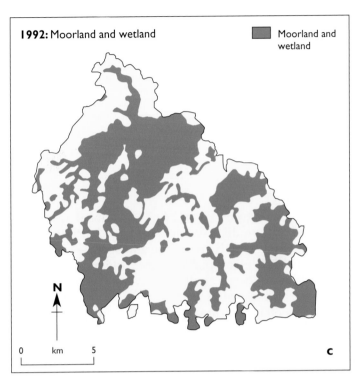

1992: Moorland and wetland

▮ Moorland and wetland

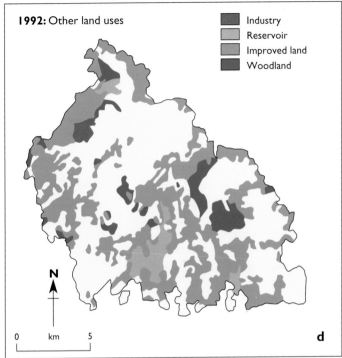

1992: Other land uses

▮ Industry
▮ Reservoir
▮ Improved land
▮ Woodland

0   km   5

c

0   km   5

d

3 From the Bodmin Moor example:
a Which parts of the Moor have experienced greatest changes of vegetation, 1946–92.
b What types of land cover/land use have replaced the moorland?
c What changes have occurred in the woodland vegetation since 1946?

4.2 The atmosphere at work

In order to understand the role of atmosphere in determining climate, ecosystems and their relationships, we need answers to two important questions:
■ What happens to solar energy once it has arrived within the atmosphere?
■ How is moisture stored, moved and released in the atmosphere?

The atmosphere is the thin envelope of gases that surrounds planet earth (Table 4.2). Over 90 per cent of the atmosphere's contents are found within 40 km of the earth's surface, and our weather systems involve processes at work in the lowest 25 km.

The earth–atmosphere energy budget

■ **Incoming solar energy**: Figure 4.5 shows that we can divide incoming short-wave solar energy into three components. The 30 per cent that is reflected and scattered does not enter the Earth–atmosphere store. Thus, it is the 70 per cent remaining which drives our weather systems and energises all life on planet earth: 20 per cent stored in the atmosphere and 50 per cent in the earth's surface layer and vegetation.

■ **Balancing the budget**: Figure 4.6 follows the pathways of the 70 per cent of total solar energy which is involved in the constant earth–atmosphere exchange. The key understanding revealed in this diagram is that each of the three components has a balanced budget: the earth's surface store gains and loses 146 units of energy; the atmosphere store gains and loses 160 units; the 70 units lost to space equate to the incoming units absorbed (stored); 50 in the Earth's surface and 20 in the atmosphere.

▼ **Table 4.2** *Composition of the atmosphere*

Permanent gases	Percentage by volume
Nitrogen (N_2)	78.08
Oxygen (O_2)	20.95
Variable gases	
Water vapour (H_2O)	0.25
Carbon dioxide (CO_2)	0.036
Ozone (O_3)	0.01

4 Analyse the right-hand side of Figure 4.6 to explain the 146 units of energy gained by the surface.

5 How much of the short-wave radiation lost by the surface to the atmosphere is returned to the surface?

Key Terms

Ozone hole: In the stratosphere, at about 20 km above the earth's surface, there is an area of ozone gas that filters out some of the incoming ultraviolet (UV) solar radiation. During the past 30 years scientists have observed a thinning of this layer, especially over Antarctica and the Arctic. Increased UV radiation can increase risks of cancer. Scientists believe that a major cause of the ozone holes is the increased output of CFCs by human activities.

The Greenhouse effect: During the twentieth century, the amount of CO_2 in the atmosphere increased steadily. CO_2 acts like a greenhouse, intercepting some outgoing infra-red long-wave radiation and retaining it within the earth's atmosphere. This causes slow global warming. The increased burning of fossil fuels is reckoned to be a major factor in this CO_2 increase, although CO_2 is only one of the greenhouse gases.

The detailed pathways show how this balance is achieved. For example, if we examine the left-hand side of the diagram we can see that the majority of energy (116 units) returns to the atmosphere as long-wave radiation. This infra-red energy is readily absorbed (stored) by the atmosphere, so that only 6 units escape to space. When water evaporates – that is, changes from a liquid to gaseous state - energy is used and stored in the water vapour (latent heat). This accounts for the surface loss of 23 units. The turbulent motion of air means that **conduction** and **convection** lift 7 units. The final 20 units are added to the atmosphere by direct absorption of incoming long-wave solar **radiation**.

Rhythms in solar energy

The solar energy regime given in Figures 4.5 and 4.6 is the global average. However, the climatic and ecosystem diversity across planet Earth tells us that the solar energy budget varies widely over time and space – ice sheets once covered much of the British Isles; all parts of the world experience 'seasons'. Temperature fluctuations occur at all time-scales. Over relatively long time-scales, the advance and retreat of ice sheets reflect variations in incoming solar energy. During 800 000 years of the Pleistocene glacial period, there have been at least ten major ice advances (glacial phases or episodes), interspersed by ice-front retreats (interglacial phases). Within these phases well-defined shorter-term climatic fluctuations are identifiable (Figure 4.7a).

Even within a single century, marked changes can occur, as the recent concern over 'global warming' demonstrates (Figure 4.7b). There is no doubt that the 1990s was the hottest decade of the twentieth century. What is debated is the extent to which this rise was a natural phenomenon, i.e. a short-term increase in incoming solar energy, or whether it was evidence of **global warming** resulting from human activities.

▼ **Figure 4.5** *The distribution of incoming solar radiation*

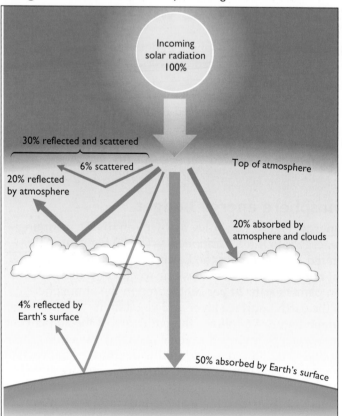

▼ **Figure 4.6** *The earth–atmosphere energy budget*

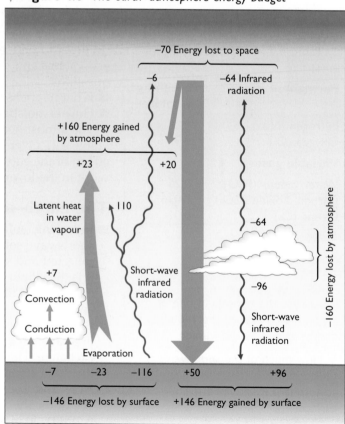

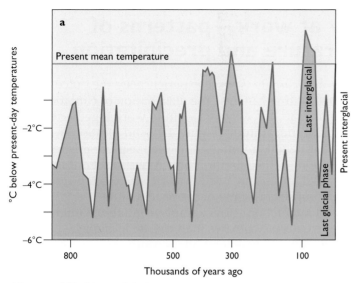

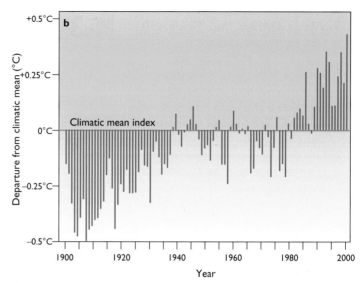

Figure 4.7 *Mean global temperature variations over time*

▲ **a** *Temperature fluctuations during the Pleistocene period (Source: Bradshaw and Weaver, 1993)*

▲ **b** *Temperature fluctuations during the twentieth century (Sources: Aguedo and Birt, 1999; US Met Survey, 2002)*

▼ **Table 4.3** *Solar energy – the diurnal rhythm*

The six-factor day model
Incoming solar radiation
Reflected solar radiation
Absorbed energy into the surface-subsurface
Sensible heat transfer
Long-wave Earth radiation
Latent heat transfer – evaporation

The four-factor night model
Absorbed energy returned to the atmosphere
Sensible heat transfer
Long-wave Earth radiation
Latent heat transfer – condensation

6 How can we tell that night-time energy loss is balanced by excess energy input during the day?

The most important spatial variation in the solar energy regime is in the energy balance of the Earth (Figure 4.8). Within 37 degrees of the Equator, more energy is received than is lost; in higher latitudes, there is an energy deficit. Regions in lower latitudes do not become progressively hotter, and higher-latitude countries, such as the British Isles, do not become colder, because energy is transferred across the globe.

The time–space relationship is illustrated by seasonal climatic rhythms. The seasonal shift in the distribution of incoming solar energy is controlled by the annual movement of the Earth around the sun and the angle of the Earth's axis in relation to the sun.

The major short-term energy rhythm is diurnal: the day–night contrast that helps to explain local microclimatic phenomena such as ground frost, road icing, fog. Energy gains cease at night but losses and transfers continue. The rhythm is explained by comparing the processes at work in the two models summarised in Table 4.3.

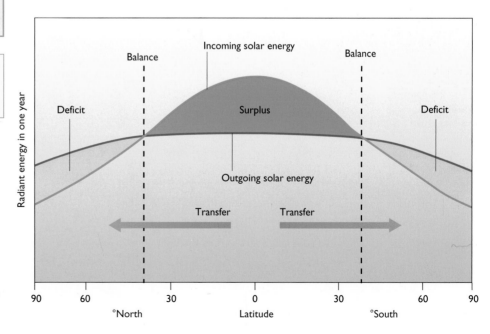

► **Figure 4.8** *Global latitudinal energy balance (Source: Ahrens, 1994)*

7 With a partner: from Figure 4.10:

a Use a software package to construct graphs for the four weather stations at the bottom. (You may wish to add the three completed graphs to your set, and so have a print-out of the seven stations.)

b Use the data for the west–east transect to assess the accuracy of these statements:

'Continental interiors have more extreme climates than coastal regions.'

'In temperate latitudes, west-coast environments are less severe than east-coast environments.'

'Altitude is less important than location in influencing climate along a given latitude.'

8 Use the data for the south–north transect to suggest and then support two general statements about the influence of latitude on climatic characteristics.

4.3 Energy at work – patterns of temperature and precipitation

The weather conditions and climate that we experience are the outcomes of atmospheric energy at work. If we analyse the distribution of climatic regions shown on a world map, three factors seem to work together to influence climatic type: latitude; altitude (Figure 4.9); and location in relation to oceans and continental landmasses. We can explore this idea in more detail by consideration of the data for North America set out in Figure 4.10 which follow west–east and south–north transects across the continent. Comparative analysis demonstrates some of the results of the latitude–altitude–location interaction – for example, continental interior locations experience more extreme climates than coastal locations. Remember, we are establishing only general relationships here: (a) each continent and ocean is unique; (b) there are landmass–ocean contrasts between the northern and southern hemispheres.

▼ **Figure 4.9** *This view of the Canadian Rockies in British Columbia demonstrates how altitude affects vegetation. The coniferous forest becomes thinner and is interspersed with short grasses as altitude increases. The upper tree line locally is about 1500 m. Above that, the period of snow cover prevents growth of vegetation.*

a N–S transect

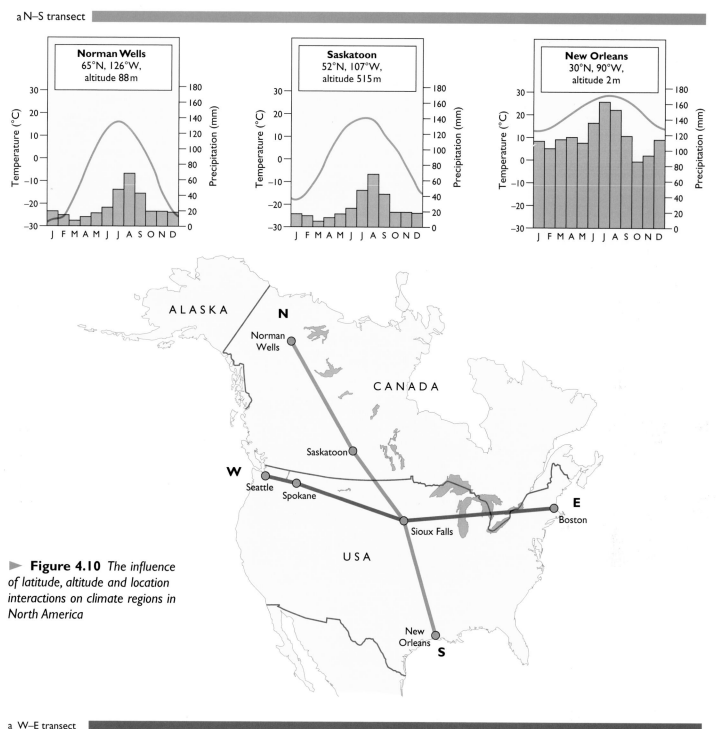

► **Figure 4.10** *The influence of latitude, altitude and location interactions on climate regions in North America*

a W–E transect

		J	F	M	A	M	J	J	A	S	O	N	D
Seattle 47.5°N, 122°W, altitude 20m	Temperature (°C)	3	6	7	10	13	16	17	18	13	11	7	4
	Precipitation (mm)	121	98	80	60	47	39	18	19	41	76	121	141
Spokane 47.5°N, 117.5°W, altitude 719m	Temperature (°C)	–3	0	4	9	14	17	21	20	15	10	3	–1
	Precipitation (mm)	53	43	31	28	33	33	15	15	23	28	53	53
Sioux Falls 43.5°N, 97°W, altitude 240m	Temperature (°C)	–10	–9	–1	7	14	20	23	22	17	10	1	–7
	Precipitation (mm)	20	20	40	50	80	102	72	70	75	50	30	25
Boston 42°N, 71°W, altitude 10m	Temperature (°C)	–2	–1	4	9	14	20	23	22	18	13	7	1
	Precipitation (mm)	91	92	94	94	83	78	72	82	79	84	107	102

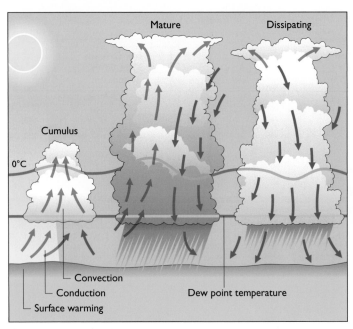

▲ **Figure 4.11** *A cumulonimbus convection cell (After: Ahrens, 1994)*

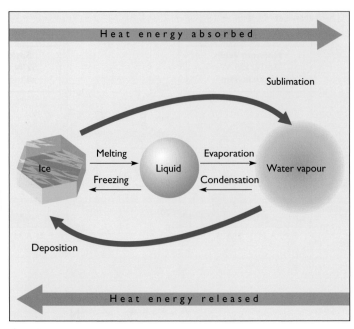

▲ **Figure 4.12** *The absorption and release of energy (Source: Ahrens, 1990)*

9 From Figure 4.12:

a Name the process involved when moisture changes from the gaseous to the liquid state.

b Does this change absorb or release energy?

c Name two processes that release energy into the atmosphere.

d Processes that absorb energy cause temperatures to fall, i.e. they take energy from the atmosphere. Name two processes that cause a drop in temperature.

10 Use Figure 4.13 to:

a Describe briefly the generalised circulation system of the northern hemisphere.

b Describe how the model helps to explain the variable weather conditions we experience in the British Isles.

Temperature controls and transfers

Heat energy is transferred by three processes:

■ **Conduction**: the direct transfer of heat from molecule to molecule in a solid or fluid.
■ **Convection**: the rising of warmed air, i.e. turbulent mass movement of a gas.
■ **Radiation**: movement by energy waves (long-wave and short-wave).

For example, a thin layer of air in contact with the Earth's surface is warmed by conduction. This warmed air rises by convection and is replaced by cooler air which is then warmed by conduction – and so on (Figure. 4.11). The absorption and release of radiation energy by solid or fluid substances influences temperatures (Figure 4.12).

Large-scale transfers of heat energy are achieved by global wind systems (about 80 per cent) and ocean currents (about 20 per cent). Winds are generated by pressure differences across the globe. For instance an atlas map of world pressure and wind systems shows SE and NE trade-wind systems blowing from sub-tropical HP centres to an equatorial low-pressure zone, known as the intertropical convergence zone (ITCZ). The location of the HP and LP centres move north and south with the seasons, following the latitude of the overhead sun and maximum surface heating. Poleward of the subtropical HP centres, wind and pressure systems known generally as the Westerlies transfer energy in a general easterly direction towards middle latitudes, that is, from regions of energy surplus to regions of energy deficit (look again at Figure 4.8). When we introduce the vertical dimension, the general hemispheric circulation can be summarised in terms of a 'three cell' model (Figure 4.13).

The jet stream (defined and located on Figure 4.13) is an important element in this model. For example, the polar jet stream of the northern hemisphere helps to explain the catastrophic floods that hit central Europe in August 2002. During summer, the normal track of this jet stream is a broad arc across the North Atlantic and southern Scandinavia (Figure 4.14).

Figure 4.13 *The 'three-cell' model of hemispheric circulation (After: Ahrens, 1999)*

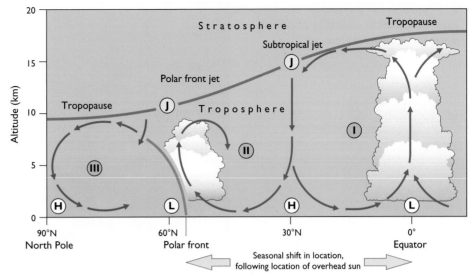

(J) Jet stream: a swiftly flowing air current in the upper troposphere – thousands of kilometres long, a few hundred kilometres in diameter and a few kilometres deep. Wind speeds generally >100km/hour. Jet stream location and strength influences weather systems and Earth surface conditions

(H) High pressure (subsiding air)   (L) Low pressure (rising air)   (I)(II)(III) Cells

Front: junction zone between major air masses

Figure 4.14 *A shift in the location of the jet stream caused heavy flooding in Vienna in August 2002.*

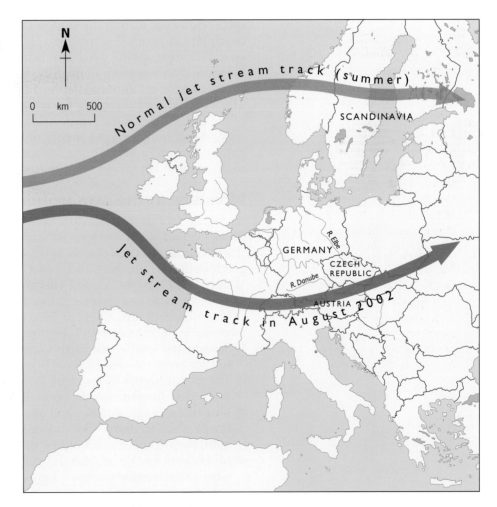

► **Figure 4.15** *Devastation in Gedersdorf, Vienna caused by abnormal jet stream activity in August 2002*

Rain-bearing depressions in the lower atmosphere generally follow this track. However, during the summer of 2002, the jet stream wave switched much farther south (Figure 4.14). This dragged the depressions across central Europe, giving persistent heavy rains and causing the worst floods for more than a century in the Elbe and Danube river basins (Fig 4.15).

Ocean currents also transfer heat energy from low to high latitudes. Thus, one factor influencing the relatively temperate nature of the climate of the British Isles is the North Atlantic Drift (Gulf Stream), the warm ocean current that transfers heat energy north-east from the Caribbean. The pattern of wind and ocean circulation is determined by (a) the distribution of oceans and landmasses and (b) the rotation of the Earth on its axis.

Moisture in the atmosphere

As Figure 4.12 shows, moisture is present in the atmosphere in all three states – liquid, solid and gas. We experience its presence as humidity – water vapour (gas); as clouds, mist or fog – water droplets (liquid); as precipitation – rain, sleet, hail, snow (Figure 4.16). As with energy, moisture is stored and transferred, and we can summarise this system in terms of the *hydrological cycle* (see chapter 2, Figure 2.1).

The following fundamentals help us to understand the part played by moisture in the character and distribution of climatic regimes and ecosystems:

■ Seventy per cent of the Earth's surface is water (oceans, lakes), with further areas covered by ice.

■ The global supply of moisture is finite, and unevenly distributed in time and space, in other words, no moisture enters from, or leaves to, space.

■ Moisture storage, transfer and availability are closely related to temperature rhythms – for example, warm air can store and hence release more moisture per unit volume than cold air.

■ Moisture is moved around the globe by the wind and pressure systems which also transfer energy – for example, the moisture-laden low-pressure systems that move on to landmasses from oceans, such as the Atlantic depressions that bring year-round rainfall to the British Isles.

■ Moisture stores and releases energy as it changes state: the evaporation process uses and stores energy for instance, and **condensation** releases this energy.

Figure 4.16 *Driving through an August afternoon thunderstorm, Wyoming, USA. Air is heated by conduction from the hot summer land surface. This triggers vigorous convection currents and the build-up of massive cumulonimbus clouds that yield sudden, torrential rainstorms.*

Key Terms

Hygroscopic nuclei: Airborne particles having an affinity for water, serving as condensation nuclei.

Absolute humidity: the amount of moisture per unit volume of air.

Relative humidity: the proportion of moisture stored in relation to the storage capacity of the air at that temperature.

Precipitation – the release of atmospheric moisture

For precipitation to occur, atmospheric water vapour must first be converted from the gaseous state. The main process involved is condensation, which normally takes place when a parcel of air becomes saturated, that is to say, it has reached its capacity to store moisture as water vapour. It is important to understand that: **A unit volume of air at a given temperature has a finite capacity to store (hold) moisture in the gaseous state.** In general, this capacity increases with temperature, and decreases as temperatures drop. Thus, we talk of humidity levels: **absolute humidity** is the amount of moisture stored per unit volume of air; **relative humidity** is the proportion of moisture stored in relation to the storage capacity of the air at that temperature. So, a relative humidity of 60 per cent means that the particular parcel of air is holding 60 per cent of the moisture it could hold as water vapour at that temperature. When relative humidity reaches 100 per cent the air becomes saturated and condensation commences (Figure 4.17a, b).

When a parcel of air is cooled, relative humidity rises. If the cooling continues, saturation point (dew point) is reached, condensation takes place, and water droplets form around tiny **hygroscopic nuclei** (for example, salt held in suspension). We see this condensation as mist, fog or clouds. When condensation takes place at the surface, deposition may occur directly on to the ground or vegetation as dew or, when temperatures are low enough, as frost. A fall in air temperature is achieved by (a) conduction from the cooling of the surface below or by air moving over a cooler surface; (b) air rising (Figure 4.18a-d). In many situations more than one mechanism may be at work, e.g. relief lifting may trigger convection, as is happening in Figure 4.18c.

Most clouds do not produce rain or snow – but why? The answer is that the water droplets and ice crystals are so small that they are kept in suspension by even the weakest updraughts. Without updraughts they fall so slowly that they evaporate in the unsaturated air below the clouds before they reach the surface. For precipitation to occur, the droplets and crystals must grow: a raindrop is approximately one hundred times larger than a typical cloud water droplet.

▲ **Figure 4.17** *Absolute and relative humidity: The amount of moisture present in the air per cubic metre is similar in both of these locations. Their absolute humidities are similar, but the relative humidities are very different.*

a *A cold day in the Trossachs, Scotland. The temperature is approximately 0°C, and the relative humidity is approaching 100%.*

b *A hot day in the Arizona desert near Phoenix. The temperature is 31°C and the relative humidity is 20%.*

11 Read the text on temperature and precipitation.
a Define the terms (i) dew point; (ii) condensation
b Explain why temperature as well as moisture is important in determining when the dew point is reached in a parcel of air.
c What do we mean when we say that desert air is 'dry'?

12 Look at Figure 4.18:
a For each situation illustrated in Figure 4.18, explain why condensation occurs.
b By means of a diagram, suggest a situation where more than one of the four cooling mechanisms is at work.

13 From Box 2:
a Describe the two main processes that permit precipitation to occur.
b What determines whether precipitation arrives as drizzle, rain, snow or hail?

Box 2 Precipitation

■ **Rain**: Precipitation in liquid form, with individual raindrops more than 0.5 mm. Drizzle has droplets 0.2–0.5 mm in diameter. The droplets may have formed in the liquid state in 'warm' clouds or as ice crystals in 'cool' or 'cold' clouds, with melting occurring during descent.

■ **Snow**: Precipitation in the form of snowflakes of complex hexagonal ice crystals. Formed in 'cold' clouds and falls through air at near or below freezing point.

■ **Sleet**: A mixture of snowflakes and rain, as a result of partial melting during descent. (In the USA sleet is defined as raindrops that have frozen to ice pellets as they fall, i.e. where cold air near the surface produces freezing.)

■ **Hail**: Ice particles that range in size from small peas to golf balls. Formed by the accretion of successive layers of ice in 'cool' convection clouds with strong updrafts. As an ice crystal falls into the lower 'warm' sector of the cloud, a thin layer of water is added – updraughts carry the particle upwards – the water layer freezes – the crystal falls once more – another water layer is added – and so on until the hailstone is heavy enough to fall to the surface. The more times the 'yo-yo' motion is repeated, the larger the hailstone becomes.

Cooling mechanisms

a (left) Radiation and conduction cooling: A mid-winter sunrise, Victoria, Australia. Heavy dew and mist indicate condensation. A surface layer of still air has been cooled to saturation point by conduction from a land surface which cooled rapidly overnight by radiation loss. This radiation cooling was assisted by clear skies (i.e. no 'blanket' of cloud) and still air. Similar cooling may occur when slow-moving air drifts across a cold surface, such as a large lake or snow-covered surface.

b (right) Convection lifting: Auckland, New Zealand. As a moist oceanic air mass moves on to the warmer land, a parcel of air is heated by conduction and radiation from the underlying warm land surface. This parcel rises and cools but remains warmer and lighter than the surrounding air and so continues to rise rapidly as a convection cell.

c (below) Relief lifting: Northern Queensland, Australia. Warm, moisture-laden tropical air moves from the ocean on to a mountainous coastline. The air has high absolute and relative humidities and so, as it is forced to rise and cool along the mountain front, saturation point is quickly reached. Condensation takes place and clouds build up rapidly, especially where convection cells develop.

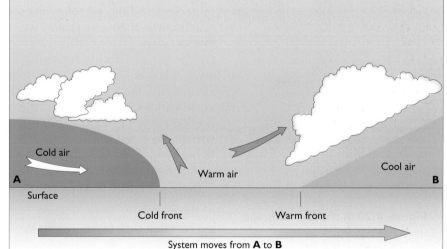

d (right) Frontal lifting: A typical situation across the British Isles. Within the circulation of a low-pressure system colder, heavier air pushes beneath a warmer, lighter air mass. The junction zones between the air masses are fronts. The relatively warm air rises along the fronts, cools, and progressive condensation takes place, producing extensive cloud development.

▲ **Figure 4.18** *Examples of cooling mechanisms. More than one process may be at work at a particular time and location.*

The processes that determine the droplet growth depend upon conditions within the clouds. In so-called 'warm' clouds, where temperatures are above freezing level, water droplets grow by the collision–coalescence process (Figure 4.19a). In 'cool' or 'cold' clouds, where temperatures in part or all of the clouds are below freezing level, ice crystals grow by what is known as the Bergeron-Findeison process (Figure 4.19b). (Remember – freezing-point temperature varies with air pressure, so it may not be 0°C.) The precipitation we receive depends upon these processes and what happens to the raindrops, snowflakes and hailstones as they fall earthwards (see Box 2).

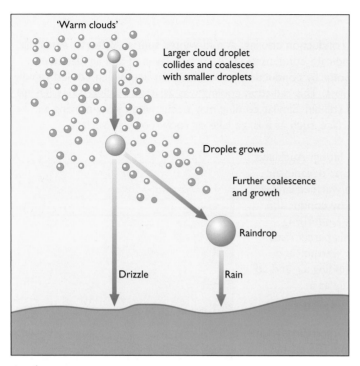

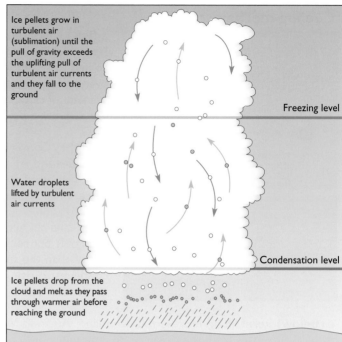

▲ ◀ **Figure 4.19** *Two theories of precipitation:* **a** *collision–coalescence theory and* **b** *the Bergeron–Findeison coalescence theory*

4.4 The availability of moisture

In understanding the role played by precipitation in the character and functioning of an ecosystem, we need to consider a set of related questions. The answers give us crucial information about moisture storage, movement and hence availability for living organisms – including humans. All ecosystems are adapted to the regimes and rhythms of energy and moisture availability. We can take one variable, such as surface type, to illustrate the significance of the questions: the natural vegetation of the hilly landscape in Borneo shown in Figure 4.20 is tropical rainforest, and there are three surface types visible:

■ **Bare soil:** Rain falls directly on to the soil surface and either runs off, is absorbed (stored) or evaporates. Only the moisture stored is available for plant growth. The proportion of incoming precipitation absorbed into the

▶ **Figure 4.20** *A terraced landscape in Borneo. The region has an equatorial climate with heavy convection rains throughout the year. Note the hilly topography and steep slopes.*

soil and regolith will depend mainly upon (a) the gradient; (b) the permeability of the soils; (c) the depth of the soil/**regolith** profile; (d) the amount of moisture already stored.

■ Forested surface: The dense forest cover intercepts a high proportion of the incoming precipitation, acts as a temporary store, and releases it gradually to the ground surface. Thus, the input to the soil/regolith store is slowed down. Some of the rainwater will evaporate from the vegetation, but ground surface evaporation will be low (see New Zealand example).

■ Managed slope: The purpose of the agricultural terraces is to store water and reduce downslope runoff. This makes a relatively high proportion of moisture available for plant use. Evaporation will vary according to the stage of crop growth, e.g. evaporation rates will be higher immediately after harvest and in the earlier stages of plant growth.

EXAMPLE: New Zealand

Vegetation/land-use/precipitation relationships

The two photographs (Figures 4.21a and b) were taken 15 km apart on the same day in 1989. One year earlier, an unusually vigorous low-pressure system had dumped up to 30 cm of rain in 48 hours across this part of North Island, New Zealand. The natural vegetation of the region is temperate rainforest. Figure 4.21a lies within the Urewera National Park and the luxuriant natural forest has been conserved. Although the stream flooded and covered the road, the forest was able to act as a huge sponge to intercept, absorb and gradually release much of the water.

The landscape of Figure 4.21b lies just outside the National Park. The forest was cleared, except for a scattering of trees, around one hundred years ago. Since then, sheep and cattle grazing have maintained a short-grass mat cover above the hillslope soils. With little protective cover, the soil/regolith could not absorb the abrupt water input. The result was extensive sheetwash, slope failure on the steeper slopes and flooding. (The new stream channel in the foreground is the result of this surge of runoff.) During the 1990s there were large-scale reafforestation schemes across these slopes.

▼ **Figure 4.21a** *Natural temperate rainforest in the Urewera National Park*

▼ **Figure 4.21b** *Ecologically degraded and eroded landscape just outside the Urewera National Park*

14 Name three variables that influence the proportion of precipitation inputs that are stored, and made available for use by vegetation. Suggest which of your three selected variables can be influenced by human activities and what the effects might be.

15 Examine the New Zealand landscapes shown in Figures 4.21a and b.
a Describe the topography and vegetation
b What evidence is there in Figure 4.21b of rapid runoff and erosion?
c Explain how this example illustrates the importance of natural vegetation in controlling the storage and throughput of water in an ecosystem.
d Suggest two methods for reducing the rate of runoff, increasing the water storage capacity and lowering the risk of further erosion across the hillslopes in Figure 4.21 (look again at Figure 4.20).

16 From Figure 4.22:
a Describe how (i) carbon and (ii) nitrogen are transferred from the dead organic matter (DOM) in the soil store to the fox.
b Illustrate the role played by soils in the functioning of ecosystems.

◄ **Figure 4.22a** *The carbon cycle (After: Huggett, 1998)*

► **Figure 4.22b** *The nitrogen cycle (After: Huggett, 1998)*

The soil store

Soils consist of a dynamic (constantly changing) mixture of mineral and organic materials, water and air. Soil is the vital link between the living and non-living components of ecosystems and performs five fundamental functions:

■ It provides physical support for plants
■ It is a home for a wide variety of living organisms
■ It receives, decomposes and stores organic matter, and stores inorganic minerals weathered from bedrock.
■ It is the store from which plants draw nutrients and water.
■ In connecting the inorganic and organic components of an ecosystem, it controls the movement of energy and nutrients via a series of cycles (Figure 4.22).

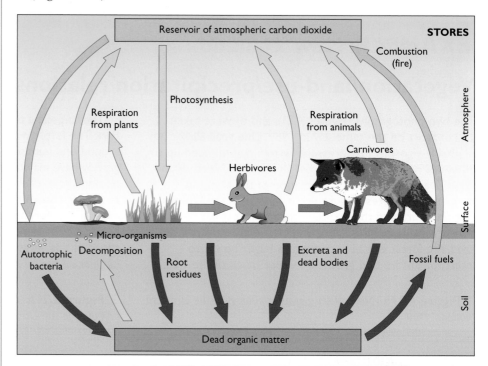

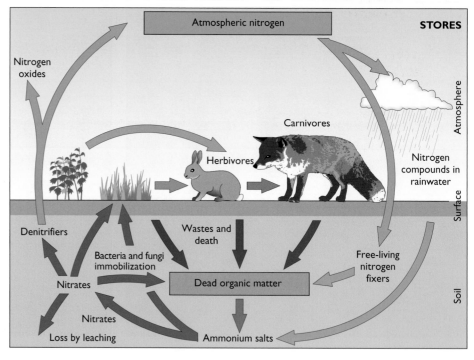

Although soil can be regarded as an open system in its own right, it is also an integrated subsystem of an ecosystem. Examples of the wide diversity of soil types are given in the study of selected ecosystems that make up the rest of this chapter. However, it is useful at this stage to highlight certain basic characteristics:

■ Soils are classified according to their vertical profiles.
■ Soils vary in their depth and maturity and take long periods to evolve.
■ Soil texture influences their ability to store, transfer and release water.
■ Soils vary in the volume and balance of inorganic and organic contents (Figure 4.23).

To sum up the soil–atmosphere–biosphere relationship:

'It is difficult to think of any characteristic of the soil or its processes that is not in some way controlled by climate. The global distribution of soils therefore closely matches the distribution of major climatic regions and world vegetation types' (Bradshaw and Weaver, 1993).

► **Figure 4.23** *Variations in nutrient stores.*
a *(right) The distribution of carbon in major biomes (after Huggett, 1996).*
b *(middle) Ecosystem contrasts in carbon stores (after Odum, 1987)*
c *(bottom) Ecosystem contrasts in nitrogen stores (gm/m^2) (after Odum, 1987)*

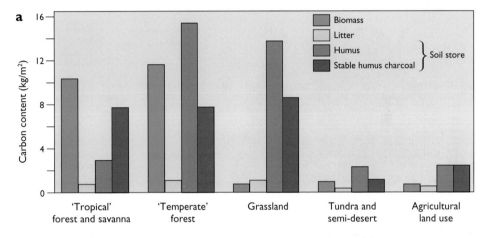

317

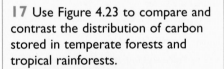

17 Use Figure 4.23 to compare and contrast the distribution of carbon stored in temperate forests and tropical rainforests.

18 From Figure 4.24:
a Explain why an ecosystem is an 'open' system.
b Construct the ecosystem example shown in Figure 4.24 in the format of a systems diagram (boxes for stores; arrows for transfers; indicate inputs and outputs).

19 One important characteristic of an ecosystem is that a significant change in one component is likely to have a 'knock-on' effect through the rest of the system. Use Figure 4.24 to illustrate this characteristic by suggesting what may happen if rainfall decreases (follow the diagram from B).

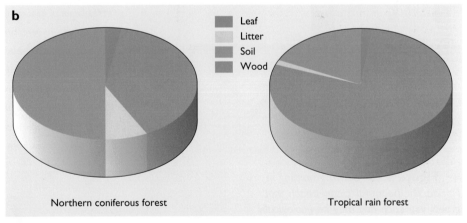

c	British 55-year pine forest	Tropical gallery forest
Leaves	12.4	52.6
Above-ground wood	18.5	41.2
Roots	18.4	28.2
Litter	40.9	3.9
Soil	730.8	85.3
% N above ground	3.0	44.0

4.5 Ecosystem fundamentals

Ecosystems, at all scales, evolve and work within the same set of principles and processes. These fundamentals underpin ecosystem study, and are illustrated in the sections and examples that make up the rest of this chapter. (The fundamentals are most easily understood by spending a little time to follow the set of diagrams, Figures. 4.24, 4.26–4.27 carefully.)

■ An ecosystem is an open system in which energy and matter move via a set of flows and cycles (Figure 4.24). A significant change in any key component may trigger changes through the system.

■ An ecosystem is dynamic – it changes over time. Ecologists have suggested two main models of this dynamism: the steady-state model, where an ecosystem evolves through a sequence of stages (**succession**) towards a **climatic climax** condition in balance with the prevailing climate; and the non-equilibrium model, where disturbances cause recurrent change and adjustments (Figure 4.26).

▼ **Figure 4.24** *An ecosystem as an open system*

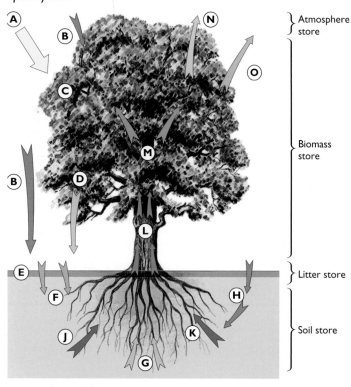

⎫ Atmosphere
⎬ store
⎭

Biomass
store

⎫ Litter store
⎬
⎭

⎫
⎬ Soil store
⎭

• Follow the flows and cycles through the diagram
• Note that this example omits additional flows created by grazing animals or human uses of plants and animals

(A) Incoming solar energy: short-wave radiation

(B) Moisture input from precipitation

(C) Photosynthesis: leaves contain the green pigment *chlorophyll*, which absorbs the incoming short-wave radiation. This sets off chemical reactions between water and carbon dioxide to produce *carbohydrates*. The carbohydrates are either stored and build up the mass of the plant, or are used to move moisture and nutrients through the plant

(D) Dead organic matter (DOM) falls to the ground

(E) The decomposition process begins in the surface or litter layer

(F) Decomposition of organic matter is completed in the upper soil profile to form *humus*

(G) Input of mineral nutrients from continued bedrock weathering

(H) Moisture percolates through the soil

(J) Uptake of soluble nutrients by soil water

(K) Roots absorb nutrient-rich water

(L) Moisture and nutrients move through the plant

(M) Respiration: the use of energy from carbohydrates to assist growth

(N) Transpiration: loss of moisture through the leaves to the atmosphere

(O) Oxygen loss from photosynthesis

► **Figure 4.25** *Beach grooming in Florida. The beautiful 'clean' beach that greets you each day may be ecologically 'dead'. Clearing the seaweed and other organic material removes the food supply for beach organisms.*

Stability and productivity

Change in environmental conditions

Intolerance | T o l e r a n c e   r a n g e | Intolerance

Increasing stress | Optimum zone | Increasing stress

Threshold | Threshold

Collapse and progressive change | Collapse and progressive change

Positive feedback | N e g a t i v e   f e e d b a c k | Positive feedback

Optimum zone:
Where there is the most efficient balance between an ecosystem and its environment

Tolerance range:
The range within which an ecosystem can adjust and adapt to changing environmental conditions. Negative feedback mechanisms work to 'damp down' the impacts of change and sustain ecosystem balance

Threshold:
A critical point at which environmental change is so severe that the ecosystem and, in particular, certain keystone species, can no longer adjust. Positive feedback – the runaway mechanism – takes over and the ecosystem collapses and undergoes fundamental change in character

▲ **Figure 4.26** *Ecosystems and environment: a general model*

20 From Figure 4.27:
a What is meant by a 'climatic climax ecosystem'?
b What effects can human activities have upon ecosystem succession?
c What do we mean by 'secondary succession'? Give an example from any photograph in this chapter.
d What types of ecosystem are depicted in the Figures 4.21a and b?

21 Use the New Zealand example to illustrate the usefulness of the model shown in Figure 4.26.

▼ **Figure 4.27** *Generalised ecosystem evolution pathways and human impacts*

- An ecosystem has an optimum set of conditions in which it functions most productively, but has a range of tolerance within which it can adjust to environmental fluctuations (Figure 4.26). Ecosystems vary in their ability to resist and to recover from disturbance and change. Tolerance and resistance may depend on certain **keystone** species, such as dominant tree species in forests or food sources in oceans (Figure 4.28).
- An ecosystem can be classified by its **biodiversity** (number of species) and its productivity (the amount of growth achieved and sustained). The primary controls are temperature and precipitation (Table 4.5).
- An ecosystem has a structure of **trophic levels** linked by a set of food chains and food webs (Figure 4.29). Each species has a place within the system and a role to play in sustaining **equilibrium**. This is called a niche. For example, in the moorland ecosystem (Figure 4.29a) foxes help to control rabbit numbers and hence reduce the risk of overgrazing from an explosion of rabbit populations. Equally, the foxes depend on the presence of a food supply such as rabbits.
- The evolution, structure and functioning of ecosystems are increasingly influenced by human activities (Figures 4.25 and 4.27).
- Sustainable ecosystem management techniques attempt to balance conservation and development values (See Canada example, page 129).

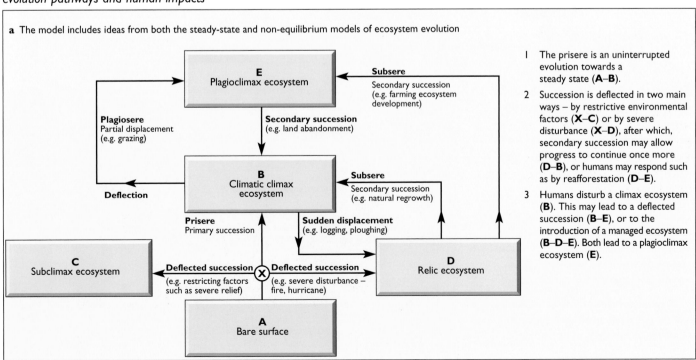

a The model includes ideas from both the steady-state and non-equilibrium models of ecosystem evolution

E Plagioclimax ecosystem

Subsere Secondary succession (e.g. farming ecosystem development)

Plagiosere Partial displacement (e.g. grazing)

Secondary succession (e.g. land abandonment)

Deflection

B Climatic climax ecosystem

Subsere Secondary succession (e.g. natural regrowth)

Prisere Primary succession

Sudden displacement (e.g. logging, ploughing)

C Subclimax ecosystem

Deflected succession (e.g. restricting factors such as severe relief)

X

Deflected succession (e.g. severe disturbance – fire, hurricane)

D Relic ecosystem

A Bare surface

1 The prisere is an uninterrupted evolution towards a steady state (**A–B**).

2 Succession is deflected in two main ways – by restrictive environmental factors (**X–C**) or by severe disturbance (**X–D**), after which, secondary succession may allow progress to continue once more (**D–B**), or humans may respond such as by reafforestation (**D–E**).

3 Humans disturb a climax ecosystem (**B**). This may lead to a deflected succession (**B–E**), or to the introduction of a managed ecosystem (**B–D–E**). Both lead to a plagioclimax ecosystem (**E**).

Table 4.4 *Net primary productivity (NPP) of terrestial biomes (gm/ m²/yr)*

Tropical rainforest	2200
Deciduous forest	1300
Boreal coniferous forest	800
Grassland – temperate	600
– savanna	900
Arctic & alpine tundra	140
Desert/semi-desert	90
Wetlands	3000

Note: Plants are the primary producers in an ecosystem. The net primary productivity (NPP) is the amount of energy stored in living tissue produced per year, after energy has been used to make the plants function and been lost in respiration. It represents the amount of growth in all the living organisms.

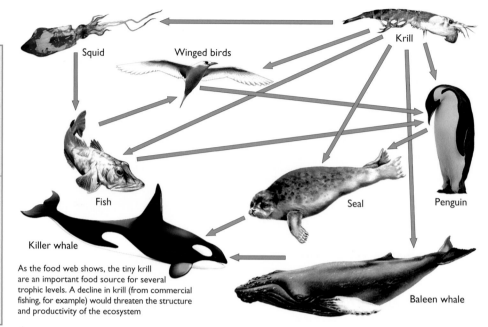

As the food web shows, the tiny krill are an important food source for several trophic levels. A decline in krill (from commercial fishing, for example) would threaten the structure and productivity of the ecosystem

Figure 4.28 *A 'keystone' species: krill in the Antarctic oceanic ecosystem (Source: Chaffey, 1997)*

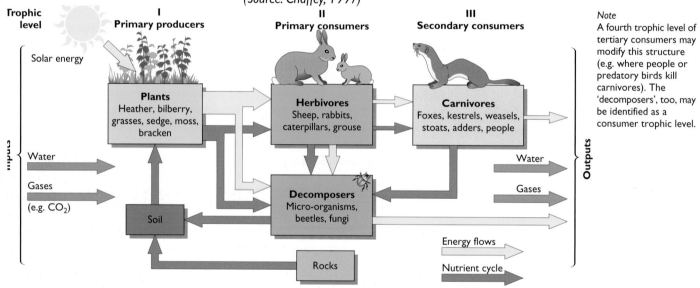

Note
A fourth trophic level of tertiary consumers may modify this structure (e.g. where people or predatory birds kill carnivores). The 'decomposers', too, may be identified as a consumer trophic level.

Figure 4.29a *A UK moorland ecosystem (after Raw and Shaw, 1997)*

Figure 4.29b *A woodland food web: Wythain Wood, Oxfordshire (after Huggett, 1998)*

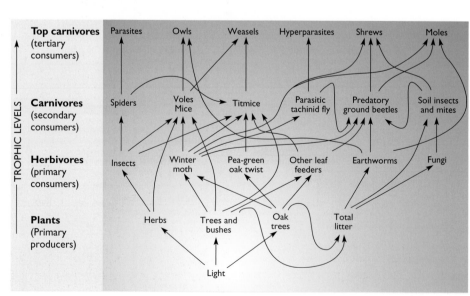

22 What do we mean by the terms (i) keystone species; (ii) trophic level; (iii) niche? Support your definitions by using the food web examples given in Figures 4.28 and 4.29.

23 Explain briefly how the data in Table 4.4 illustrate the relationship between climate and ecosystems.

EXAMPLE: Sustaining Canada's forest ecosystems

Canada is a huge country (921 million hectares). Approximately 45 per cent of its land surface is covered with some form of forest, and timber production is one of the country's major industries. Almost one milion hectares are harvested each year and the industry supports 350 000 jobs.

As more is learned about how ecosystems work, concerns over the environmental impacts have grown. For instance, each wildlife species needs a specific type of habitat and a certain amount of space, called its home range (Table 4.6). The loss of such habitat and necessary space means that the species will decline and even disappear. The following examples illustrate how modern management techniques attempt to sustain both biodiversity and economic development of Canada's forest resources.

Mountain caribou in British Columbia

There are only 1700 mountain caribou scattered across south-eastern BC. Caribou feed on the lichen that grows on mature trees in old-growth forests. Wildlife groups, government agencies and the forest industry are working together to protect the caribou habitat by minimising the impact of logging in old-growth forests and by improving the habitat in areas already logged or burned. Maintaining the caribou habitat will also benefit other wildlife species that rely on old-growth forests, such as flying squirrels, grizzly bears, pine marten and cavity-nesting birds.

Partial-cutting methods enable forest managers to retain some of the lichen-bearing trees while maintaining a constant supply of timber. A land-use strategy to classify the caribou habitat is being developed that will identify priority management zones and guidelines: primary ranges for caribou may be protected from harvesting; special management areas will be managed for both timber and caribou by using partial cutting and by establishing longer rotation periods.

▲ **Figure 4.30a** *Caribou*

▼ **Figure 4.30b** *White-tailed deer are increasingly confined by a shortage of winter habitat. Timber harvesting is now carried out in such a way as to conserve the critical winter habitat.*

Quebec's Deer Yard programme

The Deer Yard programme, developed by the Quebec government and the forest industry, provides private owners with technical and financial assistance to plan timber harvesting that will conserve the critical winter habitat of white-tailed deer. the programme also provides assistance to encourage the planting of trees that are beneficial to the white-tailed deer's habitat.

In Quebec, the white-tailed deer is at the northern limit of its range, and the availability of winter habitat is a major factor affecting its population levels. Deer require mature coniferous stands that provide shelter from the wind and cold, and allow only a thin layer of snow to accumulate on the ground. Deer also require an abundance of young, broad-leaved trees for the twigs that are their basic diet. A diversity of shelter and food within a deer yard reduces the distance that deer have to travel during rigorous winters. Maintaining these deer yards ensures the long-term stability of the deer population.

▼ **Table 4.5** *Home range needs of selected species in Canadian forests*

Species	Range (ha)
Grizzly bear	377 000
Gray wolf pack	153 000
Bobcat	11 600
Elk	943
Marten	215
Snowshoe hare	2.55

(Source: 'The State of Canada's Forests', Natural Resources Canada, Canadian Forestry Service, 1999)

▼ **Figure 4.30c** *Turtle mountain in Loch Alva, New Brunswick. The Loch Alva Protected Natural Area is a large tract of land (22 000ha) set aside from industrial development. It has no roads, no buildings, is nothing but wilderness.*

New Brunswick's land management policy

Since 1980, companies in New Brunswick have been required to prepare 25-year plans outlining the harvesting levels required to maintain a sustainable supply of timber over an 80-year period. Forecasts indicate a shortage of mature coniferous forests in 30 years that would affect the 25 bird species and four mammal species identified as dependent on this habitat.

The American marten, which is particularly dependent on mature coniferous forests, was chosen as an indicator species for this forest habitat. The analysis identified the minimum viable population level for marten as 250 adults. This population would require a total of at least 50 000 hectares of suitable habitat, in patches of no less than 500 hectares, connected by travel corridors.

Plans prepared by companies now include specific objectives for maintaining this wildlife habitat. Under the programme, each company must maintain a specific amount of mature habitat at any given time. On average, 10 per cent of the licence area is required to maintain a marten population 3.5 times greater than the minimum levels. A company's plans must also reflect the fact that the composition of the forest is not static: areas of habitat will not remain forever in one location; new areas age and become suitable as mature habitat.

4.6 The temperate forest biome: an ecosystem with strong seasonal rhythms

► **Figure 4.31** *A mature oak/beech wood, north Herefordshire. This is the natural climax vegetation of the area. Notice that only remnants survive. Even these remnants have been managed and modified over centuries.*

▲ **Figure 4.32** *Croft Castle Woods, Herefordshire, one year after clearance*

Temperate forest is the climatic climax vegetation across extensive areas of Western Europe and North America. Smaller areas are found in the southern hemisphere. However, this biome has been seriously reduced by human activities. For instance, less than 10 per cent of the land area of the UK is forested, and barely one-third of this consists of native tree species (look again at the Bodmin Moor example, pages 110–111). Under natural conditions, at least 75 per cent would be under some form of predominantly deciduous woodland (that is, species that shed their leaves in winter).

Across lowland Britain the main tree species are oak, beech, ash, lime and chestnut (Figure 4.31). A typical mature woodland may contain 10–12 main tree species per hectare, although one or two species may be dominant (oak/beech woodland). When the trees are in full leaf, the canopy may be almost continuous and more than 50 per cent of the incoming sunlight is intercepted. However, many deciduous woodlands are structured with well-developed ground and shrub layers. This diverse vegetation supports a wide range of birds and animals under natural conditions, such as fox, badger and deer (look again at Figure 4.29b).

No truly 'virgin' forests remain in the British Isles. Even designated 'ancient woodlands', such as the New Forest, have experienced centuries of use and management. Many are, therefore, **plagioclimax** ecosystems and have passed through at least one secondary or deflected succession. What we see frequently are immature woodlands, with high proportions of 'pioneer' species whose seeds are prolific, mobile and germinate readily – birch, hazel, sycamore and, in wetter sites, alder and willow (Figure 4.32). In the early years, secondary succession is rapid, but becomes more gradual over time (Figure 4.33a). A deciduous hardwood ecosystem may take 200 years to approach a steady-state climax (Figure 4.33b).

► ▼ Figure 4.33a (right) and b (below) *Succession and climax vegetation in deciduous woodlands*

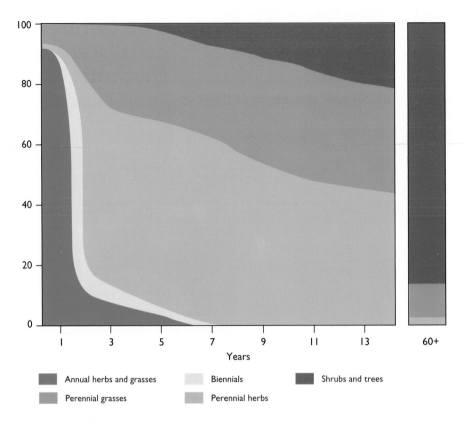

Annual herbs and grasses   **Biennials**   **Shrubs and trees**

Perennial grasses   **Perennial herbs**

The graph shows succession after land clearance for 13 years, and the vegetation structure as it approaches the climatic climax after 60 years. (The main trees will not be fully mature, but the species composition is mature.)

Main trends
• Increase in plant biomass (living matter)
• At first, rapid increase in species diversity
• Increasing complexity of structure: vegetation layers
• Quick-growing pioneer trees, mainly birch, willow and hazel, are progressively replaced by slower-growing climax species such as oak, ash, chestnut and beech
• Grasses and shrubs are gradually shaded out as trees grow and reduce daylight reception to the ground surface

24 From Figure 4.33:
a Describe the succession (i) in years 1–5 and (ii) in years 13–60
b Give the main vegetation types in each of the successional seres.

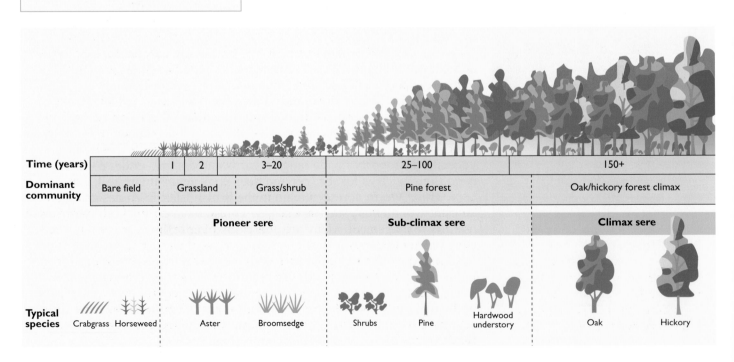

Time (years)		1	2	3–20	25–100	150+
Dominant community	Bare field		Grassland	Grass/shrub	Pine forest	Oak/hickory forest climax
		Pioneer sere			**Sub-climax sere**	**Climax sere**

| **Typical species** | Crabgrass | Horseweed | Aster | Broomsedge | Shrubs | Pine | Hardwood understory | Oak | Hickory |

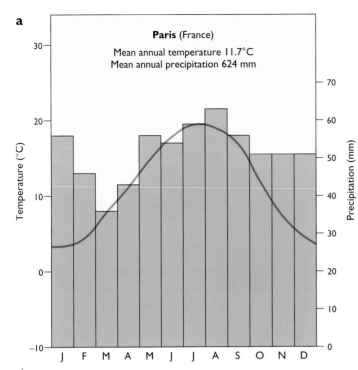

a

Paris (France)
Mean annual temperature 11.7°C
Mean annual precipitation 624 mm

b

- Moisture surplus
- Moisture deficit
- Soil moisture utilisation
- Soil moisture recharge

Potential evapotranspiration

Actual evapotranspiration

Precipitation

Potential evapotranspiration (PET) = the maximum possible amount of evapotranspiration at a given temperature, given an unlimited supply of water

Actual evapotranspiration = the amount of evapotranspiration which takes place

Figure 4.34a and b *Climate and water balance, Paris*

Average growing season in Britain

Growth

6°C

a

b

No growth

J F M A M J J A S O N D

Model to show monthly variations in growth above and below the *critical growth* threshold (6°C) for (a) lowland Britain and
(b) upland Britain

Note   These are generalised patterns – there will be local variations and fluctuations from year to year

Figure 4.35 *The average growing season in the UK (Source: Bishop and Prosser, 1990)*

25 From Figure 4.34:
a Name the wettest and driest months in Paris.
b Give the months in which there is
(i) a moisture surplus;
(ii) a moisture deficit.
c What is the main feature shown by your answers to (a) and (b)?
d Explain the relationship between temperature, moisture availability and vegetation growth.

Controlling factors

Temperature is the primary control on the deciduous regime (Figure 4.34a). In winter, with low inputs of solar energy, the tree must reduce **transpiration** and conserve moisture because it is less able to draw in nutrient-rich water from the cold soil. In frost spells, moisture in leaf tissues would freeze, expand, break up the tissue, and the plant could die. As a result the trees have adapted and remain dormant until spring. Without leaves there is no photosynthesis, no creation of carbohydrates and hence no added energy to sustain growth.

All living organisms adapt to this seasonal rhythm – for example, birds migrate, while animals hibernate, reduce their activity levels and hence their energy needs, or live on reduced food supplies and food stored during the warmer months (if you have a hazel-nut tree in your garden, it will be a race between you and the squirrels in the autumn!).

The net primary productivity (NPP) (i.e. the increase in biomass) depends upon the ability of plants – the primary producers – to react quickly as temperatures rise in spring. Species are adapted for rapid seed **germination** and leaf growth so that **photosynthesis** can begin. This maximises the length of the growing season (Figure 4.35).

Marked seasonal rhythm therefore, is a distinctive feature of this biome, with solar energy input the key controlling factor. This rhythmic variation in solar energy and temperature also influences the availability of moisture. We can use the mean data for Paris as an example (Figure 4.34a). The growing season in a normal year lasts from late March to early November, and during this period plants, especially trees, have increased moisture demands. However, the figures show that rain falls regularly during the summer. Yet when we examine the water balance (Figure 4.34b) we can see that there are periods of moisture deficiency during the summer. Higher evaporation rates and plant demands mean that more moisture is being withdrawn from the soil store than is being replaced. Fortunately, during the winter months, with lower temperatures, lower evaporation rates, reduced demands from plants, and continued precipitation, there is a moisture surplus and the soil store is recharged. This illustrates the close relationship between temperature, moisture availability and ecosystem functioning.

133

Key Terms

DOM: Dead organic matter.

Litter: Dead, partially decomposed vegetation lying on the ground surface.

Soil development

Soils that develop beneath deciduous woodlands are said to be 'fertile'. The soil store is particularly rich in organic and mineral nutrients (Table 4.6 and Figure 4.37). A clearly defined profile develops below a thick litter layer and the typical soil is classified as a forest brown earth (Figure 4.36). We can explain the richness of the soil nutrient store in terms of the temperature–moisture regime. The DOM of the litter layer is vigorously decomposed during the warmer months, and enters the upper soil horizons as dark humus. Vegetation demands are seasonal and the nutrient cycle is slower than in tropical ecosystems with year-round growth, where higher proportions of nutrients and energy are stored in the **biomass**. Moderate rainfall and good water-holding properties of the soil restrict the leaching out of soluble nutrients.

Range and tolerance

Two important features of the temperate forest biome environment are the infrequency of extreme conditions and the reliability of the seasonal rhythms. None the less, no year is 'average', and extremes do occur. Throughout their **seral** succession, from pioneer stage to maturity, temperate forest ecosystems have a broad range of tolerance (look again at Figure 4.26). They are resistant to disturbance and resilient after disturbance. Species have adaptive capabilities and coping strategies. For instance, in the hot, dry UK summers of 1995 and 1996, many tree species 'shut down' much earlier than normal: they shed their leaves, thereby reducing energy and moisture needs, although this did reduce NPP for those years. Severe winters and droughts reduce bird populations but, again, within a few years most species recover by increased breeding rates, although some specialised species such as wetland birds may remain threatened.

When such 'extremes' become more frequent, they may signal a medium-term trend, such as global warming. Members of an ecosystem modify their behaviour, and in time the ecosystem may have a different seasonal regime and rhythm (Figure 4.38). The local survival of a species depends upon (a) the extent and direction of change; (b) the speed of change; (c) the adaptive capability of the species; (d) competition from other species.

▼ **Table 4.6** *The main nutrient stores in a deciduous woodland (percentage of total)*

	Biomass	Soil	Litter
Phosphorus	8	42	50
Nitrogen	10	40	50
Calcium	7	43	50

▼ **Figure 4.36** *A typical forest brown earth developed below a deciduous woodland (Source: RSPB, 1994)*

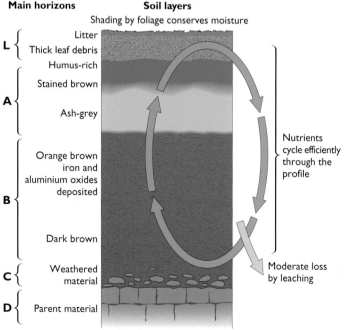

The nutrient cycle shows soluble nutrients drawn down by leaching but the soil has good water-holding properties. So, nutrients are then drawn back up the profile by plant roots and into vegetation

▼ **Figure 4.37** *Three soil stores of carbon (Source: Tivy, 1993)*

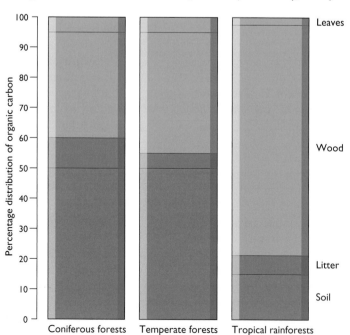

134

26 From the information in Figure 4.37 and Table 4.7, support the claim that deciduous woodland soils are 'fertile' and, when cleared, make productive agricultural land.

27 From Figure 4.38, what evidence is given that the seasonal rhythm of British climate may be changing?

▼ **Figure 4.38** *Newspaper report concerning climatic change (Source: Hereford Times, October 1999)*

WHAT'S HAPPENED TO THE WEATHER?

HOW MANY TIMES have we asked this question or heard others talk about it? In truth, we all have different views. The more senior citizens recall the severe winter of 1947, the hard frosts of 1960 and the baking hot summers of 1975 and 1976.

Since 1976, the media has spoken long and often about the depletion of the ozone level in the upper atmosphere, of global warming, of melting polar ice caps and rising sea levels; even of low-lying parts of the UK being abandoned to the sea. But how much evidence is there to support the fears? The Institute of Terrestial Ecology has been carefully monitoring our weather and recording what they believe are significant changes.

And while it is still processing the bulk of the Spring 1999 records, it has already confirmed the mean January-March temperature as 6.1 °C, compared to a mean of 4.2 °C for the decade of the 1960s.

The last decade of the twentieth century has been the warmest on record, and a whole range of different species have responded to that warmth.

For example, tree leafing in the South East has been considerably advanced: oak 10 days earlier on average today compared to ten years ago; horse chestnut 12 days earlier and lime 11 days earlier.

Migratory birds continue to arrive earlier than they did in previous decades: for example, the swallows have been arriving a week earlier than in the 1970s. Butterflies continue to make early appearances too, with some species appearing as early as March.

We don't normally expect to witness changes in our own lifetime, but these changes are very pronounced. It is too early yet to say whether earliness is actually beneficial. Certainly plants get a longer growing season and animals can produce more offspring.

There may also be problems. Habitats of endangered species may become unsuitable, pest species may make an earlier appearance too, and undesirable species from abroad could become established, such as leaf-damaging moths and mosquitoes capable of transmitting malaria.

4.7 Water as a controlling factor: wetland ecosystems

► **Figure 4.39a** *Tundra wetlands near the Arctic Circle, Alaska. The surface is frozen and snow-covered for up to eight months a year. Summer melting and poor drainage cause waterlogging, especially where permafrost underlies the surface layers.*

▲ **Figure 4.39b** *Tropical wetlands in Kakadu National Park, Northern Territory, Australia. The rainfall is highly seasonal, and the wetlands expand and shrink with this rhythm. This photograph is taken at the end of the dry season, when the billabongs (ponds) are at their lowest. Notice the vegetational gradation away from the water edge. In the wet season the billabong water extends to the woodland edge.*

Definitions of wetland

Wetlands are found across a wide latitudinal range (Figure 4.39a and b). In Britain, natural wetlands occur in upland locations and in lowlands, such as the estuaries of Essex. Wherever they occur, they possess characteristics summed up in these definitions:

'Those areas where land and water meet' (Crace, J., *Guardian Education*, September 1996).

'Any land where saturation with water is the dominant factor determining the nature of soil development and the type of plant and animal communities.'

'Areas of marsh, fen, peatland or water, whether natural or artificial, permanent, seasonal or cyclical, with water that is static or flowing, fresh, brackish or salt, including mudflats and mangrove areas' (Victoria Department of Conservation, Forests and Lands, Australia, 1988).

Key attributes

- Water as the control.
- A wide variety of water types. This helps to explain the ecological diversity of wetlands.
- Inland and coastal locations (Table 4.8).

▼ **Table 4.7** *Main wetland types by location*

Coastal	Inland
Estuary	Floodplain and bottomland
Delta	Inland delta
Intertidal flats	Lowland fens and **carr**
Sand-bar and lagoon marsh	Upland blanket bog
Mangrove swamp and forest	Raised bog
Coral reef and atoll	Tundra with and without permafrost
	Shallow pond and lake fringe
	Glaciated pothole country

28 Library research task: For each of the wetland types listed in Table 4.7, locate two examples. Wherever possible, one of your examples should be in the British Isles. (Plot your examples on maps; name the wetland and the country in which it lies.)

▲ **Figure 4.40** *Intensive agriculture in the Mississippi delta*

▲ **Figure 4.42** *Marina at Port Douglas, Queensland (Australia). The construction of this tourist development caused the removal of more than 100ha of coastal mangrove forest.*

▼ **Figure 4.41** *Listing as a Site of Special Scientific Interest (SSSI) does not necessarily save valuable wetland from development (Source: Conservation Planner (RSPB), Spring 1999).*

BATTLE FOR RAINHAM MARSHES CONTINUES

The London Borough of Havering, with support from English Partnerships, plans to build on over 200 acres of Rainham Marshes, in the Inner Thames SSSI. The RSPB, together with English Nature and the Environment Agency, feel that this is unjustified and unnecessary.

The council are currently undertaking a thorough review of options for land adjacent to the Thames. The RSPB believes the creation of a nature reserve on the whole SSSI with regeneration of the areas of genuinely derelict land next to the site is best for people and wildlife.

Value of wetlands

Several attributes combine to make wetlands highly distinctive and valued ecosystems:

■ High productivity and biodiversity of many temperate and tropical wetlands (look again at Table 4.4).
■ Gently sloping terrain slows natural drainage, but when artificially drained it may become very fertile agricultural land where climatic conditions are favourable (Figure 4.40).
■ Specialised habitats with high conservation value, such as the Okavango Delta (Botswana), the Pantanal (Brazil) and the Sundarbands (Bangladesh).
■ Resource potential and location make wetlands attractive to humans – for instance, fuel from peat extraction; recreation and tourism (birdwatching, fishing, hunting); aquaculture (fish farming); and clearance for settlement (Figure 4.41).

As a result of these various pressures, wetlands are under increasing threat. Apart from high-latitude tundra and Arctic wetlands, at least 50 per cent of the world's natural wetlands have been seriously disturbed or removed (Figure 4.41).

In Australia, the state of Victoria lost 35 per cent of its wetlands, including more than half of the inland wetlands. Along the Queensland coast, the explosive growth of tourism is causing accelerated destruction of the coastal mangove forest (Figure 4.42). The mangroves are a vital breeding and nursery habitat for many aquatic species, and provide a protective barrier against tropical storms.

▲ **Figure 4.43** *Wetland adaptation: Tupelo and cypress trees in standing water, with air roots, Mississippi Delta, Louisiana*

29 Give two reasons why wetlands (i) have high conservation value and (ii) are threatened by human activities.

30 Explain why some high-latitude wetlands are found in regions of low mean annual precipitation.

▼ **Figure 4.44** *Climate graphs for a) Fairbanks, Alaska and b) Miami, Florida*

Water and specialisation

All living components of wetland ecosystems are adapted to living in standing water or saturated conditions, either permanently or periodically. The periodic rhythms may be seasonal or daily (e.g. tidal). Ecosystems are delicately adjusted to these rhythms, dependent upon how long the surface is exposed above water – for instance, the vegetational sequence across a saltmarsh (see chapter 3).

One crucial adaptation is to ensure the supply of air to the plant. For instance, plants may develop stems with structures containing spaces for air between the cells. This allows them to float and draw air downwards to the roots. Other plants work in reverse: they have 'air roots' which grow up through the water (Figure 4.43). Animals and birds adapt their feeding habits and their build to this environment – for example, the long-legs of wading birds and long, shaped beaks for probing mud and sand.

We must remember, of course, that temperature is also a control. Look again at Figure 4.39a: the tundra is a 'wetland' only during the brief summer months. Animals (e.g. caribou) and birds (e.g. wildfowl) migrate, spending only those months there when food and attractive habitat are available. Such regions have low mean annual precipitation but rapid spring thaw, poor drainage and low, highly seasonal evaporation rates, which mean that much of the available moisture remains (Figure 4.44a). Even subtropical and tropical wetlands, with year-round high temperatures, experience seasonal moisture rhythms (Figure 4.44b). Look again at Figure 4.39b.

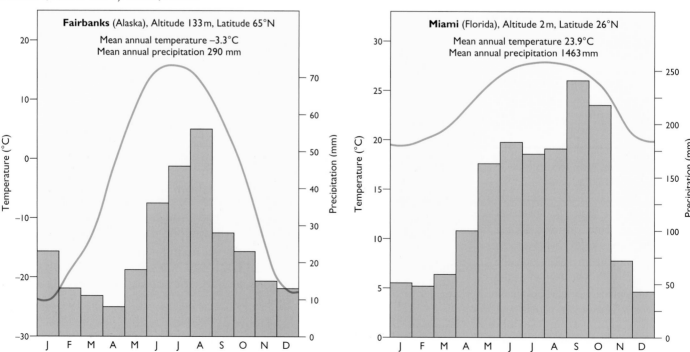

EXAMPLE: Everglades, Florida (USA)

The future for wetlands – the case of the Everglades

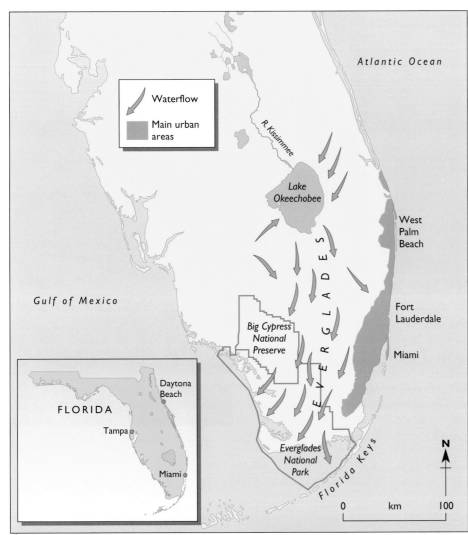

The ecosystem

The Everglades ecosystem has been described as 'a river of grass' and is the largest continuous wetland in North America. The 'river' is a very shallow (less than 15 cm deep) flow of fresh water 80 km wide. It runs 200 km south from Lake Okeechobee over the low, flat limestone platform of South Florida (Figure 4.45a). The vegetation and wildlife are delicately adjusted to seasonal rises and falls of around 1 m in the water-table. This produces three main types of plant associations and habitat (Figure 4.45b).

The largest areas are covered by saw grass 'prairie' that frequently burns during the dry season and is the main source of organic nutrients in this highly productive ecosystem (Figure 4.45c). The hardwood hammocks develop on small, drier sites, while the cypress trees are adapted to semi-permanent inundation (Figure 4.45d). As the water approaches the Gulf of Mexico, conditions become more saline, and mangrove forests form a southern fringe to the Everglades (Figure 4.45e). These wetlands have high conservation value, indicated by the designation of a National Preserve and National Park (Figure 4.45a).

Figure 4.45a *The Everglades, Florida*

Figure 4.45b *Plant associations in the Everglades (Source: Bradshaw and Weaver, 1987)*

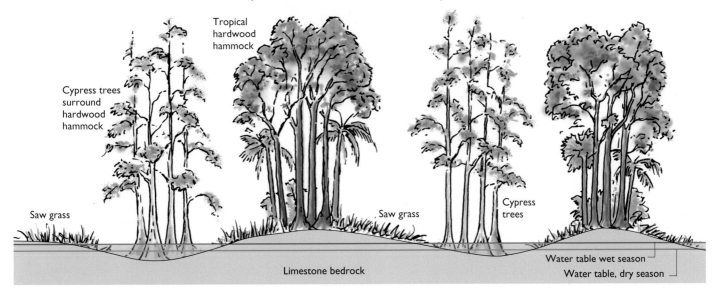

31 Use the Everglades example to discuss these ideas:
- An ecosystem is an open system.
- Wetlands are fragile, specialised ecosystems with a narrow range of tolerance to disturbance.

▲ **Figure 4.45c** *Saw grass 'prairie' and hardwood hammocks in the dry season*

► **Figure 4.45d** *The young cypress tree and the grasses are growing directly on the limestone platform, with a very thin veneer of organic material. This surface is covered with shallow, nutrient-rich water for up to nine months each year. Recently the water-table has been falling.*

▼ **Figure 4.45e** *Dense mangrove forest with its distinctive raised root systems in the brackish water along the southern edge of the Everglades.*

Threats and answers

Since 1950, 50 per cent of the wetlands have been drained for intensive irrigated agriculture and urban settlement. These activities take off large volumes of water and return a variety of pollutants into the wetland system. The flow of fresh water into the Everglades National Park has been reduced by up to 50 per cent. This also reduces the supply of nutrients to the mangroves which are a rich breeding and feeding habitat for many marine species, because the Everglades are an open system. Even small fluctuations in water levels affect the plant associations and wildlife. Ecologists and the National Park managers claim that the ecosystem is collapsing because of the reduced water supply and increasing levels of pollutants – there has been a 90 per cent reduction in bird numbers. So serious is the threat that in 1999 the US President approved a massive reclamation programme (Figure 4.46). The aim is to restore hydrology and the ecosystem of the Everglades to as near to natural conditions as possible. Above all, the throughput of unpolluted water must be sustained.

▼ **Figure 4.46** *A plan to save Florida's Everglades from further destruction*

SAVING THE EVERGLADES

MORE THAN 50 INCHES of rain falls each year on the Florida Everglades but still one of the world's great ecosystems is dying of thirst.

Yesterday the Clinton administration made its first steps towards reversing the devastation wrought on this unique national park on the southern tip of the USA by the need for land for commercial development.

A $7.8bn rescue plan could become the world's biggest environmental project if it is approved by congress. The project would create a new plumbing system for the area, returning much of it to its natural swampland state. Until the early 1900s, summer rains and the overflow from Lake Okeechobee created the unique ecosystem of the Everglades. All that began to change when settlers, taking the place of native Americans, began to drain the land for commercial development.

The process was accelerated in the 1940s with the creation of a series of canals and dykes to prevent floods, to harness water for the cities and to create more land for homes. By the early 1970s, this had become a 1700-mile network.

Under the 4,000-page-long blueprint drawn up by the Army Corps of Engineers, it is intended that more than 240 miles of canals and dams would be removed, and wells would be constructed to retain much of the 1.7bn gallons of water which currently drains into the Atlantic or the Gulf of Mexico each day. The blueprint includes the construction of reservoirs to serve the six million people who live along Florida's coast as well as spare capacity to deal with the estimated six million more people who are expected to move into the area in the next 50 years.

4.8 Disturbance and stability in ecosystems – the example of fire

All ecosystems experience disturbance – also called **perturbation**. Such disturbances vary in: character – a sudden, violent 'pulse' or a progressive series of 'ramps'; scale – the fall of one tree that opens a gap in a forest canopy or the dying off of a species locally; intensity – hurricanes that sweep across Florida's Everglades; and frequency – the '100-year storm' that felled millions of trees across south-east England in 1987. Many ecosystems show a remarkable capability to recover from apparently catastrophic 'wipe-out' events (Figures 4.47 and 4.48).

▲ **Figure 4.47a** *Mt St Helens 1982. During the violent eruption of 1980 (see Figure 1.16, page 20), a blast of hot gas, moving at high speeds, devastated forests up to 15 km from the mountain (a). Yet 13 years later, pioneer vegetation and young coniferous trees were re-establishing themselves (b).*

▲ **Figure 4.47b** *Mt St Helens 1993*

One important agent of disturbance in a variety of ecosystems is fire. This is not necessarily a negative feature because, in many ecosystems, fire is a control component, occurring seasonally or periodically. The main source of natural fires is lightning. Ecosystems adapt to the recurrence of fire and may even rely on it. For instance, some tree species require the heat from fire to open up their seeds for germination; in wet season–dry season climates, burning late in the dry season clears dead vegetation and supplies nutrient-rich ash for the commencement of the wet season (Figure 4.48a and b). Animal and bird populations are dominated by more mobile species, or species that can survive below the surface, as heat from normal 'surface' fires is not conducted deep into the soil. The fire agent therefore, sustains the productivity of these fire-tolerant ecosystems, although the impacts depend upon the heat intensity of the fires.

As the Kakadu ecosystem (Figure 4.48) shows, the 'surface' fires clear perennial surface vegetation without damaging the main perennial species. However, there may be occasional catastrophic fires with much greater heat energy. These become 'crown' fires where all living matter is totally destroyed and the ecosystem restarts from a bare surface (Figure 4.49a-d).

▼ **Figure 4.48 a and b** *Kakadu National Park, Australia: photograph a shows freshly burnt land while photograph b shows fresh vegetation growing after a fire. The ash provides vital nutrients.*

a

Ground fire: Occurs in subsurface organic materials, e.g. peat or humus below a forest. Spreads slowly, but can kill tree roots

b

Surface fire: Runs rapidly across ground-level vegetation – grasses, low shrubs, surface litter. Recovery usually fairly quick, e.g. the Kakadu example, Figure 4.48

c

Dependent crown fire: Heat and flames from surface ignite tree crowns. Fire in crowns moves at same speed as the surface fire. High temperatures, up to 850°C. Common in savannas. Recovery usual in a few years

d

Running crown: Moves rapidly, and occurs with high winds and dry conditions. Very high temperatures (up to 1000°C) A 'fire storm' is an extreme version, where heat causes trees ahead of the main fire to explode into flame. Recovery, 10–100 years

▲ **Figure 4.49 a-d** *Types of wildfire (After: Alexander, 1993)*

The second source of fire is people. Many traditional societies use fire as a method of land management. For example, tropical savannas are frequently plagioclimax ecosystems, where tree species are subdued by a combination of natural and anthropogenic (human cause) fires (Figure 4.50). In the UK, fire is an essential part of moorland management: the burning off of old, 'woody' heather helps the growth of younger, more nutritious plants (primary producers) that support more birds and animals (secondary consumers). Such controlled burning is limited in extent and in heat energy, and is environmentally sustainable, generally at subclimax or plagioclimax levels.

There is evidence that during the second half of the twentieth century, catastrophic fires became more frequent. These seem to be the result of a combination of climatic change or meteorological extremes and intensifying human manipulation of ecosystems, as the great Indonesian fires of 1997–98 illustrate (see example on the next page). Yet, such disasters also demonstrate the resilience of fire-tolerant ecosystems to even high-energy crown fires (see example on the Yellowstone fires of 1988, page 146).

▼ **Figure 4.50** *Fire and the Brazilian cerrado (savanna), north-east Brazil (Source: Mistry, 1998)*

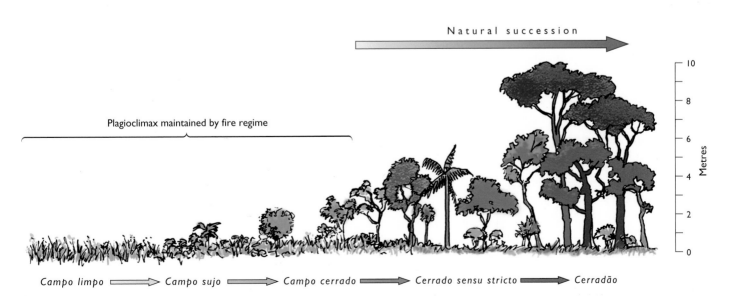

Natural succession

Plagioclimax maintained by fire regime

Metres
10
8
6
4
2
0

Campo limpo ⟹ *Campo sujo* ⟹ *Campo cerrado* ⟹ *Cerrado sensu stricto* ⟹ *Cerradão*

EXAMPLE: Indonesia

Large-scale fires in the tropical rainforests of Indonesia, 1997–98

Background

Between September 1997 and June 1998, a series of destructive fires swept across about 10 million hectares in Sumatra and Borneo. [On the island of Borneo, Kalimantan is part of Indonesia while Sabah is a state of Malaysia.] Up to one-half was primary or mature secondary tropical rainforest (TRF). The rest was mainly plantation, grazing and farmland. The huge, persistent fires spread atmospheric haze over extensive regions for virtually a year (Figure 4.51a). Incoming solar energy was reduced by up to 25 per cent and daytime temperatures were as much as 6°C lower than normal.

TRF ecosystems thrive in consistently warm, humid conditions, and there is a year-round moisture budget surplus (Figure 4.51b). None the less, there is a seasonal rainfall rhythm, and a relatively dry season (Figure 4.51c). As a result, wildfires (natural fires) do occur, although fire is not a normal control element in the TRF biome. For many centuries, local peoples have used fire as part of their 'slash-and-burn' agricultural systems, so maintaining plagioclimax ecosystems in settled areas. The 1997–98 fires however, were unusually intense, prolonged and widespread and have had what many scientists believe may be catastrophic impacts.

Impacts

The primary TRFs of Sumatra and Borneo have especially high conservation value as they are the oldest and most diverse in the world. Even before these fires, more than 50 per cent had been disturbed or removed. Only 6 per cent of the remaining primary forest lies within protected areas such as National Parks, and some of this was destroyed in the great fires. The fires were crown fires, which burned not only the ground vegetation and litter but also killed up to 90 per cent of trees in the impacted areas. As a result, seed supply for regeneration is scarce, recovery may be slow and the

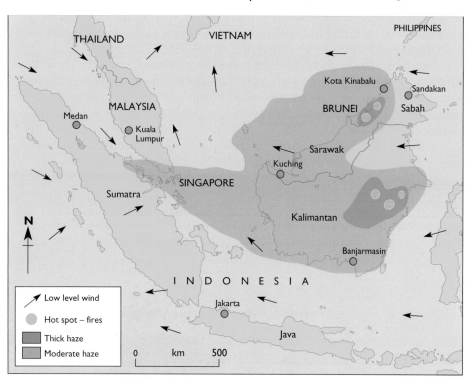

Key Terms

El Niño event: A periodic warming of the eastern Pacific Ocean waters off the coast of South America that disturbs global weather patterns.

◀ **Figure 4.51a** *Fires and atmospheric pollution in Indonesia, April 1998 (Source: Jim, Geography 1999).*

▼ **Figure 4.51b** *The water budget, Singapore*

▶ **Figure 4.51c** *Climate statistics for Jakarta (Indonesia)*

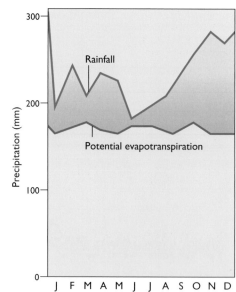

Jakarta, Indonesia Latitude 6°S Altitude 8m	J	F	M	A	M	J	J	A	S	O	N	D
Temperature (°C)	26.1	26.1	27.2	27.2	27.2	27.2	26.7	27.2	27.2	27.2	26.7	26.7
Precipitation (mm)	310	310	200	150	125	100	65	45	70	115	145	240

Mean yearly temperature 26.7°C

Total yearly precipitation 1875mm

secondary succession may not include the entire original species composition. There is evidence, too, that riverine and coastal ecosystems are being affected by increased sedimentary transport as a result of increased erosion and runoff.

In their undisturbed state, TRFs are highly complex, competitive ecosystems. Their components are generally specialised, occupy precise niches and are sensitive to disturbance. Despite the warm, moist conditions that favour rapid growth, recovery may be slow, particularly towards the original structure and composition. It may be hampered, for example, by seed shortages, the local extinction of less mobile animals, and the destruction of soil organisms that normally speed up soil regeneration (Figure 4.51d).

Causes

Meteorological fluctuations: During 1997–98 one of the strongest **El Niño** situations of the twentieth century developed. This periodic warming of the eastern Pacific Ocean waters off the coast of South America disturbs global weather patterns. Across the western tropical Pacific, including Indonesia, the principal effect is drought: during 1997 and 1998, the rainfall was up to 50 per cent below normal (Figure 4.51g). This extended period of moisture deficiency caused unusual drying out (desiccation) of the ecosystem, and seriously increased fire risk (Figure 4.51h). El Niño years in 1982–83, 1986, 1991, and 1994 were also dry years with increased fire events.

▼ **Figure 4.51e** *Ecosystem recovery*

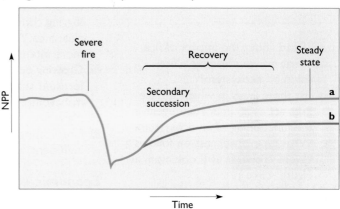

▼ **Figure 4.51f** *Short- and medium-term impacts of prolonged atmospheric pollution*

▼ **Figure 4.51d** *Fire impacts – one scenario*

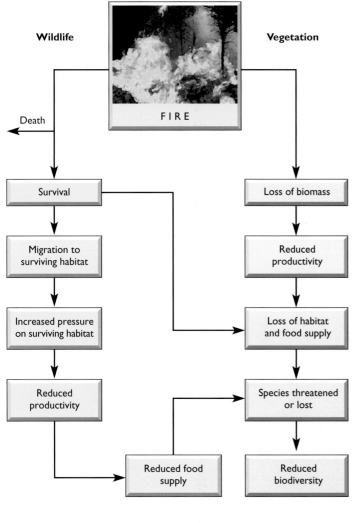

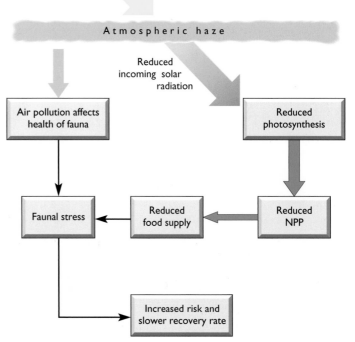

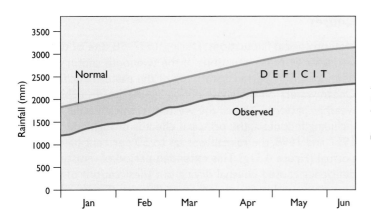

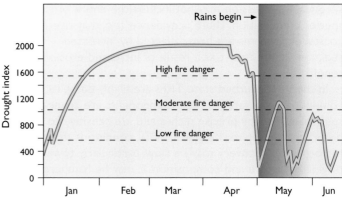

▲ **Figure 4.51g** *Fire risk and drought (Source: Geography, 1999)*

---

32 Read about the impact of fire:
a What do we mean by a 'fire-tolerant' ecosystem?
b List three ways in which surface fires act as an important agent in the functioning of an ecosystem.
c What type of succession follows a severe crown fire? (Look again at Figure 4.51e.)
d List the factors that made the Indonesian fires of 1997–98 so serious.

Human activities

■ Despite warnings of fire risk, shifting cultivators, plantation managers and logging companies continued with their seasonal 'burn' policies for land clearance. What they did not realise – or accept – was that the fires would be more intense and often impossible to control.
■ Growing population pressure means that many new settlers are moving into forested areas. These recent arrivals often have little experience in how to manage fire in a shifting cultivation system.
■ Many fires were in remote, inaccessible regions.
■ Fire-fighting organisation by central and local government agencies was inefficient, and equipment and skilled people were lacking.

Economic conditions

During 1997–98, the Indonesian and Malaysian governments were struggling with economic collapse and political unrest. This meant that they gave lower priority to fires in remote regions far from the main cities.

EXAMPLE: Yellowstone, USA

The great Yellowstone fires of 1988

Yellowstone, in the north-west corner of Wyoming, USA, is the world's oldest National Park. The main reasons for its designation were the geothermal geyser basins together with a diverse climax vegetation dominated by coniferous forest interspersed with grassland that supported abundant wildlife (Figure 4.52a). During the summer of 1988 a series of wildfires, ignited by lightning strikes, swept across approximately one-half of the park (Figure 4.52b and c). The extract on the next page from a 1992 National Park Service report sums up the situation:

◄ **Figure 4.52a** *Geysers, forest, grasslands and buffalo (North American bison): key attributes of Yellowstone's ecosystem*

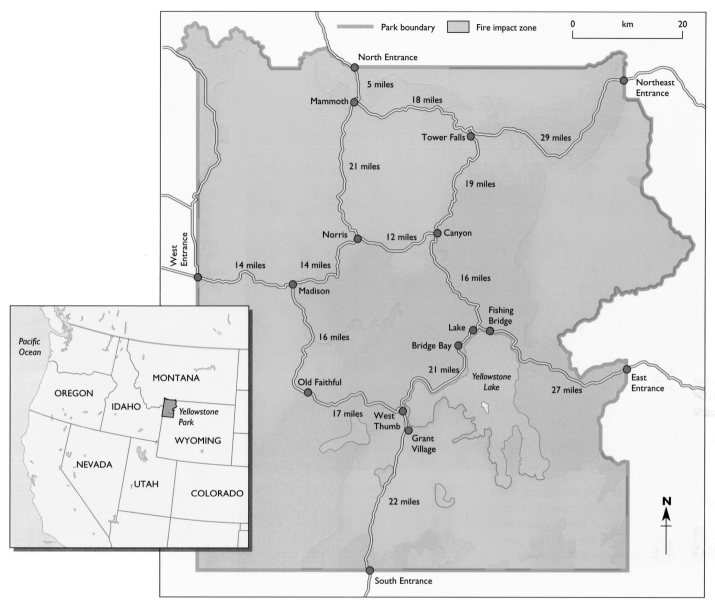

▲ Figure 4.52b *Fire impact on Yellowstone National Park, 15 September 1988*

'Scientific evidence indicates that natural fires have been a part of Yellowstone's environment for thousands of years. Fire has played a role in the formation of soil, vegetation and wildlife patterns. Large fires have burned at average intervals of 20–25 years in the grasslands, at intervals of 200–400 years in the conifer forests.

During the summer of 1988, an extended drought, along with high temperatures, strong winds, and large amounts of fuel (dead wood and litter), set the stage for the largest fires in Yellowstone since the 1700s. Despite extensive control efforts, the crown fires were extinguished only by autumn's precipitation. In most places soil heating was not severe and has served to make nutrients more available. In a few areas, soil temperatures were high enough to burn most organic matter and the return of vegetation will be slow.

Most surprising has been the rapid regrowth of aspen trees and wildflowers, fed by the abundance of nutrients, rain and sunlight. One problem is that elk browse on young aspen and may prevent the young trees maturing. Lodgepole pine, one of the main tree species, is well adapted to fire. The cones store large numbers of seeds that are released as heat

33 Use the Yellowstone example to (i) illustrate the resilience of ecosystems, (ii) show how environmental managers are increasingly using techniques that copy natural processes and the way natural ecosystems behave.

from the fire melts the resin that binds the scales of the cones. Already, young lodgepole seedlings are flourishing (Figure 4.52d).

This seems, therefore, to have been an extreme example of natural wildfires. However, the extract mentions 'large amounts of fuel'. Until 1976 the National Park managers had a policy of fire suppression and a thick surface layer of inflammable dead organic material (DOM) slowly built up. By the mid-1970s the managers understood the role played by fire and introduced a policy of allowing natural fires to burn. But by 1988, extensive areas still retained the unusually dense layer of DOM, and in the hot, dry, windy conditions, the fires rapidly ran out of control. Today, the park managers have adopted a policy of 'prescribed burns'. Each year they select areas where they ignite controlled surface fires. They are, in fact, imitating natural ecosystem processes. In 2002, prescribed burning and forest thinning became official government policy for forest management throughout the United States.

▲ **Figure 4.52c** *This photograph, taken in 1988, shows how the crown fires varied in intensity. Patches of trees survived while over extensive areas the lodgepole pines were killed, leaving their dead trunks standing.*

▶ **Figure 4.52d** *This photograph shows the recovery of lodgepole pine after seven years (1995). The dead trunks (snags) are gradually falling and will slowly decompose, providing a steady supply of nutrients.*

4.9 Ecosystems and economics

The Human Environments chapters in this book (chapters 5–8) show that we are making ever-increasing demands on global natural resources. As populations grow and countries strive to improve their standards of living, this trend will continue. However, if development is to be sustainable, policies need to take into account both the economic and environmental value of ecosystems. For instance, in 2002 a group of scientists published a report on the economic benefits of ecosystem conservation (Figure 4.53).

▼ **Figure 4.53** *The economic benefits of ecosystem conservation (Source: The Guardian, 9 August 2002)*

WORLD'S WEALTH STILL RELIES ON NATURE

Preservation of the world's remaining wilderness could be the ultimate bargain. Scientists and economistst calculate that forests, wetlands and other natural ecosystems are worth far more to human economies than the farms or buildings that could replace them.

They report in the US journal Science that the wilderness converted to human use each year costs economies $250bn a year, every year.

Put another way, it would cost the world $450bn to extend and effectively protect threatened areas of temperate and tropical forest, mangrove swamps, coral reefs and so on. But in return, these global reserves would supply humans with at least $4400bn in 'goods and services'.

Humans depend on insects to pollinate crops, on forests to recycle carbon dioxide, slow erosion and prevent floods, on estuarine fswamps as fish hatcheries and to buffer towns from storms and tidal surges. Ultimately, natural ecosystems provide humans with food, water, air, shelter, fuel, clothing and medicines. In 1997 economists tried to put a price on the things nature supplies, and arrived at a total of £33 trillion a year. This year, with backing from the Royal Society for the Protection of Birds and the Department for the Environment, Food and Rural affairs, British and US scientists did their sums again.

They surveyed 300 case studies of what happened when the natural environment was converted to human use, and chose five for closer analysis. These included the intensive logging of a Malaysian forest, a Cameroon forest converted to a small-scale agriculture and commercial plantation, a mangrove swamp in Thailand turned to shrimp farming, a Canadian marsh drained for agriculture, and a Philippine coral reef dynamited for fishing. In each case the value of the natural ecosystem – as storm and flood protection, for sustainable hunting and tourism, or to soak up carbon dioxide – outweighed the returns from human use. The Malaysian forest would have been 14 per cent more valuable left standing. The Canadian marsh would have returned 60 per cent if left alone for hunting, trapping and fishing.

Summary

Atmosphere and ecosystems are closely interrelated open systems.

Human activities are having increasingly important impacts on processes in the atmosphere and ecosystems.

There are well-developed spatial and temporal rhythms in the distribution of solar energy and therefore, in regional temperatures.

Moisture is available in the atmosphere in three forms (gas, liquid, solid) and reaches the Earth's surface as precipitation (rain, snow, hail).

Atmospheric moisture and precipitation vary over time and space at a diversity of scales.

Soils are a mixture of mineral and organic materials, water and air, that move between the lithosphere, biosphere and atmosphere in a series of cycles.

An ecosystem consists of organic and inorganic components.

Ecosystems evolve through seral stages of succession towards a condition of dynamic equilibrium with their environment. This optimum state is called the climatic climax.

All ecosystems have adaptive capability, but vary in their resistance to and tolerance of disturbance.

HUMAN ENVIRONMENTS

The following chapters focus on people – where they live, their work and their impacts on their surroundings. Individual chapters follow module themes central to Geography syllabuses, but it is important to bear in mind throughout that the themes are interrelated: they function as a system. For instance, population change (chapter 5) has a direct relationship with the growth of cities (chapter 6). In turn, the dynamics of economic activities (chapter 8) has a direct influence on urban location, form and functions (chapter 6). Furthermore, there is an intensifying interaction between rural environments (chapter 7) and urban environments (chapter 6). Resource demands and development (chapter 8) are also an integral element of population dynamics (chapter 5).

Thus, we need to cross-reference the themes of the individual chapters to develop balanced attitudes and values concerning people and their relationship to planet Earth. By linking the chapter themes we can arm ourselves with enough material to make decisions, seek explanations, forecast changes and assess policy options, all of which contribute to a better understanding of the processes at work in human geography.

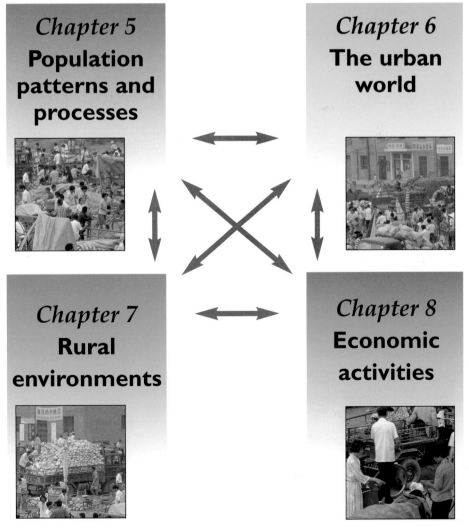

Chapter 5
Population patterns and processes

Chapter 6
The urban world

Chapter 7
Rural environments

Chapter 8
Economic activities

Chapter 5 · Population patterns and processes

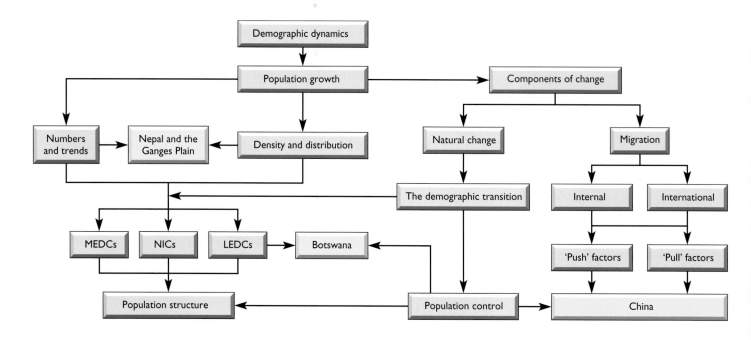

5.1 Introduction

On 1 January 1900, there were approximately 2500 million people on planet Earth. Today there are more than 6000 million (6 billion). During the 1990s, the world's population grew by about 97 million people each year (Figure 5.1a and b). The enormous growth was one of the outstanding features of the twentieth century. This 'population explosion' has become a high profile global issue (Figure 5.2) and is an influence on a wide range of topical issues like the pressure on world resources, global warming and world poverty (Table 5.1).

► **Table 5.1** *Human impact on planet Earth (Source: Simmons 1998)*

World habitat and human disturbance				
	Total area (km²)	Percentage undisturbed	Percentage partially disturbed	Percentage human dominated
World total	162.0	51.9	24.2	23.9
World total minus rock, ice and barren land	134.9	27.0	36.7	36.3

Criteria
Undisturbed: *Primary vegetation; very low population density (under 10 persons/km²), even lower in tundra and semi-arid regions.*
Partially disturbed: *Shifting or extensive agriculture; secondary but naturally or regenerating vegetation; overgrazing; logging.*
Human dominated: *Permanent agriculture or urban settlement; desertification or other permanent degradation; primary vegetation removed.*

Key Terms

Demography: The study of population.

Population Geography: The study of the spatial and temporal dynamics of population.

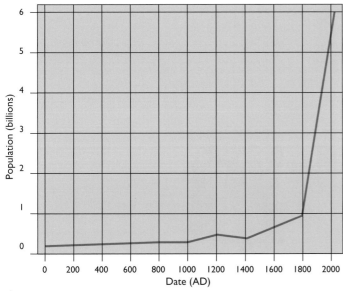

Figure 5.1a *World population growth 1–2000 (Source: Central TV Enterprises, 1993)*

▶ **Figure 5.1b** *World population growth 1000–2150 (Source: BBC Education, 1993)*

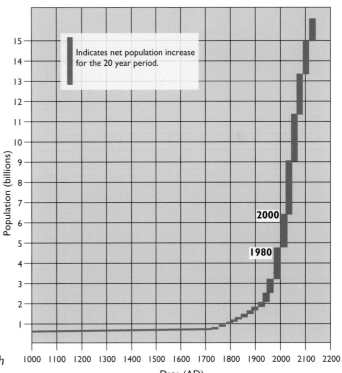

▶ **Figure 5.2** *The global population explosion (Source: The Guardian, 30 July 1999)*

1 Use the information in Table 5.1 to illustrate, in one sentence, why the population explosion has become such an important issue.

2 From Figure 5.1 both parts:
a In approximately which year did world population reach
(i) 1 billion; (ii) 5 billion?
b Which 20-year period records the most rapid population growth?
c Do the graphs support this statement: 'during the twentieth century, world population growth accelerated but during the twenty-first century it will slow down'? Back up your answer with figures.

3 What are the two most important pieces of information given in the newspaper article of Figure 5.2? Compare your answer with those of others in your group.

Make room for 6 billion

ON OR AROUND 12 October 1999, a very important baby will be born somewhere in the world. The baby's arrival is not in itself big news, since three are born every second, but this will mark world population reaching six billion. The five billionth baby isn't even a teenager yet, having been born in 1987. In took all of human history until 1800 for the population to reach its first billion; the second took only until 1930. A mere 69 years later, six billion are crowding the planet. Today, the world's population is twice what it was in 1960. Because of declining fertility, the UN Population Division was forced to make some dramatic revisions to its projections in late 1998. Instead of increasing by 80 million a year, population was increasing by 78 million. Originally, it had been calculated that we would hit six billion on 16 June, but the date had to be moved forward. Dr Nafis Sadik, executive director of the UN Population Fund (UNPFA), called this 'very encouraging news', though she noted that 97 per cent of population growth is occurring in LEDCs, where health services and family planning remain scarce. By 2050, the UN projects that MEDCs will have 1.6 billion people, slightly fewer than today, while LEDCs will have doubled from 4.2 billion in 1995 to 8.2 billion in 2050.

Yet we must be careful not to focus only on the negative effects of the population explosion. The continued pressure on resources has challenged human inventiveness. From this have come the huge advances in agricultural and industrial productivity and the creation of new industries and jobs. Population growth and rising expectations have encouraged individuals, businesses, governments and international agencies to seek to improve the quality of life and to find sustainable ways of using global resources.

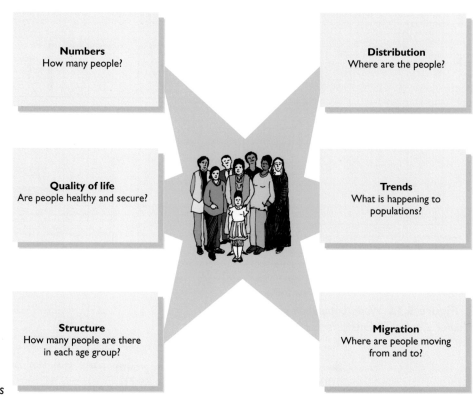

Numbers
How many people?

Distribution
Where are the people?

Quality of life
Are people healthy and secure?

Trends
What is happening to populations?

Structure
How many people are there in each age group?

Migration
Where are people moving from and to?

▶ **Figure 5.3** *Demographic characteristics*

The central theme of this chapter is **population dynamics**. By 'dynamics' we mean change over time and space. Section 5.2 deals with changing distributions and densities – where people are. Following sections focus on how populations change – the shifting balance between births, deaths, migration and age groups. As you develop your understandings of population dynamics, you will find three levels of question emerging. The first level seeks mainly descriptive answers (Figure 5.3). The second level looks for explanation and causation. For instance, what processes help to explain why the population of France is growing at less than 0.5 per cent a year, while Nigeria's growth rate is 3.0 per cent a year. The third level raises 'should' and 'how' issues that involve values, priorities and decisions: 'Is family planning a good thing, for whom, and who should decide?' The materials in this chapter give you practice in assessing the usefulness of data and information for answering these levels of questions.

5.2 Numbers and trends

We are constantly bombarded with population statistics at global, continental and national scales. For instance, a TV programme tells us that the population of the continent of Africa has reached 800 million; a newspaper headline shouts 'Make room for 6 billion!', and so on. However, we need to use demographic data with care.

■ The data may not be accurate. Most countries take **censuses** every ten years, but they vary in the efficiency of the counts.

■ The data quickly becomes out of date. Even websites may be using roughly estimated updates.

■ Data may vary from one census to another and from one country to another in the way it is collected. This makes comparisons risky.

Yet without up-to-date, accurate figures to tell us about population characteristics and trends, it is difficult to understand issues, to draw sensible conclusions and so to make decisions and policies – such as how many houses, schools and hospitals will be needed in the future.

4 With a partner: For your own city or county, identify three current or recent issues in which accurate, up-to-date population statistics would be valuable.

154

The global scale

The graphs of Figure 5.1 provide data to answer several key questions. For instance, 'How many people were there in the world at particular times during the twentieth century?', and 'Is global population growth accelerating or slowing down?' This information, however, gives us only the most general understanding of population dynamics, because people are spread unevenly across the continents (Figure 5.4). Thus, we need to interrogate (to ask questions of) statistics that give information on population distribution and growth rates. Table 5.2 provides such data at the continental scale. (Antarctica is excluded because it has no permanent human population; Asia is sub-divided into four large-scale regions because of its size and population.)

▼ **Figure 5.4** *World population density and distribution*

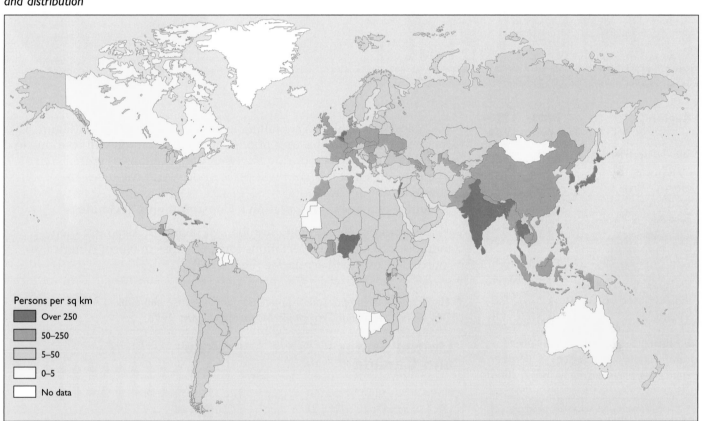

Persons per sq km
- Over 250
- 50–250
- 5–50
- 0–5
- No data

▼ **Table 5.2** *Population totals by continents, 1950 and 2001 (million)*

Continent	Area (million sq.km)	1950	2001
Africa	30.3	224.0	813.0
Latin America	20.5	166.0	527.0
North America	21.5	172.0	317.0
Asia *	31.8	1402.0	3721.0
Europe	23.0	547.0	726.0
Oceania	8.5	12.5	31.0

*Asia	Area (million sq.km)	1950	2001
Eastern Asia	11.8	671	1498
South-central Asia	10.8	499	1504
South-eastern Asia	4.5	182	532
Western Asia	4.7	50	187

(Source: Social Trends, 32, 2002)

5 From Table 5.2:
a For 1950 and 2001, name the continents with (i) the highest and (ii) the lowest population densities.
b Name the continents with (i) the highest absolute numerical growth and (ii) the highest proportional growth rate during the 1950–2001 period. (Use the four regions of Asia.)
c Select suitable forms of graph or pie chart to depict
(i) population totals for 1950 and 2001
(ii) proportional growth by continents.
d Use the data to assess the accuracy of these statements:
■ During the second half of the twentieth century, population growth was most rapid in less developed regions of the world.
■ There is little relationship between population change and size of area.
e Use your information to prepare a brief report with the title: 'Peopling the planet, 1950–2001'.

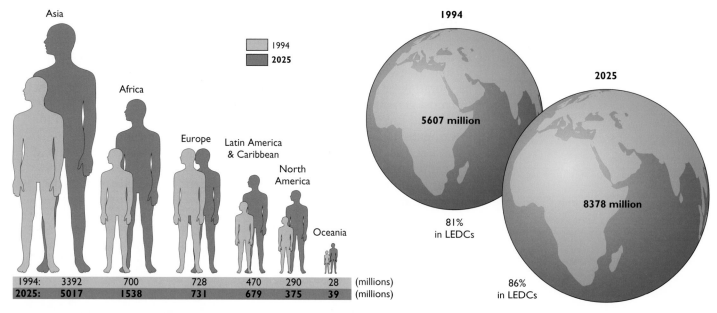

Region	1980–1990	1990–1999
Sub-Saharan Africa	5.0	2.6
Middle East & North Africa	5.0	2.3
South Asia	3.7	1.9
East Asia & Pacific	2.6	1.2
Latin America	3.3	1.7
East Europe & former USSR	1.1	0.3
Developing countries	3.2	1.7
Industrialised countries	1.2	0.6
WORLD	2.9	1.4

▲ **Figure 5.5** *Trends in population growth rate by world regions, % per year (Source: Barrett, Geography, [85}2, 2000)*

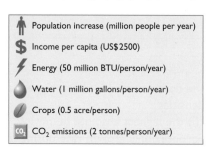

Population growth rates are falling in all major world regions (Figure 5.5). Nonetheless, the global balance of population distribution will continue to shift to **LEDCs** and **NICs** during the twenty-first century (Figure 5.6).

The pressures that this shift will place upon world resources are summarised in this equation:

Resource demands = Population x Consumption x Technology

For example, as the growing numbers of people in Asia raise their standards of living and expectations towards those of North America, so resource demands will accelerate (Figure 5.6).

▼ **Figure 5.6** *A comparison of population and resource demands in MEDCs and LEDCs (Source: National Geographic Magazine, August 1998)*

United States and Canada

The USA and Canada use about twice as much energy per capita as Europeans, 10 times more than Asians and 20 times more than Africans. The USA alone uses more than one-quarter of the world's oil. By one estimate, the average American will consume 62,000 pounds of animal products, 55,000 pounds of plant foods, 770 tons of minerals and the energy equivalent to 4 000 barrels of petroleum.

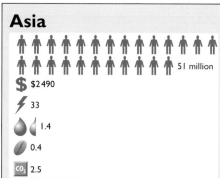

Asia

As Asia's economy grows, so does its consumption levels. China's economy alone grew ten-fold between 1953 and 1989, and its energy consumption grew 18-fold. China's use of coal, which emits more carbon dioxide than any other fuel, is projected to double by 2015. Already, China is second only to the USA in CO_2 emissions; by 2020 it's likely to be ahead of the USA. Developing Asian nations still use far less energy per capita than industrialised nations but the gap is narrowing.

6 Look at Figure 5.5:

a What are the main contrasts in population growth shown by the data in Figure 5.5?

b Summarise the contrasts in current resource use in MEDCs and LEDCs as suggested in Figure 5.6.

c Use the information in Figures 5.5 and 5.6 to discuss the assertion that the major environmental issues facing the world in the twenty-first century will focus on LEDCs.

▼ **Table 5.3** *Changing population growth rates of the five poorest LEDCs on the World Development Report's ranking list*

Percentage average annual population growth		
Country	**1985–90**	**1997**
Rwanda	3.0	2.6
Mozambique	1.6	2.2
Ethiopia	3.1	1.6
Tanzania	3.2	2.9
Burundi	2.9	3.0
Country	**Rate of per capita GDP change**	
	1985–90	**1990–97**
Rwanda	+ 2.1	– 14.0
Mozambique	– 0.2	+ 6.0
Ethiopia	+ 2.3	+ 1.0
Tanzania	+ 3.4	+ 3.0
Burundi	+ 4.0	– 1.4

7 Look at the figures in Table 5.3. If each country maintains the current rates of population increase, approximately how long will it take for the population of each to double?

8 Use the information in Figure 5.7 to illustrate the statement that population growth rate declines as level of development rises.
(Note the high growth rate in Jordan during the mid 1990s was due to in-migration from neighbouring countries experiencing political instability and war.)

The national scale

Generalisations at continental and macro-regional scales have limited accuracy and usefulness. To examine population dynamics with greater precision and to be able to answer more specific questions, we need data for smaller area units. The figures in Table 5.2 tell us that between 1950 and 2001, the population of Latin America grew by 361 million but we know nothing about where the growth was greatest and where population pressures and problems are emerging.

The most commonly available figures are from individual countries, collected by national censuses. In Britain, population censuses have been taken every ten years since 1801, with the exception of 1941, during the Second World War. Not all countries collect census data as regularly.

Another valuable source of demographic and economic information for individual countries is the World Development Report, published annually by the World Bank. The report uses a set of **development indicators** (such as average income levels; **GDP** per capita; balance of payments; educational and health indices) to produce a three-class ranking system: low-, middle- and high-income countries. The examples in Figure 5.7, taken from the Report, indicate a clear relationship between population growth and the level of **development**. These examples show that low-income LEDCs have relatively high population growth rates whereas high-income MEDCs have low rates of population growth. NICs are frequently in the middle-income class but generally show rapidly declining growth rates.

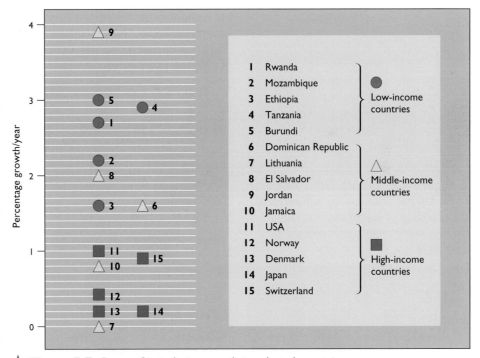

▲ **Figure 5.7** *Rates of population growth in selected countries*

The mean annual population growth rates for LEDCs as a group is falling from 2.7 per cent between 1980 and 1990, to 2.4 per cent between 1990 and 1997, although the change is uneven (Table 5.3). Nonetheless, these are still high growth rates which raise serious issues for governments and international agencies. With an annual growth rate of 1 per cent, the population will double in 67 years; at 2 per cent in 35 years; at 3 per cent in 24 years and at 4 per cent, the population will double in just 17 years. Clearly, providing homes, schools, health facilities, and jobs for the additional people, in such a short time, presents a great problem for governments. This is especially true for LEDCs where population growth

Key Terms

Population distribution: The geographical patterns created by where people live.

....

Population density: The number of people per unit area, usually expressed as persons /sq.km.

....

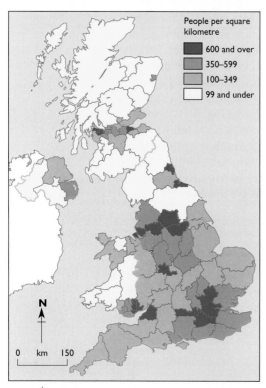

People per square kilometre

- 600 and over
- 350–599
- 100–349
- 99 and under

0 km 150

▲ **Figure 5.8** *The population distribution of the UK has been strongly influenced by the processes of industrialisation and accompanying urbanisation over the past 250 years.*

rates frequently exceed economic growth rates. Few LEDCs sustain annual economic growth rates of 2 per cent a year (Table 5.3). Over the last ten years, LEDCs recorded an average annual increase in per capita GDP of only 1.4 per cent. Furthermore, LEDCs often lack the necessary financial, administrative and industrial structures for rapid and large-scale provision. In contrast, NICs can achieve high economic growth rates even in uncertain economic conditions, for example, despite a financial collapse in 1997, Malaysia was recording an economic growth rate of 8 per cent in 1999.

Densities and distributions

Population distribution is the geographical pattern produced by spatial variations in **population density**. At the national scale, densities vary widely (Table 5.4). Equally significantly, these average densities hide considerable internal variations that produce well-defined distribution patterns (Figure 5.8, Table 5.4). These patterns are the outcome of the complex interaction between cultural, environmental, economic and technological factors and processes working over long periods of time (see the Nepal and the Ganges Plain example). For instance, Canada is the world's largest country but has one of the lowest population densities. Yet this low overall density statistic hides a highly distinctive distributional pattern: more than 90 per cent of the 29.5 million people live in an East–West strip along Canada's southern border. This pattern is explained by the interaction of a set of natural, cultural and economic factors – the harsh climates of the more northerly part of Canada; the westward movement from the initial European immigrant settlements of the St Lawrence Basin; the proximity of the USA to this East–West corridor (Figure 5.9). Each country exhibits its own, distinctive distributional pattern of varying population densities.

▼ **Table 5.4** *Population distribution in the UK (million)*

Area	1951	2001	2021 (forecast)
England	41.2	49.1	54.3
Wales	2.6	2.9	3.1
Scotland	5.1	5.1	5.0
N. Ireland	1.4	1.7	1.8

(Source: Social Trends, 32, 2002, ONS, 2002)

9 From Figure 5.8 and with the help of an atlas:
a Name the main urban centres of population.
b Name the counties of south-east England with moderate population densities.
c **Group research project:** Select one major conurbation outside London, and identify the main economic factors that have influenced the population growth.

▼ **Table 5.5** *Population density variations for selected countries*

Country	Area (thousands/sq.km)	Population (millions)	Density (persons/sq.km)
Singapore	1	3.0	3000
Bangladesh	144	118.0	820
Netherlands	37	15.5	420
UK	245	59.3	242
China	9561	1200.0	125
Brazil	8512	159.5	19
Canada	9976	29.5	3

◄ **Figure 5.9** *Calgary, Alberta owes its growth to its location on a road and rail route through the Canadian Rockies, the agricultural productivity of the Prairies and the mineral wealth of the region.*

EXAMPLE: Nepal and the Ganges Plain (India)

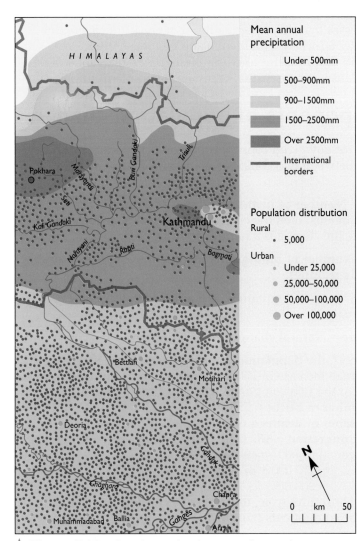

Figure 5.10a *Population distribution and annual precipitation*

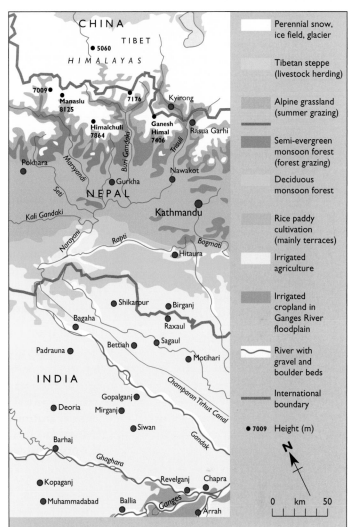

Figure 5.10b *Land-use (natural and cultivated)*

The two maps (Figure 5.10a and Figure 5.10b) cover a North–South transect from the Himalayan crest, across the mountainous terrain of Nepal to the broad, flat plain of the River Ganges and its tributaries. Figure 5.10a shows a well-defined population distribution pattern, with two zones of relatively high population densities. Analysis of the two maps reveals the relationship between this pattern and the physical, climatic and hydrologic environment. Over many centuries, communities have developed farming techniques, like the building of terraces and irrigation canals, which are well-adapted to this environment. Today the region supports a dense network of villages and towns.

10 From the Nepal–Ganges example:
a Describe the pattern of population distribution.
b Sketch a North–South cross-section of the region.
Add the:
■ Names of the countries and the national boundaries
■ Names of physical features and the rivers
■ Main land-use zones
■ Zones of high population densities.
c For each of the three environmental characteristics – physical, climatic, hydrological – give one example of its influence upon population density and distribution.

159

Key Terms

Population change: The changes in population over a given period, expressed in absolute figures or proportional figures (rate of change).

....

11 Look at the following figures for the UK during the 1981–91 decade:
Mean annual number of births = 757 000
Mean annual number of deaths = 655 000
Mean annual number of immigrants = 250 000
Mean annual number of emigrants = 210 000

Now calculate the mean population change.

Population distribution patterns are constantly changing, and section 5.4 introduces the main factors which influence these shifts. However, as we have learned from the study of the UK population (Figure 5.8), 'distribution' includes where people live, i.e. settlements. Thus, a number of the forces and processes that drive population dynamics are covered in some detail in chapter 6 (urban settlements) and chapter 7 (rural settlements). You will find it useful to cross-reference the relevant sections of these chapters in answering questions on population distributions.

5.3 Components of population change

Population change for any given country or region is the outcome of the balance within and between two components natural change and net migration.

Natural change = Number of births – Number of deaths
When births exceed deaths, then the change is a natural *increase*; when deaths exceed births, the change is a natural *decrease*.

Net migration = Number of immigrants (people moving in) – Number of emigrants (people leaving)

The balance between these two equations is expressed in terms of the following simple equation:

Population change = Natural change + Net migration

Between 1991 and 1997, the population of the UK grew from 57.8 million to 59.0 million, an average increase of 200 000 people a year. We can derive the relative importance of the two components of change from this data:
Mean annual number of births = 753 000
Mean annual number of deaths = 640 000
Mean annual net migration = +87 000
Population change: (753 000 – 640 000) + (87 000)
= 113 000 + 87 000
= 200 000

Even within such a short time-scale, there may be considerable variation in contribution from the components (Figure 5.11).

▼ **Figure 5.11** *Components of population change, England and Wales, 1991–96*

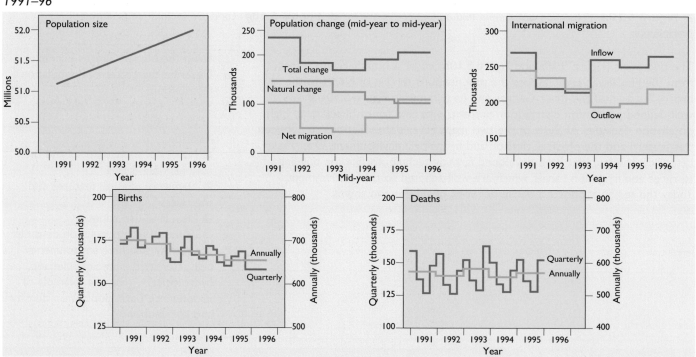

Key Terms

Birth rate: The number of live births in a given time (usually one year), expressed as a rate/1000 population.

Death rate: The number of deaths in a given time (usually one year), expressed as a rate/1000 population.

Natural change: The balance between births and deaths, expressed numerically or as a rate/1000 population.

Demographic Transition: The sequence of change in birth rates and death rates and the changing relationship between them in a country or region over time. The transition is from a condition of high birth rate/high death rate/low natural change, to one of low birth rate/low but slowly rising death rate/static or negative natural change.

Migration: The permanent or semi-permanent movement from one place of residence to another. This movement may be voluntary or involuntary, international or internal.

Population structure: The number or the proportion of the total population in a particular age group, usually subdivided by gender and presented as an age–sex pyramid.

Births and deaths are usually expressed as *rates*/1000 population, and so we refer to natural increase *rates* and natural decrease *rates*. Thus, from Table 5.6, we can state that in 1997 the natural increase rate for France was 3.3/1000 population (12.4 – 9.1). As later sections in this chapter make clear, it is important that governments and international agencies have accurate information on the balance between births and deaths, between immigration and emigration, and the shifting role played by each component of population change.

▼ **Table 5.6** *Population change in France, 1987–97. How does this suggest the slowing population growth of France?*

Year	Population (million)	Births (thousands) (rate/(1000)		Deaths (thousands) (rate/(1000)		Natural Increase (thousands)	Net Migration (thousands)	Total Change (thousands)
1987	55.8	768	13.8	527	9.4	+240	+44	+284
1992	57.4	744	13.9	522	9.1	+222	+90	+312
1997	58.6	725	12.4	534	9.1	+191	+40	+231

So far, the data and calculations have been mainly descriptive – they tell us what is happening. The following sections focus on *how* and *why*, examining the processes causing fluctuations in each component over time. For instance, how natural increase rates decline as development advances and why migration tends to occur in surges. It is useful at this point to scan the key terms to familiarise yourself with terms that are used in these sections.

The demographic revolution or transition

The data we have studied so far suggests that as a country develops, industrialises and becomes more 'advanced' its rates of population growth decline. Today, MEDCs such as the UK, France and the USA have low population growth rates. In contrast, most LEDCs have higher rates of population growth (look again at Figure 5.7). NICs and 'middle income' countries tend to have intermediate population growth rates.

Net migration balance may be a significant factor at certain periods (see section 5.6), but over the longer term, population change is controlled by the relationship between births and deaths, i.e. natural increase or decrease rates. There seems to be a common sequence of change in this birth rate/death rate relationship as countries progress through the development process. This sequence has been expressed as a model, known as the Demographic Transition Model (Figure 5.12). The key to understanding this model is that the wider the gap between the birth rates and death rates, the more rapid population growth becomes. (The model ignores the direct effects of migration on population change.)

The model is commonly used in two ways. First, it can be used to follow the course of demographic change for a country over time. For example, the UK was in Stage 1 until around 1760 when the Industrial Revolution began. Over the next one hundred years or so, the economy and society was transformed to make the UK an urbanised, industrial country (Stage 2). By the late nineteenth century, birth rates were falling rapidly (Stage 3), and since the 1930s both birth rates and death rates have been low (Stage 4).

The second way in which the model may be applied is to locate countries at a particular time. For instance, nowadays, very few countries remain in Stage 1, but many LEDCs are in Stage 2 (e.g. Peru, Sri Lanka). NICs and rapidly developing countries such as China and South Africa are in Stage 3, while many MEDCs are at the end of Stage 4 (e.g. the UK, the USA).

Figure 5.12 *The Demographic Transition Model*

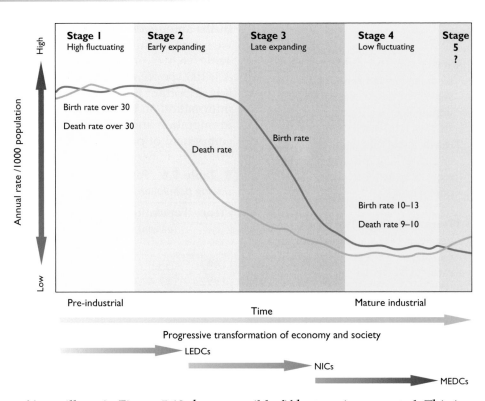

You will see in Figure 5.12 that a possible fifth stage is suggested. This is because in an increasing number of countries, annual numbers of deaths are beginning to exceed births. Table 5.7 includes some examples of these countries in Europe. This is one result of the 'ageing' of populations: as the proportion of the total population in older age groups increases, so death rates will inevitably begin to rise (see section 5.5 on population structure). When this happens, unless birth rates rise and or there is an increase in immigration, population totals will fall overall. Official forecasts estimate that the UK will enter Stage 5 within the next 30 years.

▼ **Table 5.7** *Birth rate/death rate relationships for selected European countries, 1997*

Country	Population (million)	Birth rate (/1000 population)	Death rate (/1000 population)
Ireland	3.7	14.1	9.6
France	58.5	12.4	9.1
Yugoslavia	10.5	12.5	10.6
UK	58.9	12.3	10.7
Poland	38.6	10.7	9.9
Austria	8.1	10.4	10.0
Spain	39.3	9.1	9.0
Sweden	8.8	10.2	10.5
Germany	82.0	9.9	10.4
Lithuania	3.7	10.2	11.1
Russia	147.7	8.9	14.1
Bulgaria	8.3	7.7	14.5

We need to ask two basic questions about the Demographic Transition Model: 'Does it work?' and 'Is it useful?' Like all models, it is very generalised, and may not apply in detail to all countries. However, if we accept its general structure, it is a useful device for forecasting what is likely to happen to a country's population. Thus, suppose data for Peru allows us to place the country in Stage 2 in a particular year, we can then use the model to forecast that in the next few decades, falling birth rates (in Stage 3) should be the main focus of population policy (see China example).

12 From Figure 5.12:
a During which stage of the demographic transition is population growing most rapidly?
b Why does population growth (numerical and proportional) increase rapidly during Stage 2?
c Why does the rate of population growth slow down during Stage 3?
d Why might numerical increases be maintained during much of Stage 3, even though proportional increase rates are declining?
e What is the main similarity between Stages 1 and 4?

13 From Table 5.7:
a Calculate the rate of natural change (+ or –) for each country in 1997.
b Which countries support the idea that a Stage 5 is beginning to develop in the Demographic Transition Model?
c Give a brief description of the characteristics that a Stage 5 would have in the Demographic Transition Model.

One important feature of the graph is the time-scale of the horizontal (x) axis. For Western European countries such as the UK and France, the Stage 1-4 time-scale has been approximately 200 years. The lengthy periods of Stages 2 and 3, when population growth was most rapid, were also periods of rapid economic growth and social change. In contrast, since the 1950s, many LEDCs have moved suddenly into Stage 2, but without parallel economic development: population growth rates have outstripped economic growth rates. As a result, population-control policies have focused on an urgent reduction in birth-rates (see section 5.4).

For LEDCs, the time-scale of the horizontal axis is shortened – Stage 1–3 spans less than 50 years. Of course, some countries, including NICs such as Taiwan, Thailand and Malaysia, have succeeded in achieving parallel economic development as they pass through the stages of demographic transition in just a few decades (Figure 5.13). These countries have been able to financially support their growing population.

► **Figure 5.13** *Thailand is one of South-East Asia's NICs, and has experienced rapid economic and population growth. The population (60 million) has doubled in the last 30 years. The population of the capital, Bangkok, is forecast to reach10 million early this century.*

EXAMPLE: China

Analysing population change in China

Jianfa Shen is a geographer in a Chinese university, and summarises what is happening to China's population:

'In 1950 there were 550 million people in China. By 1995 the total had reached 1200 million. During this time, our national fertility rate fell from 6 to 2. This is just below replacement level, but as the graph (Figure 5.14c) shows, our population will continue to grow until about the year 2040. In the first four decades of the twenty-first century, China will have to cope with another 400 million people!

At first it seems odd: we have a low fertility rate, strong government policies on family planning, and yet population is continuing to grow. The answer lies in the way population structure changes over time. Fertility rates only began to drop sharply after the government introduced its 'one-child per family' policy in 1978. The large numbers of people born between 1950 and 1978, are now having children. Only when this generation grows old and dies from around 2040, will the overall population begin to decline. Because we are living longer than in the past, this decline will take longer. Today our average life expectancy is 70 years; in 1950 it was 60 years.

One of the most important features over the next 40 years therefore, will be the ageing of our population structure. An increasing proportion of Chinese people will be in the older age groups.'

	1994	1999
Number of women of childbearing age (million)	326	543
Number of married women of childbearing age (million)	230	249
Percentage of late marriage*	58.2	60.12

*The legal age of marriage is 20 years for women
A late marriage is considered to be after the age of 24

◄ **Figure 5.14a** *The 'one-child' policy aims to limit each married couple to a single child. This is a severe restriction in China, where children and the family are a central part of the culture.*

► **Figure 5.14b** *China today has a youthful population structure. Until this generation has reached the end of their child-bearing years, the population will continue to rise.*

14 From the China example:

a Describe the population dynamics of China in terms of the Demographic Transition Model.

b Explain the population trends shown on the graph (Figure 5.14c). Write brief statements from the point of view of either:
   ■ A government official supporting the 'one-child policy'.
   or
   ■ A young married couple objecting to the policy.

▼ **Figure 5.14c** *Population graph for China, 1950–2070*

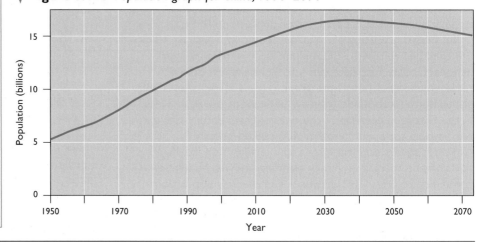

Key Terms

Life expectancy: The average length of life from birth of individuals in a population.

Infant mortality rate (IMR): The number of children who die before their first birthday.

Total fertility rate (TFR): The average number of children born to a woman in a particular population.

Replacement-level fertility (RLF): The fertility rate at which each generation has only enough children to replace themselves in a particular population.

15 Use the information in Table 5.8a, b to support the idea that there is a relationship between the development process and the Demographic Transition Model of population change.

5.4 Factors influencing natural change

The demographic transition is one outcome of the development process. There is a clear relationship between social, health and economic progress and the stages of the model (Figure 5.12). Evidence from all continents indicates that as people become better educated, healthier, and have improved job and income prospects, they live longer, fewer die young, and women have fewer children (Tables 5.8a,b). Thus, both birth rates and death rates decline.

▼ **Table 5.8a** *Natural change and development indicators, 1995*

	10 poorest countries (LEDCs)	10 richest countries (MEDCs)
Life expectancy (years)	48	76
Infant mortality rate (IMR)	100	6
Total fertility rate (TFR)	6.5	1.6

▼ **Table 5.8b** *Quality of life indicators for selected countries, 2000*

Country	Pop (million)	Annual pop growth (%)	Infant mortality (/1000 live births)	Life expectancy (yrs)	
				Male	Female
UK	59.8	0.2	7	75	80
Canada	31.1	1.0	6	76	82
Sierra Leone	4.9	3.0	170	36	39
Bangladesh	129.2	1.7	79	58	58
India	1013.7	1.6	72	62	63

(Source: Social Trends, 32, 2002)

Increases in life expectancy create an initial fall in death rates. However, in the longer term, as the proportion of people in older age groups increases, so death rates may rise again. This is one feature of Stage 5 in the Demographic Transition Model.

Many governments have policies for controlling population growth and for improving the quality of life. Such policies aim to increase **life expectancy**, while reducing IMR (**infant mortality rate**) and TFR (**total fertility rate**), so that population growth rates are lower than economic growth rates. There is no doubt that, despite continuing contrasts, IMR and TFR are falling (Figure 5.15). The IMR is a sensitive indicator of living conditions, and can be influenced fairly rapidly by the availability of clean water, improved diet, local clinics, etc. The TFR strongly reflects the longer-term population dynamics, such as how many children a woman bears, and how many live through adulthood. In forecasting future trends, the RLF (**replacement-level fertility**) becomes a crucial benchmark. For instance, if in a particular population, 50 per cent of births are female, each woman will need to have two children. On average there should be an equal number of children of each gender born. However, as some of these daughters will die while they are very young, the RLF is normally given as 2.1 or 2.2. For a population to replace itself over time, the TFR must stay at

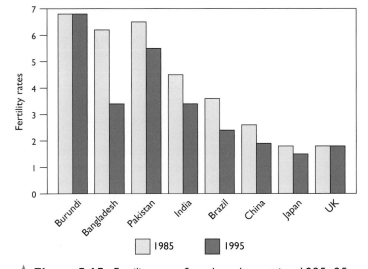

▲ **Figure 5.15** *Fertility rates for selected countries, 1985–95*

16 From Figure 5.15:
a Which of the countries have a TFR below RLF?
b What is likely to happen to population totals in these countries if these TFRs continue?

17 Using the data of Table 5.9, choose an appropriate statistical technique to test the hypothesis that female education is a significant factor in influencing fertility rates.

▼ **Table 5.9** *Relationship between female literacy and fertility rates*

Country	TFR	Female literacy (%)
UK	1.8	99.0
USA	2.1	99.0
Japan	1.4	99.0
Rep. of Korea	1.3	96.8
Spain	1.2	97.1
Mexico	2.7	86.7
Malaysia	3.4	77.5
Saudi Arabia	6.4	47.6
South Africa	4.0	81.2
Indonesia	2.5	77.1
China	1.0	70.9
Kenya	5.5	52.3
Bangladesh	2.9	24.3
Uganda	7.1	48.7
Burkina Faso	7.2	8.6

(Source: Human Development Report)

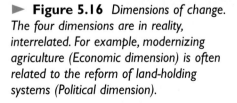

► **Figure 5.16** *Dimensions of change. The four dimensions are in reality, interrelated. For example, modernizing agriculture (Economic dimension) is often related to the reform of land-holding systems (Political dimension).*

18 Give two reasons why the introduction of family planning and birth control into a society can be a 'sensitive' issue.

or above 2.1 or 2.2. This helps to explain the natural decreases which are occurring in several of the MEDCs in Table 5.7 (page 162), where birth rates are far too low to attain RLF. This process is also the logic behind the slogan for the population control programme in Indonesia – 'Two is enough'.

Issues in population control

The successful introduction of population-control policies and family-planning projects is a difficult and sensitive process. It involves the interaction of four overlapping dimensions of change (Figure 5.16) and requires a combination of 'top down' and 'bottom up' approaches. Thus, in examining any particular country, we should ask these key questions:

■ Are there appropriate, well-funded and managed policies and projects by the government and international agencies?
■ Is there a 'readiness' in communities for change, based on education (Table 5.9), awareness of the benefits, and a cultural acceptance of family planning and smaller families?

Socio-cultural
■ A shift in social attitudes and cultural traditions.
■ Adaptation of religious rules and taboos.
■ Development of women's rights.
■ Improved levels of education, especially for women.
■ Rising material expectations.

Medical/scientific
■ Increasing range of birth-control techniques.
■ Education to improve understanding of the benefits and methods of family planning.
■ Provision of community-based health clinics with trained staff, in addition to modern hospitals.
■ Increased availability of acceptable contraception devices.

Economic
■ Increasing job opportunities in a range of enterprises beyond agriculture.
■ Modernising agriculture.
■ Increasing proportions of women with paid jobs.
■ Introducing a money-based economy, including savings and loan systems.

Political
■ Consistent, effective government policies on population control and family planning.
■ Reform of land-holding systems to give more people access to land and hence to food production.
■ Infrastructure improvements such as roads, water supply, electricity.
■ Acceptance by governments of assistance from international agencies.
■ Maintenance of political stability that allows positive change to take place.

For example, Kerala, a state in southern India, has the highest life expectancy (71 years), the lowest IMR (26/1000) and a fertility rate 50 per cent lower than the national average. Literacy rates, especially for women, are higher than elsewhere in India, based on education and health programmes introduced by Christian missionaries in the nineteenth century. Furthermore, unlike much of India, women in Kerala can inherit land, and so have a greater equality with men. The desire for improved health and population control has meant that communities have made demands on the State Government to create local health-care systems. During the 1990s, the State Government was spending 10 per cent of its budget on health and medical services, the highest in India.

Evidence suggests, therefore, that: 'the countries that have reduced their birth rates fastest are not always those with the highest incomes. They are countries that gave priority to health, education and women's rights. What brings fertility down is primarily social development, especially for women' (Harrison, 1993) (see Botswana example).

Such progress is not easy to achieve, as there may be tensions between the four dimensions of change, which helps to account for the wide diversity of fertility rates across the world. For instance, religion remains a powerful factor influencing social change. In Latin America the dominant religion is Catholicism. The Catholic Church opposes birth control, but in several of these countries contraception is legal (e.g. Mexico and Brazil). Families, therefore, must resolve the conflict between their civil rights and their religious teachings. In the Muslim world too, where Church and State are often closely linked, there are frequent tensions concerning women's rights and family planning. However, control in Islam is less centralised than in Catholic countries, and local Imams (religious leaders) have considerable powers to interpret the Koran. In Tunisia and Turkey, both mainly Islamic countries, fertility rates are (rapidly) falling. Tunisia had a rate of 5.2 in 1980, which had fallen to 3.0 in 1995. Turkey had a fertility rate of 4.3 in 1980, but, 3.1 in 1995. It would seem, therefore, that in Catholic and Muslim countries where fertility rates are falling significantly, families, and particularly women themselves, are making decisions based on self-interest rather than on religious grounds. In China, the strong communist government has imposed a one-child policy that has been reluctantly accepted by the Chinese people (see China example, pages 163–164). Countries with less powerful central governments, e.g. India, have failed to enforce these policies. Indonesia has had its 'Two is enough' policy since 1970. The level of enforcement has varied, but a national network of local health clinics has been developed which has helped the TFR to fall from 4.3 in 1980 to 2.7 in 1995 (Figure 5.18).

In other instances, change may be enforced, through environmental pressures such as food shortages resulting from prolonged drought (see Box 1). When famine and disease combine, rising mortality and shortened life expectancy become major factors in population trends, for example, the catastrophic combination of prolonged drought and the AIDS epidemic across southern and central Africa (Figure 5.17).

Box 1 Main responses to drought and famine in Northern Ethiopia

- Changed attitudes towards early marriage
- Changed attitudes to having large families
- Increased acceptance of family planning services
- Reductions in fertility
- Outmigration, especially by young people who would be having children.

▼ **Figure 5.17** *Changes in selected African countries with high HIV prevalence, 1970–2000*

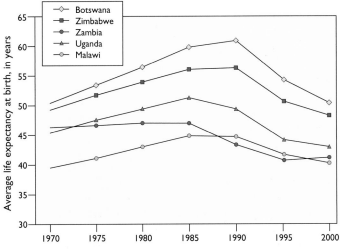

▶ **Figure 5.18** *Education and support are important in encouraging change in traditional attitudes and in developing understanding of the benefits of family planning.*

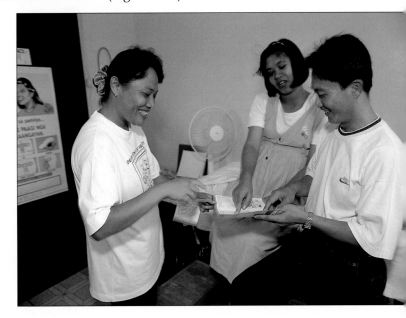

EXAMPLE: Botswana

Social change and fertility in Botswana

On the 'World Development Report' ranking list, Botswana is classed as a 'lower-middle-income economy'. In 1997, the GDP/capita was $US3000. Although it is a LEDC, its economic growth rate during the 1985–95 decade was high, at 6.6 per cent a year. This has been based largely on income from the diamond mines and the growth of tourism which in turn has helped to generate jobs in other service industries. Despite rapid urban growth (8 per cent a year in this period), 65 per cent of the population still lives in rural areas and is largely dependent upon agriculture. The figures show that although population growth is still around 3.0 per cent a year, the country is experiencing the demographic transition, moving rapidly into Stage 3.

▼ **Table 5.10** *Demographic trends in Botswana, 1985 and 1997*

Demographic trends	1985	1997
Population: 1.5 million		
Average annual growth rate (percentage)	3.5	3.1
Life expectancy (years)	62	70
IMR (/1000 live births)	63	33
TFR (/1000)	6.5	4.9
Adult literacy (percentage of population)	20	30
Access to safe water (percentage of population)	56	70
Access to health facilities (percentage of population)	65	85

19 From the Botswana example:

a Use the data to illustrate the process summarised in the Demographic Transition Model (Figure 5.12).

b Summarise the main factors that have influenced the decline in the TFR in Botswana. To what extent does the Botswana example support the idea that the primary cause of fertility decline is social change?

c **Group project:** What are the benefits and costs of introducing programmes of family planning and birth control? (Think in terms of economic, socio-cultural and environmental factors.) Who should decide whether such programmes are introduced? Should they be compulsory? What methods should be adopted? (You will find it useful to carry out a library/website search to broaden your background knowledge.)

Mary Ngwenya is a Regional Officer with the Botswana Tourist Board, and is studying for a post-graduate degree at a British university:

'The changes date from the 1940s, when the young men began to move away to work in the mines in South Africa and Zambia, and later in distant parts of Botswana. They sent money back and sometimes came home, but by the 1970s, up to 25 per cent of the men aged 12–35 years were away. This affected marriages and family life, and has changed the lives of women in particular.

Today, over 50 per cent of women have jobs and one-third have completed secondary education – this compares with only 25 per cent of men. Someone said to me recently – 'women in Botswana are educating themselves out of the marriage market'. There's some truth in this – 40 per cent of professional jobs are now held by women and many of us are not content with a less educated man. I have two sons who have the same father, but I have never married, and my mother looks after the boys when I am away. Today, over 50 per cent of women aged 15–49 – our childbearing years – have never married, and around one in every two children is born outside formal marriages.

'The combination of the spread of women's independence and Botswana's strong family-planning programme had brought fertility rates down. As younger women become educated and take jobs, they tend to be older when they have their first child, to practise birth control, and to have fewer children. The motivation for family limitation is stronger among women who work. This is especially true for the growing number of unmarried women. As economic expectations rise, men too are coming to see the benefits of fewer children.'

5.5 Population structure

One important factor influencing population change is the number and proportion of people in various age groups. This division of a population by age and gender is called the population structure, and is usually illustrated as **age–sex pyramids**. Pyramids vary in the way they are drawn: the age and gender groups may be numerical or proportional and the age groups vary in range and detail (Figure 5.19 a–d). Notice too, the variations in structure. LEDCs have high numbers and proportions in the younger age groups, and are said to have youthful age-sex structures. In contrast, MEDCs have ageing structures, i.e. relatively high numbers in older age groups.

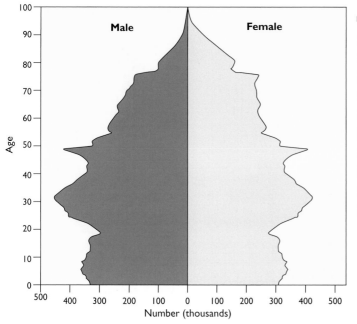

▲ **Figure 5.19a** *Population pyramid, UK (an MEDC), 1997*

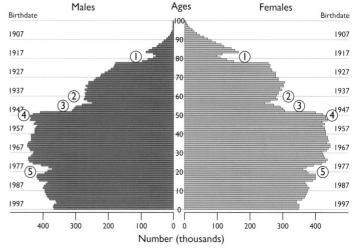

1 Lack of births due to the First World War.
2 Years when people born during the First World War reach fertility.
3 Lack of births due to Second World War.
4 The Baby Boom
5 Fertility rates drop below two children per woman.

▲ **Figure 5.19b** *Population pyramid, France (an MEDC), 1997*

Numbers (hundred thousands)

▲ **Figure 5.19c** *Population pyramid, Malaysia (an NIC), 1995*

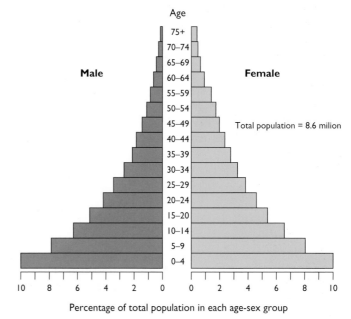

Total population = 8.6 milion

Percentage of total population in each age-sex group

▲ **Figure 5.19d** *Population pyramid, Malawi (an LEDC), 1995*

20 From Figure 5.19 a–d:

a What evidence is there that fertility rates and birth rates in the UK and France are falling?

b Give two effects of the Second World War (1939–45). (Think of the number of 50-year-olds in the UK in 1997.)

c To what extent do the population structures of the UK and France indicate that they are in the final stages of the demographic transition process?

d Suggest the likely main differences in the population structures of the UK and France in the year 2017.
Give your answer in terms of these age groups: 0–19; 20–39; 40–59; 60+.

e Use the examples of the UK and France to illustrate what is meant by an 'ageing' population structure.

f Are there any significant differences between the population structures of Malaysia, an NIC, and Malawi, an LEDC?

g Use the examples to illustrate what is meant by a 'youthful' population structure.

h Use the data of the four pyramid graphs to illustrate the characteristics of the demographic transition process.

21 From Figure 5.20:

a Which age groups dominate the migrants into and from South Africa?

b Suggest one important difference between the age–sex structure of immigrants and emigrants, and what effect this may have on future population structures.

The main purpose of population structure information is for forecasting. Thus, if we know how many young children there are aged 0–4 years, then we can estimate the number of school places needed in the next decade; if we know the numbers age 15–19 years, then we can forecast the numbers who will be entering the job market. Over time, each age cohort moves upwards through the pyramid like a ripple or wave, as the pyramid for France shows (Figure 5.19b). Of course, not all people now in the 15–19 age cohort will move into the 20–24 group: some will emigrate; some will die. Forecasting needs to take these losses in all age groups into account. Losses may be balanced by immigration, especially of young adults, who tend to dominate most types of voluntary migration. The mobility of young adults is especially important because they are entering the child-bearing period of their lives and so will influence future population trends and structures (Figure 5.20).

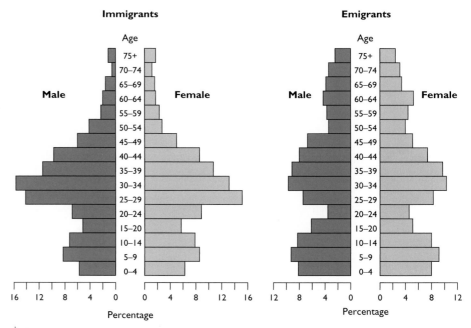

▲ **Figure 5.20** *Age–sex structure of migrants in South Africa, 1985–95*

The proportion of a particular age cohort who die while in a particular age group is called the **age-specific death rate**. Note too, that particular episodes, such as wars, have long-lasting effects on population structures. The effects may be direct, i.e. numbers killed, and indirect, i.e. fewer people in certain age cohorts to have children.

Population structure and age–sex pyramids summarise a great deal of information concerning population dynamics which has been the main theme of this chapter. Numbers and proportions in specific age/gender groups reflect past processes, e.g. wars, migrations, changing IMR and TFR, they illustrate the present condition, and they indicate future situations, for example, numbers of children available to become parents. There are clear relationships between population structures and the demographic transition process. For example, LEDCs in the earlier stages of the demographic transition exhibit youthful population structures. The IMR and TFR are falling but remain high enough to sustain high rates of natural increase; life expectancy is rising but still moderate; and up to 50 per cent of the population may be under 15 years of age, e.g. Mexico, Ethiopia, Indonesia. In contrast, MEDCs, many of whom are entering Stage 5 of the demographic transition process, have ageing population structures. This is the result of persistently low IMR; TFR below replacement levels; and long life expectancy, e.g. France, Japan (Figures 5.21, 5.22). Without significant net immigration, the populations of these countries will decline (Table 5.12).

JUST 500 JAPANESE?

Even as the six billionth human is born, it's time to forget fears about the world being overpopulated. By the end of the next millennium, Tokyo will be a ghost town, and Japan will be empty. The country's population will be just 500 by the year 3000, and just one by 3500. When that person dies the Japanese nation will be no more.

These apocalyptic predictions aren't the rantings of a doomsday cult, or of an academic out to gain some publicity, but of the Japanese government itself.

Of course, many things can change in 1000 years. But what is frightening about the forecast is that it's a mathematical certainty if Japanese women carry on having an average of just 1.4 children and if Japan doesn't change its immigration policy. If it continues as they are the Japanese will die out. It's just a question of when.

And so will we.

Britain is one of 61 countries that are not having enough babies to replace their populations, according to the United Nations. For a population to remain stable, women need to have, on average, 2.1 babies each. One in four is opting to have none at all.

In all countries of the European Union, fertility is now so low that populations are set to decline – if they haven't already. Spanish women are having just 1.5 babies each, which is the lowest fertility rate in the world. In some parts of Spain the average has dropped below one.

▲ **Figure 5.22** *Japan 2000*

▲ **Figure 5.21** *An extreme scenario of declining birth rates Japan 3000?*

22 How does the information in Figure 5.21 and Table 5.11 help to explain why most of the global population growth during the twenty-first century will occur in countries of the developing world?

23 If the trends shown in Table 5.11 are continued, what is likely to happen to the population structures of LEDCs?

▼ **Table 5.11** *Trends in fertility and population growth rates*

a Fertility

	Birth rate/ 1000 pop.		TFR	
	1980	2000	1980	2000
LEDCs	34	24	4.8	2.9
MEDCs	15	11	1.8	1.6
World	30	21	4.0	2.6

b Growth rates (% p.a.)

	1980–1990	1990–2000
LEDCs	3.2	1.6
MEDCs	1.2	0.6
World	2.9	1.4

(Source: UNICEF, 2001)

5.6 The migration component of population change

Migration influences population growth and the geographical distribution of population. Its effects are more immediate than the slow-acting shifts in birth rates and death rates. As this section will show, migration is a complex process, but its meaning is straightforward: migration occurs when people move from one place to another on a semi-permanent or permanent basis, i.e. they acquire a new address. Official definitions of migrants are normally based on moving to a new place for at least one year.

In studying migration, we must try to answer some fundamental questions:
- How many migrants are there?
- Where do they move from and to?
- Why do they move and what factors influence their movements?
- What trends in migration are taking place?
- What policies are in place to control migration?
- What effects does migration have on the origin and destination regions?

▼ **Figure 5.23** *Push and pull factors influencing migration*

	Push	Pull
Economic	Unemployment, poverty	Job opportunities, better earnings
Social/cultural	Religious and political persecution, ethnic rivalries	Freedom of religious, political and social beliefs
Quality of life	Poor housing, health and education	Improved housing, health and education
Environmental	Poor environments and natural disasters	Pleasant and safe environments

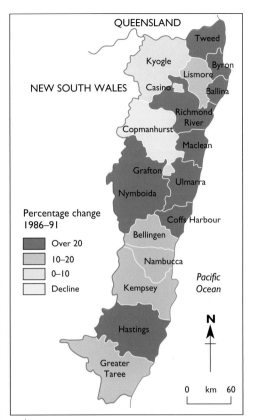

Migration can be split into two distinct types: internal migration (movements within a country) and international migration (movements between countries).

In answering the question 'Why do people move?', whether internally or internationally, we need to distinguish between two main categories. First, there is migration that is to some extent voluntary – people make a decision to move. Second, there is compulsory or involuntary migration, where individuals and communities are forced to move. It is useful to think too, of a third category which we may call obligation migration, where family or friendship ties or job requirements may trigger migration.

It is important to remember that voluntary migration, both internal and international, involves varying degrees of freedom and choice. The motivation to move arises from some dissatisfaction with the present situation, and the idea that life would be better somewhere else. Our decisions are based upon perceived quality of life (Figure 5.24). This process has been expressed in terms of **'push' factors** that encourage people to move from a region or country, and **'pull' factors** that attract people to a region or country (Figure 5.23). As the examples in the following sections illustrate, each situation varies in the degree of freedom of choice an individual believes they have in making the decision to move, whether internally or internationally.

▲ **Figure 5.24** *The north coast of New South Wales, is rapidly growing. Between 1986 and 1991 the population grew by 81 000, and 52 per cent of this was due to internal migration. The motive for this migration is an improved quality of life.*

5.7 International migration

24 Use the data in Figure 5.25 to complete a table showing the following items, for five contrasting years across the 1972–96 period:
  Natural change
  Net migration
  Total change
  Migration as a percentage of total change
  (Use + and – symbols as appropriate.)

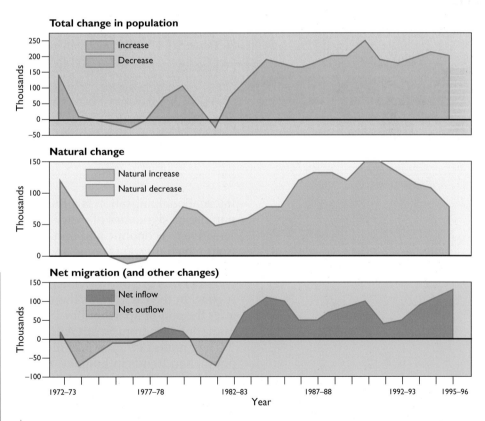

▲ **Figure 5.25** *Population change in England and Wales 1972–73 to 1995–96, and its components*

▼ **Table 5.13** *Net migration balance for selected European countries, 1990–95 (Average annual totals in thousands)*

German	+563
Italy	+109
France	+59
Ireland	–2
Portugal	–13

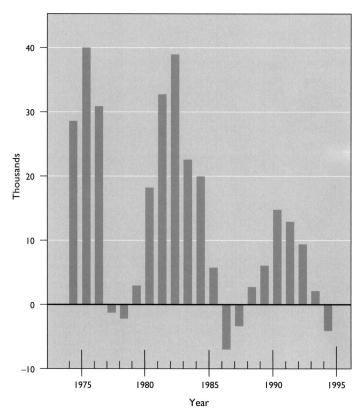

▲ **Figure 5.26** *Population change in South Africa, 1974–94*

302

▼ **Table 5.14** *UK international migration, 1996–2000, thousands per year*

Regions	In	Out	Net
New Commonwealth	58.5	22.4	+36.1
European Union	89.2	72.9	+16.3
Old Commonwealth	75.5	62.3	+13.1
All countries	321.5	232.5	+89.0

(Source: Social Trends, 32, 2002)

International migration is a fluctuating and significant component in the population dynamics of every country (Table 5.13). Migration patterns are sensitive to economic and social conditions inside a country and in regions from which migrants come and to which they go. The figures for South Africa, 1974–94 illustrate this clearly (Figure 5.26). The late 1970s and 1980s were periods of political violence and instability during the apartheid regime. In the early 1990s there was uncertainty about what would happen after the end of **apartheid** in 1994. In all three periods there was a net migration loss from South Africa.

The graphs of Figure 5.25 illustrate the influence of migration on recent population trends in England and Wales. For example, during the 1993–96 period, growth from natural increase fell, while **immigration** rose until it accounted for over 50 per cent of population growth.

All countries experience a constant ebb and flow of international migration, and fluctuations in migration balance. For instance, during the 1990s the UK was gaining an average of 250 000 immigrants a year, and losing an average of 210 000 emigrants (Figure 5.27). Between 1996 and 2000, the net gain averaged 89 000 a year (Table 5.14). Immigration is forecast to account for 70 per cent of UK population growth during the decade 2002–2011.

Key Terms

Immigration: The movement of people into a country or region.
....

Emmigration: The movement of people from a country or region.
....

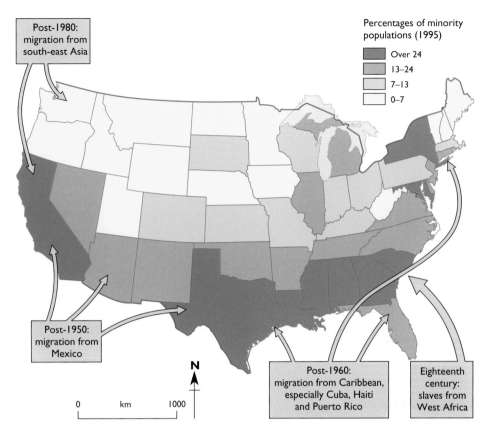

Percentages of minority populations (1995)

- Over 24
- 13–24
- 7–13
- 0–7

Post-1980: migration from south-east Asia

Post-1950: migration from Mexico

N

0   km   1000

Post-1960: migration from Caribbean, especially Cuba, Haiti and Puerto Rico

Eighteenth century: slaves from West Africa

▲ **Figure 5.28a** *Minority (non-white) population and migration surges into the USA*

About one-third of immigrants initially settled in London. By 2000, there were 1.3 million foreign residents living in the UK, compared with 2.3 million in France, and 4.2 million in Germany. Remember that this UK figure does not include immigrants with British passports (such as those from remaining British colonies). International migration flows are complex, but the three maps (Figures 5.28a, b, c) illustrate five basic ideas which provide a foundation for understanding and explaining the migration flows:

- **Emigration** from and immigration to particular countries occurs in a series of surges over time.
- Some countries and regions remain attractive immigrant destinations, while others are regular sources of emigrants.
- Emigrant and immigrant flows have distinct geographical (spatial) patterns.
- Migration flows are most intense between relatively adjacent countries as it easier to travel from one to another.
- Modern migration patterns are dominated by flows from poorer regions (LEDCs) to richer regions (MEDCs), because people move in search of well-paid jobs and a better quality of life.

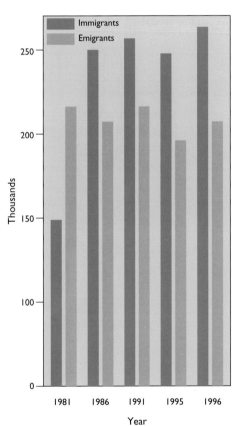

Immigrants
Emigrants

Thousands

1981 1986 1991 1995 1996

Year

▲ **Figure 5.27** *UK International migration trends, 1981–96*

25 From Figure 5.27 use the years 1981 and 1996 to illustrate how migration influences population change.

▶ **Figure 5.28b** *Main origins of migrants to Germany, 1997*

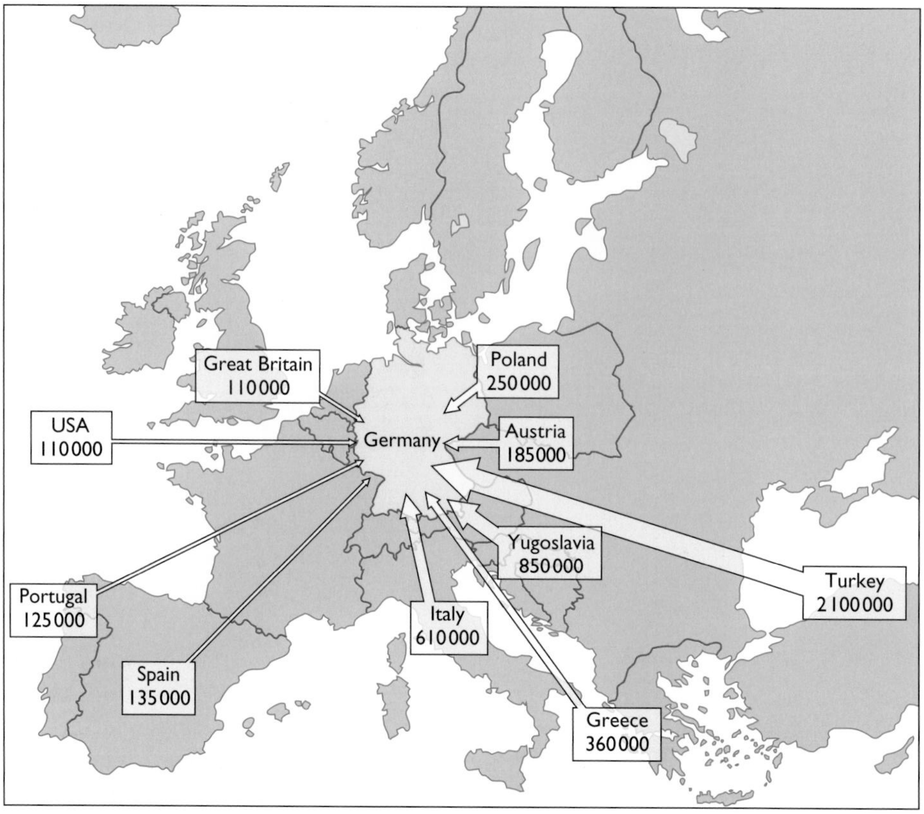

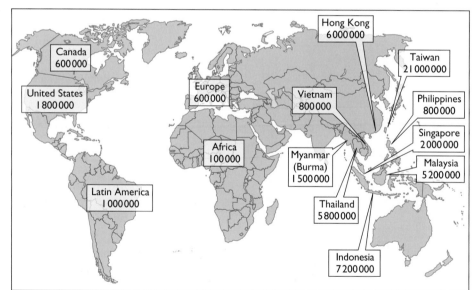

302

▶ **Figure 5.28c** *The distribution of ethnic Chinese living outside mainland China*

---

26 Use the three maps of Figure 5.28a, Figure 5.28b and Figure 5.28c to illustrate and assess the accuracy of the five basic ideas on international migration flows in the bulleted list (page 175).

Short-term migration

Not all migrants move with the intention of settling permanently. Many **economic migrants** are short-term rather than long-term migrants. They move to take jobs and send part of their earnings back to their families. The majority of Turks who work in Germany (Figure 5.28b) intend to return to their families. Fifty per cent of the labour force in Oman are short-term economic migrants, and include doctors, nurses and engineers as well as servants and labourers. In Hong Kong there are at least 150 000 Philippino women working as domestic servants and 'nannies' (Figure 5.29). Large numbers of mineworkers in South Africa are also attracted from neighbouring countries by the relatively high wages.

▶ **Figure 5.29** *Sunday morning in downtown Hong Kong: thousands of Philippino maids and nannies gather in groups to enjoy their day off, chatting and picnicking.*

27 Suggest two reasons why citizens of a country may oppose continuing immigration, and two reasons why governments have encouraged immigration.

Much short-term migration is controlled by the immigration policies of the receiving countries through a limited-term work-permit and visa system. In Germany, short-term economic workers are called the 'gastarbeiter' (guest workers). The Philippino maids in Hong Kong must reapply for work permits each year and must also obtain permits to enable them to return home for holidays. These entry restrictions mean that a proportion of immigrants in any country are illegal. For example, in California (USA), up to 40 per cent of the workers of Mexican origin are believed to be illegal immigrants. Yet the attractions of the new 'home' can be very strong, and increasing numbers are applying for long-term residence permits and citizenship. This has become a major issue in several countries, including France and Germany (Figure 5.30).

▼ **Figure 5.30** *Proposed changes to immigration laws in Germany*

Whose fatherland?

IF ALL GOES ACCORDING TO plan, hundreds of thousands of young Turks will soon be declared German nationals – with no fuss and no renouncing their original Turkish citizenship. A proposed new law, backed by German Chancellor Gerhard Schroder's centre-left government, would ease the process for obtaining citizenship in a country that has a long history of excluding those whose ethnic background is not German. Those who would benefit would be as many as 4 million of the country's 7.3 million immigrants – including 2.1 million Turks who began coming to Germany in the 1960s as guest workers and stayed to build homes, families and lives, not to mention the German economy.

At present, most of Germany's immigrant population is concentrated in urban enclaves, where it receives the material benefits of Europe's most bountiful economy, but not the rights and protections of citizenship. Under a 1913 law based on the legal principle of jus sanguinis, or 'blood law', parentage and ethnicity determine German nationality, not place of birth. A descendant of Germans stranded in Russia since the eighteenth century is considered a German, while a Slav born in Munich is not.

The new law would introduce a series of liberalisations: a second-generation foreigner born in Germany would automatically be a citizen if at least one parent came there before the age of 14; and immigrants could apply for naturalisation after eight years of residence instead of 15 years if they can show knowledge of German language and politics; the partner in a German-foreign marriage would readily become a citizen in just three years. In all three cases the new German citizen would not have to give up their original Turkish (or any other) nationality.

(Source: Time, 25 January 1999)

Involuntary international migration

The main type of involuntary (compulsory) migrants are known as refugees and are forced to move by cultural rivalries, political conflicts, war and natural disasters (see Figure 5.31 and Box 2). United Nations estimates suggest that there are more than 20 million refugees in the world (Table 5.14).

Slavery is another form of involuntary migration which is fortunately rare today (e.g. the eighteenth-century slave trade between Africa and the USA).

It is often difficult to distinguish between 'refugees' (asylum seekers who fear for their lives) and 'economic migrants (people who move to find jobs and better living conditions). When political unrest and economic problems combine, waves of migration often result, for example, from Kosovo in 1999–2000 and from Afghanistan in 2001–2002 (Figure 5.32). The recent rise in asylum seekers trying to enter the UK is a vivid example (Table 5.16), and creates policy issues for the government (Figure 5.33). Many refugees may wish to return home when conditions improve.

▼ **Table 5.14** *Global refugee totals, 1997 (million)*

People displaced and living outside their home country	13.2
People displaced but still living in their home country	4.9
People displaced and recently returned to their home country	1.3
Other categories	1.4
Total	**20.8**

▼ **Figure 5.31** *Refugees from Kosovo fled to makeshift camps like this one in Macedonia.*

► Figure 5.32 *The Tampa is on its first visit back to Australia after rescuing 400 asylum seekers who were initially refused admission in September 2001. The asylum seekers were confined to living on the boat for a long time.*

Box 2 Involuntary migration

■ 1995–96: More than 2 million people were displaced by tribal rivalries between Hutus and Tutsis in Ruanda, central Africa.

■ 1996–98: Two-thirds of the population of Montserrat were forced to leave the Caribbean island by recurring volcanic eruptions.

■ 1999: Up to 2 million people were forced out of Kosovo as a result of their campaign for independence from Yugoslavia.

▼ Table 5.15 *Refugees/asylum seekers into the UK, 1997 and 2000*

Origin	1997	2000
Europe	8 000	23 000
Asia	7 500	23 000
Africa	10 000	18 000
Middle East	2 500	14 500
Americas	3 000	2 000

(Source: Social Trends, 32, 2002.)

► Figure 5.33 *The UK – a safe haven for Kosovan refugees?*

28 With a partner: Monitor (keep a check on) the news media – including websites – for a two-week period. Keep a record of any involuntary migrations in the news:
- ■ Who are the migrants and what are their approximate numbers?
- ■ Where are they moving from and to?
- ■ Why are they moving?
- ■ Plot the movements on a world map, naming the countries.

UK policy for Kosovo refugees

SOME 60 PER CENT OF THOSE Kosovan Albanians who were airlifted out of camps in Macedonia were family reunion cases. Those who had immediate family members who had been through the asylum determination process in the UK were granted permission to enter, while others have been given Exceptional Leave to Remain on humanitarian grounds for one year. This means that they are entitled to benefits, in line with UK nationals, and are eligible for employment. The government has guaranteed that they will not curtail people's leave to remain and if those on ELR want to apply for refugee status then their cases will go through the usual asylum determination process.

(Source: Forced Migration Monitor, 5 August 1999)

▼ **Table 5.17** *Rural–urban flows for selected countries, 1980–95*

Percentage of total population living in urban settlements		
	1980	1995
Mozambique	13	34
Tanzania	15	25
Ghana	31	38
Indonesia	22	36
Malaysia	42	55
India	23	29
Brazil	66	78
Mexico	66	76
Nicaragua	53	64

5.8 Internal migration

Economic reasons and quality-of-life motivations dominate migration decisions. For instance, one of the outstanding features of modern migration patterns is the massive flow from rural to urban areas. In 1950, barely one in four people lived in towns and cities, but demographers forecast that early in the twenty-first century, one-half of the world's population will live in urban settlements (see chapter 6). The main causes are increasing rural unemployment and poverty, and rising expectations ('push' factors) interacting with the perception that more jobs, higher wages and attractive life-styles are to be found in cities ('pull' factors). These forces are particularly powerful in LEDCs and NICs (Table 5.17).

In contrast, in a growing number of MEDCs, increasingly powerful urban–rural flows have developed in recent decades. This process, known as **counter-urbanisation** is brought about by families seeking an improved quality of life in rural and semi-rural environments.

Internal population shifts tend to be particularly vigorous within LEDCs and NICs experiencing rapid economic and social change. For example, peninsular Malaysia had a population of approximately 15 million in 1995. In the mid-1990s some 750 000 people a year were migrating internally, and in 10 of the 12 states, migrants made up 4–6 per cent of the total population. In China in 1995, at least 105 million people were living in a different province from that in which they were born. In MEDCs too, people are constantly on the move. In the USA, approximately one in every ten households moves address each year. The population of the UK is becoming increasingly mobile, and there are constant regional flows (Table 5.19). The pattern and scale of these movements from less prosperous northern regions to the economically booming regions of southern England has become a high profile environmental and economic policy issue (Figure 5.34).

Even within growth regions, migration processes and patterns are complex. For example, during 2000, London attracted 163 000 people from other parts of the UK (Table 5.18). However, 232 000 people left London. Of this total, approximately 63 per cent moved to the South-East and East regions of England (rows 2 and 3, column a).

▼ **Table 5.18** *Internal migration within SE England during 2000, thousands*

	Origin			
	(a) London	(b) South-East	(c) East	(d) UK total
Destination				
1 London	–	53	29	163
2 South-East	88	–	28	224
3 East	59	28	–	146
4 Total	232	210	125	–

(Source: Social Trends, 32, 2002)

29 From Table 5.17:
a For the countries in Table 5.17, what was the average percentage of the population living in urban settlements in 1980, and 1995?
b What factor other than migration influenced the urban growth shown in Table 5.17?

30 From Table 5.18
a What was London's gain or loss as a result of internal migration during 2000?
b How many people moved from the South-East and East regions of England into London?
c How many people moved from London to the South-East and East regions of England?
d Did the South-East and East regions gain or lose population as a result of internal migration?
e During 2000, London had a net gain of 18 600 people aged 16–24, but other age groups declined, especially the 35–44 age group. Suggest reasons for this pattern.

NORTHERNERS HEED SOUTH'S SIREN CALL

Our Social Services Correspondent reports

Almost a quarter of a million people have fled the old industrial communities of the north and midlands since 1991, official figures revealed indicate.

While the populations of traditional manufacturing areas such as Tyneside, Merseyside and the Black Country are shrinking, parts of southern and eastern England are experiencing huge influxes of job and retirement migrants.

The number of people living in Cambridgeshire has risen almost 10 per cent in just seven years. Within the boom county, numbers in the East Cambridgeshire district – based around Ely – have soared by almost 20 per cent.

The latest figures reflect the scale of the population drift from north to south that has outstripped planners' expectations and has led to growing numbers of houses being abandoned in northern cities and towns.

Councils in the south-east are meanwhile searching for space to build 1 million extra homes that the government forecasts will be needed over the next few years. Over the last eight years London has grown by 4.3 per cent compared to a national average of just 2.6 per cent. Some of the biggest population increases have been in districts outside London such as Milton Keynes at 13.4 per cent and Worcester at 12.8 per cent.

▲ **Figure 5.34** *Regional migration in England from the North to the South-East*

▼ **Table 5.18** *Internal migration patterns in England 1986 and 1996*

a Region	**In-migration (thousands)**		**Out-migration (thousands)**	
	1986	**1996**	**1986**	**1996**
North East	36	39	46	45
North West	90	100	101	103
Merseyside	22	25	37	31
Yorkshire/Humberside	79	91	91	98
East Midlands	102	102	85	94
West Midlands	87	91	95	101
Eastern	145	139	128	121
London	183	168	232	213
South East	243	228	204	199
South West	149	139	103	110
b Migration between components of the UK (thousands)				
England	116	111	101	105
Wales	55	55	50	53
Scotland	44	47	58	54
N.Ireland	9	11	15	12

(Source: Regional Trends, 1998)

31 From Table 5.19:

a For 1996, make two ranking lists for regions in terms of their in-migration and out-migration flows (rank largest totals as 1).

b Use your lists to assess the accuracy of this statement:
  ■ Regions with largest in-migration flows also have the largest out-migration.

c Use an outline map of the ten regions of England to construct a simple choropleth map showing
  ■ regions with net gains
  ■ regions with net losses.

d From your map and Table 5.18(b), summarise the migration patterns between the various parts of the UK.

e To what extent do the figures for 1986 and 1996 suggest a consistent pattern of internal migration flows in the UK? (Use Table 5.18a and b.)

Step-wise migration

Many studies of migration show that much long-term migration takes place in two or more steps (Figure 5.35). Thus, people leaving rural areas in Scotland may be attracted first to Glasgow or Edinburgh; a second move then takes them to London or abroad. In Brazil, millions of landless families from southern states have moved into the favelas (slum quarters) of Rio de Janeiro or Sao Paulo. Many then move further away, to Amazonia, to clear plots of land in the rainforests, often as part of government resettlement schemes. In China, step-wide migration has been part of government policy which sets strong controls on population movements (see Migration in China example).

301

303

300

181

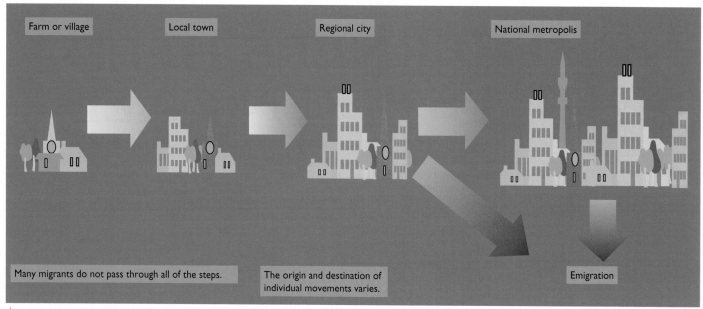

Figure 5.35 *Step-wise migration showing movement from rural areas to urban areas*

EXAMPLE: Migration in China

▼ **Figure 5.36** *Around one in four of these people in Guilin, a regional city in south-east China, are part of China's 'floating population'.*

Government policy and step-wise migration in China

Phase 1: The Communist government in China has tried to keep a firm control on internal migration. In 1958 it introduced the household registration scheme known as the hukou system. A household, and the individuals within it, were registered in a particular settlement or district. This gave them proof of identity (an ID card), housing, welfare and educational rights, but it also controlled their movements. The primary aim of the policy was to restrict rural–urban migration. If people moved, they lost their hukou rights and entitlements.

Phase 2: From 1980, China began to introduce a market economy, shifting from a command economy. Capitalist systems need mobile supplies of labour. So, in order to bring economic diversification, and to reduce rural overpopulation, people were allowed to leave the land, but not their district. These short-distance migrants were encouraged to set up small businesses in villages and towns. However, during the 1980s, millions of people did move longer distances from the smaller towns and the rural areas to larger cities, often a provincial capital. They became known as 'the floating population' as they were migrants without hukou registration rights (Figure 5.36). By the early 1990s this floating population had reached at least 100 million.

Phase 3: In the late 1980s, to encourage more rapid industrialisation, a scheme of 'temporary residence permits' was introduced. This allows people to move to cities, but does not give them full hukou rights. Since then there have been huge migratory surges to the east-coast cities such as Shanghai, Hong Kong and Guangzhou (see chapter 6 Shenzhen and Shanghai examples).

32 From the section on step-wise migration:

a Define the term 'step-wise migration'.

b Suggest reasons which encourage step-wise migration

c How can government policies influence internal migration flows?

33 Group discussion: What are the advantages and disadvantages of uncontrolled internal population movements?

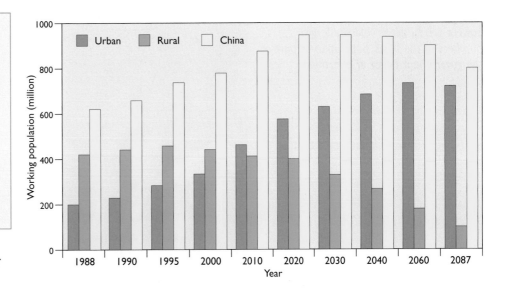

► **Figure 5.37** *The transformation of China's population (Source: Shen, 1998)*

Resettlement policies and migration

Throughout the second half of the twentieth century, there were widespread programmes of resettlement and colonisation, where government policies generated large-scale compulsory internal migration. The governments gave social, economic and ideological reasons to justify these forced movements of people. In China, for instance, the government is anxious to colonise the far western province of Xinxiang, in order to improve political stability and security. So, during the last decade, at least 2 million people were resettled from other provinces. Furthermore, the Chinese government claim that the massive Three Gorges project of dams and reservoirs is essential to provide HEP for industry and flood control for millions of people living in the Yangtse valley (Figure 5.38). But the project will displace more than 1 million people, many of whom will be resettled in Xinxiang.

► **Figure 5.38** *The Three Gorges Dam, China, under construction. The massive dam across the mighty R. Yangtse was completed in 2003. As the reservoir level rises upstream, at least one million people are being displaced and resettled.*

► **Figure 5.39** *Huge townships such as Soweto were built to relocate large numbers of the black population as part of the government policy of apartheid.*

Present-day population distribution patterns may also be the result of political ideologies. In the USSR, the communist regime, which was in power from 1919 to 1990, forced the resettlement of millions of people to Siberia. In South Africa, one significant influence upon the current population distribution has been the government policy of apartheid. Apartheid meant 'equal but separate'. The central idea was that the black and white populations of the country should live separately, but within a single economic system, and so racial conflicts would be avoided. Thus, the black majority population were allocated self-governing 'homelands' within South Africa. Millions of people were relocated from white areas to the homelands, which were often in regions of low productivity, or to black townships around the major cities (Figure 5.39).

Apartheid was abolished in 1994 when all citizens gained the right to vote, but population redistribution will occur only slowly.

Clearly many colonisation and resettlement programmes that involve involuntary migration bring up issues of human rights. UN agencies and humanitarian organisations, such as Amnesty International, campaign against such injustices. However, it is difficult to define what rights external agencies and other countries have to intervene in the internal affairs of an individual nation, and under what circumstances they can do so.

34 Group discussion: Does a government have the right to force people to move, and if so, under what circumstances, and how should it be accomplished?

5.9 Summing up the migration system

It is clear that the patterns and processes of migration are complex. Lee has suggested a model which summarises this complexity in terms of an origin–constraints–destination model (Figure 5.40).

The potential migrant weighs the advantages (+) and the disadvantages (−) of their present home (origin) against the perceived advantages (+) and disadvantages (−) of a new home (destination). This origin–destination evaluation is influenced by a set of external factors or constraints which must be overcome. Only when the 'pull' factors of the destination and the 'push' factors of the origin are strong enough to overcome the constraints, will migration take place.

► **Figure 3.40** *Lee's model of migration. The decision to migrate involves a spectrum of degrees of freedom (how much choice you have). Even with involuntary, forced migration, individuals and communities may either move passively, or take the decision to fight for the right to stay.*

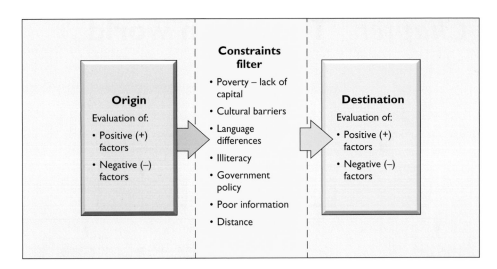

Summary

World population reached 6 billion in 1999, and will continue to grow through the first half of the twenty-first century.

Over 90 per cent of future growth will occur Less Economically Developed Countries; population totals in many MEDCs are likely to decline.

Population change is the outcome of the balance between natural change and net migration.

There is a relationship between population dynamics and level of development: as development progresses, a country passes through a demographic transition.

Rates of natural change are influenced by factors such as Infant Mortality Rate, Total Fertility Rate, Age-specific Death rate and Life Expectancy.

Population structure is determined by the numbers/proportions in specific age/gender groups. There is a clear relationship between population structure and the Demographic Transition Model.

Migration is a significant component of population change, and may be internal or international; voluntary or compulsory.

Migration is sensitive to social, economic and political processes and conditions.

Chapter 6 The urban world

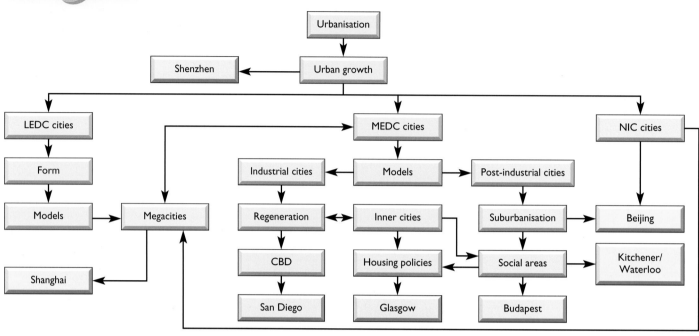

Urbanisation → Urban growth

Shenzhen ← Urban growth

LEDC cities → Form → Models → Megacities
Shanghai ← Megacities

MEDC cities → Models
Industrial cities ← Models → Post-industrial cities
Regeneration ↔ Inner cities
CBD → San Diego
Inner cities → Housing policies → Glasgow
Housing policies ← Social areas
Suburbanisation → Social areas → Budapest

NIC cities → Beijing
Kitchener/Waterloo

6.1 Introduction

Box 1

- During the 1850s, Britain became the first country in the world where a majority of people were living in towns and cities.
- In 1950, only 25 per cent of the world's population lived in urban settlements.
- In 1980–90 the urban population of the world was growing at 4.5 per cent per year.
- In 2000 at least one in two people were living in urban areas.
- Demographers forecast that by 2025, more than 60 per cent of the world's population will be urban.

One of the outstanding features of the twentieth century, in every country in the world, is that an increasing proportion of the population became urban. There have been towns and cities for more than three thousand years, but until 150 years ago, all societies were predominantly rural (Box 1). We call this process **urbanisation**: the progressive transformation from a rural to a predominantly urban society and way of life. The scale and speed of this urbanisation process has brought major costs as well as benefits.

▶ **Figure 6.1** *This is a residential area of Hong Kong. In many large urban areas people live crammed together in high-rise apartment blocks such as these.*

Key Terms

Urbanisation: the gradual transformation of a rural society to and urban society.
....

Concentration: the clustering of people and businesses.
....

Dispersion: the spreading out of people and businesses.
....

Redistribution: the constantly changing spatial organisation of people and businesses.
....

Five broad questions form a framework for this chapter:

■ Does urbanisation follow a common path in different countries?
■ What are the functions of cities and how are they organised?
■ How do cities grow over time?
■ What are the major issues facing cities and how can they be resolved?
■ How can quality of life be improved in modern towns and cities?

In examining these questions we shall be dealing with the interaction of three sets of important geographical forces: **concentration**, **dispersion** and **redistribution** (see Key Terms).

For example, in Hong Kong (Figure 6.1), powerful concentration forces produce some of the highest population densities in the world. In cities such as Chicago, (Figure 6.2) dispersion forces dominate, producing decentralised cities at very low densities. We need to understand what causes these contrasts and what impacts they have upon people's lives, e.g. who makes the decisions and how this affects quality of life for the various groups and communities.

Remember that a city is complex. It has many functions and its form may change. For instance, evolution takes place by **extension** (growth) and by internal change (**intensification** and **replacement**). So, a suburban housing area itself may extend from a town. Over time the development changes – perhaps there is some infilling (intensification), modernisation, e.g. addition of garages, extra rooms, etc. and finally replacement as buildings become out-dated (see section 6.7).

▶ **Figure 6.2** *The city of Chicago is relatively low-rise and spreads over a large area of land. The network of freeways and high levels of car ownership encourage such low density dispersion. Note the predominance of detached houses.*

1 For a region with which you are familiar, select one settlement you define as urban and one as rural. List the three main factors influencing your choice.

2 Group exercise: Define the terms town and village. Think carefully of the criteria you will use to make your definitions.

6.2 What do we mean by urban?

Definitions of urban settlements must be placed alongside definitions of rural settlements (see chapter 7). Traditional definitions of rural settlements emphasise the dominance of agriculture as a way of life, although this characteristic may not be so clear-cut today in a case such as commuter villages. In contrast, urban is generally used to refer to relatively large concentrations of people engaged in a diverse range of economic activities, where agriculture is not dominant, and where relatively high-order facilities, like hospitals, offices and theatres, are clustered. Thus, we talk of 'an urban way of life' or 'urban economies'. In most countries, urban and rural settlements are defined in terms of size, function, density, and administrative designation. In reality, the outer limits of urban settlements are today increasingly blurred. Indeed, many geographers believe that 'over much of the world, the distinction between what is urban and what is rural has lost any real meaning' (Carter, 1995).

6.3 Urbanisation and urban growth

Urbanisation

In **MEDCs** and **NICs** (Newly Industrialised Countries), urbanisation has progressed roughly parallel with industrialisation. The Industrial Revolution of the nineteenth century energised the urbanisation of countries in Western Europe and North America. By 1900, at least 70 per cent of people in countries such as Britain, France, and the USA, lived and worked in towns and cities. Remember that 'New World' countries such as the USA and Australia, were settled by waves of immigrants, most of whom stayed in urban areas.

By the mid-twentieth century, the urbanisation process was largely complete in MEDCs. At the end of the twentieth century, urbanisation levels in most MEDCs exceeded 75 per cent, and are expected to increase slowly (Table 6.1). Most NICs and LEDCs are still experiencing urbanisation rates in excess of economic growth rates (Figure 6.3) and today, a number have over half of their populations living in urban areas (Table 6.2). Indeed, several of the most urbanised countries are in South America (Figure 6.4).

Urban growth

As urbanisation progresses, the number of cities increases, the average size of cities grows, and population tends to become concentrated in a small number of very large **urban agglomerations**. In Japan, 43 per cent of the population live in three urban agglomerations (Figure 6.5). In the UK, the 1991 census recorded four **conurbations** with more than 1 million people: Greater London 7.65 million; West Midlands 2.30 million (see Box 2); Greater Manchester 2.27 million; West Yorkshire 1.45 million. The phenomenon is found throughout the world, in MEDCs, NICs and LEDCs.

Urbanisation and urban growth in a particular country tend to occur in one or more major surges, after which the rates slow down. These surges of transformation are usually driven by a combination of demographic, economic and political forces as is seen in the current explosive urban growth in China (see Shenzhen example).

311

3 From Table 6.1:

a Choose appropriate graph types to summarise the figures shown (A software package will be useful).

b Use the graphs and figures to compare changes in urbanisation for the more developed and less developed regions of the world.

c In which regions are urban growth and urbanisation likely to progress fastest during the period 1994–2025?

4 From Figure 6.3:

a Name three countries with the highest annual urban growth rates.

b Summarise the global pattern shown.

5 From Figure 6.4, name three MEDCs and three NIC/LEDCs with over 80 per cent of their population classified as 'urban'.

6 Using the information available, list what you consider to be the two most important features of current urban change.

▼ **Table 6.1** *Urban populations, by region, 1970–2025*

Region	Urban population (millions)			Urban share (percentage)		
	1970	1994	2025	1970	1994	2025
More developed regions	677	868	1040	67.5	74.7	84.0
Australia–New Zealand	13	18	26	84.4	84.9	89.1
Europe	423	532	598	64.4	73.3	83.2
Japan	74	97	103	71.2	77.5	84.9
North America	167	221	313	73.8	76.1	84.8
Less developed regions	676	1653	4025	25.1	37.0	57.0
Africa	84	240	804	23.0	33.4	53.8
Asia (excl. Japan)	428	1062	2615	21.0	32.4	54.0
Latin America	163	349	601	57.4	73.7	84.7
Oceania (excl. Australia, New Zealand)	1	2	5	18.0	24.0	40.0

(Source: Pacione, Geography [86]3, July 2001)

▼ Table 6.2 *Percentage of the population living in urban areas of selected countries*

Country	Urban population (per cent)	
	1987	1997
Israel	89	90
Cuba	72	76
Brazil	70	76
Russia	73	73
Nicaragua	58	64
Turkey	55	63
Ecuador	53	61
Morocco	46	52
Botswana	22	48
Haiti	28	33
Malawi	13	20
Burundi	5	6

7 Look at Table 6.2.
a Name the two countries which are urbanising most rapidly.
b Use an atlas to locate the countries listed in the table. Support the statement that LEDCs are experiencing a surge of urbanisation.
c Is it true to claim that the rate of urbanisation varies widely? Use the data in the table to support your answer.

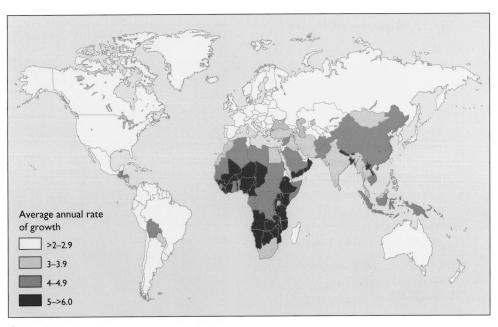

▲ **Figure 6.3** *Average annual urban growth rates, 1995–2000*

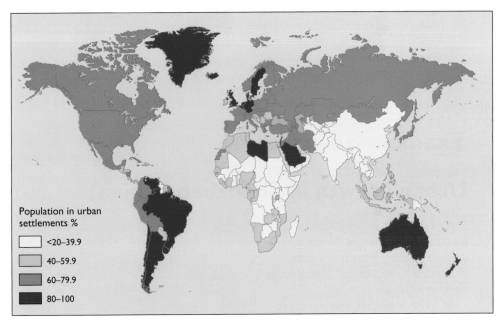

▲ **Figure 6.4** *Percentage of total population classified as urban in 2000*

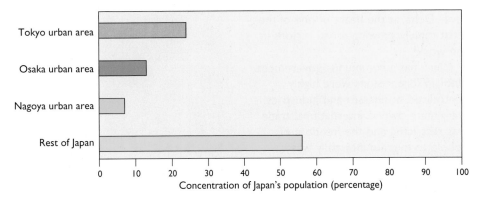

▲ **Figure 6.5** *Nearly half of Japan's population is concentrated in three urban areas.*

Box 2 Beware of figures

■ Population totals depend on the unit area being used. For example, the population of Tokyo used to construct the graph in Figure 6.5 is 30 million (24 per cent of Japan's population). However, the population of the administrative city of Tokyo is 8 million. The 30 million total is the total for an urban agglomeration that includes a number of cities, including Yokohama, which has itself, a population of over 3 million.

In the UK, totals given for London in official publications, vary from 7 to 10 million people. In the USA, the city of Los Angeles makes up less than half of the 14 million people in the Los Angeles Standard Metropolitan Area (SMA), which is used in the US census figures. The message, therefore, is to check wherever you can what area is being included in the figures.

▲ **Figure 6.6** *A satellite image of Tokyo (built-up areas are in blue and green). Tokyo is a large urban agglomeration including a number of smaller cities. Note the infilling of Tokyo Bay to create more space.*

Example: Shenzhen

Urban growth in Shenzhen (China)

The power of the combination of political and economic forces to generate urban growth is vividly illustrated by the city of Shenzhen in southern China.

In 1970, Shenzhen was a market town and a port of approximately 60 000 people. In 2000, it was an industrial city with a population approaching 1 million. It lies in the Pearl River Delta, at the heart of one of the most rapidly growing urban regions in the world.

China has a communist government. Until 1978, decisions were highly centralised, businesses and industries were state-owned, international trade was restricted and the freedom of people to migrate internally was severely controlled. Since 1978, the Chinese government has introduced more liberal policies in order to encourage industrialisation and modernisation.

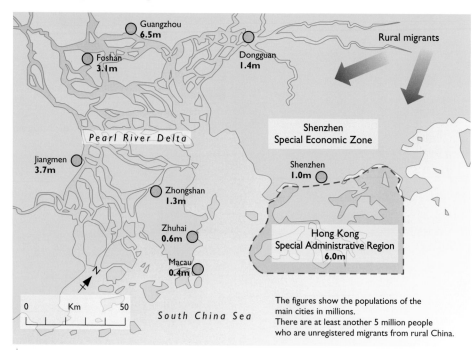

▲ **Figure 6.7** *The location of Shenzhen is ideal for trade and business with its neighbours*

8 Summarise the changes that have taken place in Shenzhen and list the factors that have encouraged its explosive growth.

The location of Shenzhen and the other towns of the Pearl River Delta suddenly became very important. They were close to Hong Kong and to major trading routes. In 1980, the Chinese government designated Shenzhen as a Special Economic Zone (SEZ), allowing greater freedom for international trade (Figure 6.7). Hong Kong is today part of China, but until 1997 it was a British protectorate. In the 1980s, it was becoming one of the world's leading financial, industrial, and trading centres, despite its tiny area. SEZ status meant that Hong Kong developers and companies could invest and do business in neighbouring Shenzhen, and that import–export trade could move freely between Hong Kong and Shenzhen. Today, Shenzhen and the other Pearl River Delta cities have become the main industrial and trading links between China and the rest of the world. (Remember that China has a population of over 1 billion people, about 20 per cent of the world's population, so the explosive urban growth of Shenzhen is not surprising.)

A further boost is that labour costs in Shenzhen are much lower than in Hong Kong. Thus, many Hong Kong manufacturers and processors now sub-contract their production to Shenzhen companies, e.g. electrical and electronic equipment. This export-processing activity is the basis of Shenzhen's most important industries.

A final energising force for growth has been a change in housing policy by the Chinese government. Until the mid-1980s, most housing was provided by the government at very low rents. This was expensive to do and there were housing shortages. Since then, private developers have been allowed to build, and families have been encouraged to buy their homes (Figure 6.8). The result, not surprisingly, has been an acceleration of building. In Guangdong province in 1985, 196 000 houses were completed, and in 1995 there were 960 000 completions. In Shenzhen in 1985, 75 per cent of people rented local authority flats whereas in 1995, 60 per cent of people owned their own homes. The population of Shenzhen doubled between 1985 and 1995.

▶ **Figure 6.8** *These new flats and houses in Shenzhen are privately owned by families.*

6.4 The changing form of cities in MEDCs

Three hypotheses (ideas) provide a useful basis for the analysis of changing urban forms and morphologies of urban settlements in MEDCs such as the UK, France, Germany, the USA and Canada:

- That industrial cities evolved primarily as a result of concentration forces.
- That the evolution of post-industrial cities is driven primarily by dispersion forces.
- That solutions to the main problems of industrial and post-industrial cities require the effective use of concentration, dispersion and redistribution forces.

In a general sense, by industrial cities we mean those that evolved during the growth surge generated by the Industrial Revolution of the nineteenth century (Table 6.3).

Their structures were controlled by the predominance of manufacturing industry, steam-transport technology and a relatively immobile workforce. Industrial cities were **compact cities** with high built and living densities, where the main aspects of life were in close proximity (Figure 6.9b).

By 'post-industrial cities' we mean those being created by economic structures increasingly dominated by service industries, modern communications systems and a mobile workforce (Figure 6.9a). These are **spread cities**, dominated by low-density developments, where the main components of life are spatially separated. Some experts believe that sustainable cities require a return to some of the qualities of the compact city, where the different aspects of our lives are more spatially integrated. Yet even in highly urbanised countries such as the USA, where the spread city form is dominant, there are wide regional variations in urban impact (Figure 6.10).

▼ **Figure 6.9a** *The spread city.*
Zoning of activities leads to the reliance on the private car.

▼ **Figure 6.9b** *The compact city.*
Compact nodes reduce travel and allow walking and cycling.

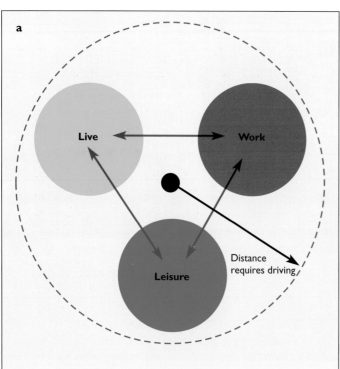

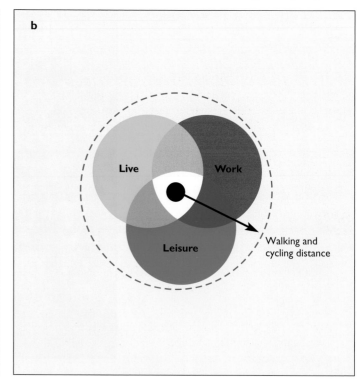

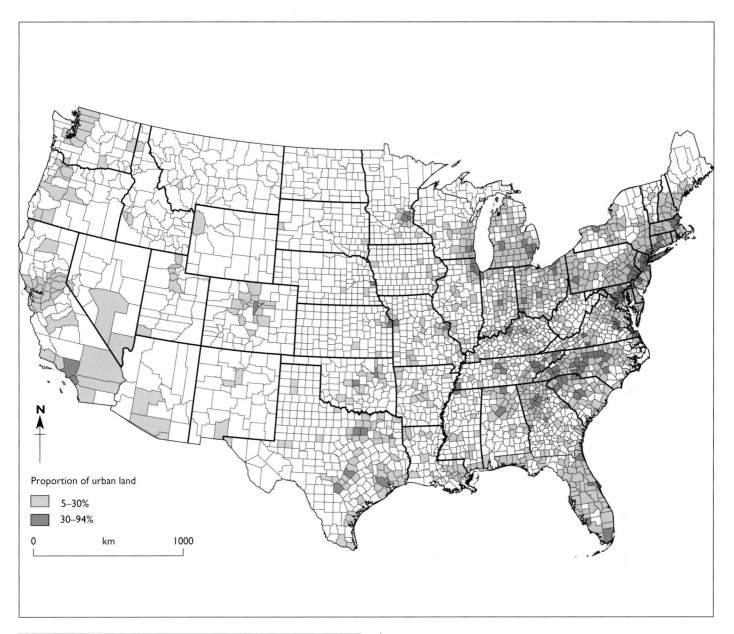

Proportion of urban land

☐ 5–30%

☐ 30–94%

0   km   1000

▲ **Figure 6.10** *Proportion of urban land-use in USA, 1997*

▼ **Table 6.3** *The nineteenth-century urban growth surge in the UK*

City	Population totals		
	1801	**1851**	**1901**
Birmingham	71 000	265 000	765 000
Manchester	75 000	336 000	645 000
Leeds	53 000	172 000	429 000

9 The simple models of Figures 6.9a,b show two types of city form. Make a list for the compact city and a list for the spread city, and show the advantages and disadvantages of each type. Divide your lists into three categories: economic, social, environmental. (You may find it useful to compare your lists with those of other members of your class or set.)

10 Many cities contain elements of both the compact and spread form. Illustrate this idea by giving examples from a city with which you are familiar.

11 From Figure 6.10:

a Name five states with high levels of urban impact in terms of land developed for urban land uses.

b To what extent does the pattern shown for the USA support the idea that urban populations are becoming increasingly concentrated in a small number of huge agglomerations?

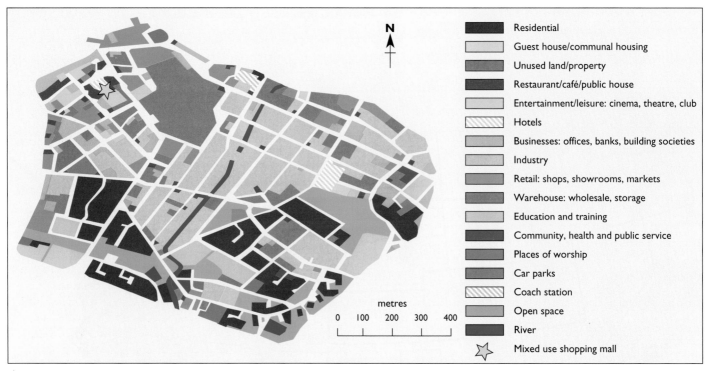

▲ **Figure 6.11** *This land-use map of a section of inner Birmingham shows the complex diversity of types of land use in a city.*

6.5 Urban models and the search for order and pattern

Industrial cities

As the land-use map of Figure 6.11 shows, urban environments are a complex mix of land uses. Yet, within this apparent chaos, patterns are discernible. For instance, how many of the following general statements apply to cities with which you are familiar?

▼ **Figure 6.12a** *Burgess's concentric zone model (1924) shows a central CBD surrounded by rings of different land use.*

▼ **Figure 6.12b** *Hoyt's sector model (1939) still has a central CBD but divides land use into sectors radiating out from the centre.*

▼ **Figure 6.12c** *Mann's model (1976) was based on British cities. It combines elements from both Burgess's concentric zone model and Hoyt's sector model.*

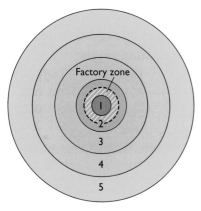

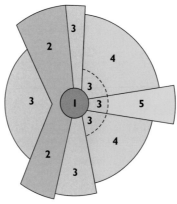

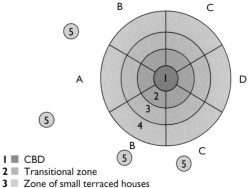

1 ■ Central Business District (CBD)
2 ■ Transition zone (wholesale light manufacturing)
3 ■ Low-class residential zone (working men's homes)
4 ■ Medium-class residential zone
5 ■ High-class residential zone (commuters' homes)

1 ■ Central Business District (CBD)
2 ■ Wholesale light manufacturing
3 ■ Low-class residential
4 ■ Medium-class residential
5 ■ High-class residential

1 ■ CBD
2 ■ Transitional zone
3 ■ Zone of small terraced houses in sectors C and D; larger bye-law housing in sector B; large old houses in sector A
4 ■ Post-1918 residential areas, with post-1945 housing on the periphery
5 ■ Commuting-distance 'dormitory' towns

A Middle-class sector
B Lower middle-class sector
C Working-class sector (and main sector of council estates)
D Industry and lowest working-class sector

Legend (Figure 6.11):
- Residential
- Guest house/communal housing
- Unused land/property
- Restaurant/café/public house
- Entertainment/leisure: cinema, theatre, club
- Hotels
- Businesses: offices, banks, building societies
- Industry
- Retail: shops, showrooms, markets
- Warehouse: wholesale, storage
- Education and training
- Community, health and public service
- Places of worship
- Car parks
- Coach station
- Open space
- River
- Mixed use shopping mall

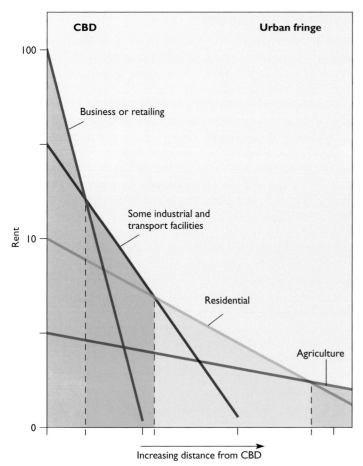

- There is a central business district dominated by professional and commercial businesses.
- There is a high-density zone of mixed land uses and varying ages surrounding the central area (Figure 6.11).
- Main transport routes radiate from the city centre.
- Strips of commercial development occur along main roads.
- Residential areas distinguished by age, scale, character, value and status are scattered through the city.
- As distance increases from the city centre, land uses are increasingly segregated, e.g. industrial areas are separated from housing estates.

From such common characteristics, several general models of urban growth and form have been proposed. For example, the Burgess model shows the city as a series of concentric zones around a central core (Figure 6.12a)). The Hoyt model suggests a set of sectors radiating from the core (Figure 6.12b)). Both Hoyt and Burgess were Americans who based their analyses on US cities such as Chicago in the 1920s. In 1965, Mann combined the concentric and sector approaches and applied them to British cities (Figure 6.12c)). Notice that the three main criteria used in constructing this model are age, land-use type, and socio-economic status.

Land value and urban form

These models are descriptive, and do not *explain* patterns. One popular explanatory approach is based on land value. This idea suggests that land value varies according to the attractiveness of the site. For many businesses this attractiveness is related to accessibility to customers, resources and transport. There is most competition for the most attractive locations. As a result, land values, often measured in terms of rental value per unit area, are highest at such attractive and accessible locations. Land uses sort themselves out by ability and willingness to pay – how much people can afford, or feel they can afford. This has been summarised as the **bid-rent model** (Figure 6.13). Remember too that land use varies because of the amount of land needed to undertake an activity (Figure 6.14). Using accessibility as a basic criterion, a complex land-value surface may evolve across the city with distinct high value suburban commercial and business nuclei.

▲ **Figure 6.13** *The bid-rent curve measures rental value of land and its uses. It compares the attractiveness to businesses of different city locations.*

---

12 From the models in Figure 6.12:

a Suggest two elements of the urban models of Burgess and Hoyt (Figure 6.12a,b) that are still common today in British cities.

b Look at Mann's model for British cities (Figure 6.12c): Describe the character of sector A in terms of the three criteria of **age, land use type** and **socio-economic status**.

13 How can the bid-rent model (Figure 6.13) be used to help explain the distribution of housing in the city? (Remember, factors other than land value may be involved.)

▶ **Figure 6.14** *An office block can function efficiently as a tall, high-density structure, yielding high rents from a relatively small site. In contrast, factories like these in Wrexham, need horizontal organisation and a large amount of space.*

The current usefulness of models

These models, as with all models, are highly simplified and generalised versions of reality. Each city is unique, and has a different history. Thus, the nineteenth-century growth of Birmingham was based largely on small workshop industries and fragmented land ownership. This produced a fragmented pattern of development. In contrast, in the cotton and woollen towns of Lancashire and Yorkshire, a simpler morphology was produced by a series of large mills, each surrounded by closely packed streets of workers' homes.

Furthermore, the classic models are now regarded as out-of-date and inadequate for the study of modern cities. For example, they ignore the role of government in setting out social and planning legislation and policies. They also pre-date the powerful dispersion forces of cars, of modern communication systems and of the rising affluence of large proportions of the population. Perhaps the main use for the classic models, therefore, is as frames of reference which place today's urban problems into their historical context. These models can be used to provide us with a background to urban topics such as the inner-city areas, suburban estates and traffic congestion (see sections 6.7 and 6.10).

▼ **Figure 6.15** *The conurbation of Birmingham and the Black Country is multicentric, unlike the monocentric models of Burgess, Hoyt and Mann.*

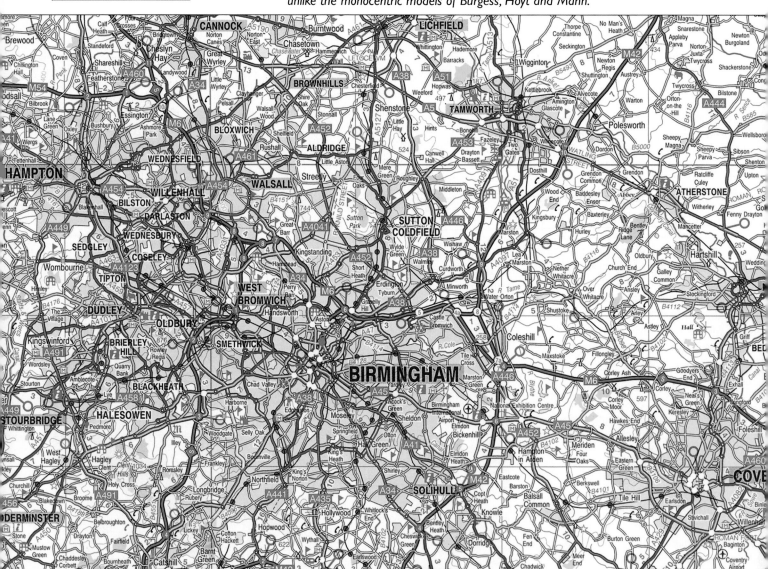

14 Define the terms monocentric and multicentric urban forms.

15 Use Figure 6.15 to illustrate what we mean by a multicentric city or conurbation. You may support your answer with an outline plan on tracing paper, showing:
- the outline of the conurbation built-up area (pale blue on the map)
- the location and name of the main towns in the conurbation (in bold type on the map)
- the dimensions of the conurbation in kilometres, e.g. NW–SE; NE–SW.

16 Give examples of what is meant by the claim that the distinction between urban and rural environments is becoming increasingly blurred. (Use Figures 6.16a,b to help you.)

Finally, models such as those of Burgess, Hoyt and Mann portray the city as a geographically separate unit. In reality, sub-regional industrialisation often saw the growth of a series of towns. As they expanded they merged to form a conurbation, e.g. the West Midlands conurbation which stretches from Birmingham to Wolverhampton. Conurbation structure is **multicentric** not **monocentric** as suggested by the models (Figure 6.15).

Post-industrial cities

During the second half of the twentieth century, urban environments were experiencing two major directions of change: massive extension and progressive internal reorganisation. The extension processes are producing two new forms of city. First there is the emergence of a broad zone of discontinuous development beyond the core city, to create what has been called a spread city. The south east of England, with London as the core city, is a large-scale example of this concept. This physical and functional continuum is blurring the distinction between what are urban and what are rural environments (Figure 6.16a and b).

The second outcome of extension is the growth of huge, sprawling cities, developed at very low densities. This urban form is evolving most clearly in North America, and may involve growth around an existing core city like Los Angeles, or may be self-generating like Las Vegas or Phoenix (Figure 6.17).

▼ **Figure 6.16a** The 'spread city' is the result of dispersion and decentralisation forces and processes.

▼ **Figure 6.16b** The 'regional city'. This structure illustrates the progressive extension of urban influence and functions, and the blurring of distinctions between urban and rural environments.

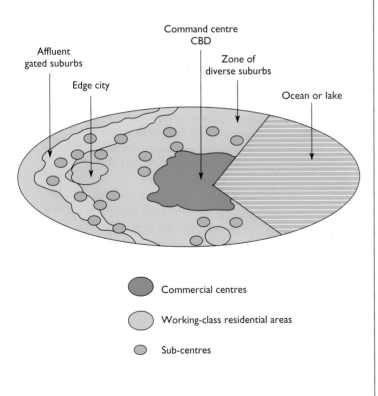

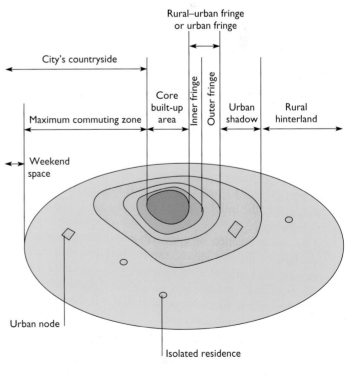

Figure 6.17 *Phoenix, Arizona has been called the '100 mile nowhere city'. Driving the freeway from the south you can cover 100 miles before leaving the northern city limits.*

These new urban forms are multi-nuclear, the nucleations being a series of large-scale office, business, service and shopping complexes set within extensive residential developments. This vigorous preference for the suburban lifestyle is causing major regional shifts in urban development, and making urban regeneration programmes in older cities difficult (Figure 6.18).

The primary aims of internal redistribution processes and policies are to change and regenerate the **morphologies** (internal structures) of

Figure 6.18 *Suburban life in the USA is steadily increasing*

THE SPRAWLING SUBURBS

They are new, affluent and mainly white, and they represent the changing face of the United States. A generation of American mini-cities is rising in the suburbs, changing the country's political, economic and social landscape.

New figures from the US census bureau reveal that America's prosperous suburban cities, fuelled by high-tech economic growth and centred on shopping malls, will become the heartland of the country during the next millennium.

As the population of the north and east falls or remains steady, the fastest growing cities in the US are increasingly self-contained, middle-class suburban com–munities, found mainly in the sunbelts of the south and west.

'The dominant trend in American life today is not the revival of the old cities but the across-the-board decentralisation of work and life to the new sprawling suburbs,' said Bruce Katz, who chairs the Centre of Urban and Metropolitan Policy at the Brooking Institution in Washington.

Four of the 25 fastest-growing cities in America – Chandler, Scottsdale, Glendale and Mesa – are the offshoots of just one city, Phoenix. The Arizona metropolis is now the seventh largest city in the US.

Robert Lang, an urban studies specialist with a community development charity, the Fannie Mae Foundation, said, 'What we're calling cities now, we used to call suburbs. These places are dominated by single-family homes, office parks and shopping malls.'

All across America the 'burbs' are the new focus of city life. Once the typical city dwelling was a multi-storey working-class apartment block. In the new cities it is a large detached house with a double garage and a garden.

Malls, accessible only by car, have taken the place of street-corner shops and cafés, and as inner-city industry has declined and the new high-tech sector has boomed, jobs and services are increasingly moving out to the suburbs too.

But the great unstated dimension of this suburban surge is race. 'Race permeates everything in this country, and the growth of the suburbs is a white phenomenon,' Mr Katz said. 'We work together, but we don't live together.'

In many of the eastern and older cities, the move to the suburbs during the 1960s and 1970s was characterised by 'white flight' – provoked by the fear of the black populations in the inner cities. In the 1990s this is still a factor, but driven now by the Hispanic move to the inner cities of the south and west.

(Source: The Guardian 3 March 1999)

industrial cities to fit the needs of **post-industrial cities** and societies, including coping with pressures arising from continued urban extension. Many high-profile urban issues and problems arise from these aims. Here are three examples:

- **Traffic congestion**: the radial transport network of many industrial cities pre-dates the dominance of the motor vehicle and the progressive separation of home and workplace (Figure 6.19).
- **Industrial change**: the shift in emphasis from manufacturing to service industries brings fundamental changes in the character and location of economic activity, in the accommodation and infrastructural needs, and in workforce demands (see chapter 8).
- **Social change**: people's preferences and expectations for residential environments change over time. This creates demand for extra land, upgrading of older properties, shifts in where families live, and for the greater equality of choice and opportunity (see section 6.7).

Factors influencing urban expansion

▼ **Table 6.4** *Processes encouraging urban extension*

Processes	Examples
Economic	■ Many economic activities have become increasingly footloose, i.e. less closely tied to specific locations. This locational flexibility applies to secondary and especially to tertiary and quaternary industries. ■ It is generally cheaper to develop on peripheral 'greenfield' sites than on urban 'brownfield' sites. ■ Increasing affluence of large sections of a population gives greater choice of residential location.
Technological	■ Modern electronic communication systems facilitate the movement of information. ■ The increasing dominance and versatility of the motor vehicle increases mobility but also requires freedom of movement. ■ The construction industry has responded to the needs of modern businesses, producing new layouts and building designs.
Social	■ An accelerating preference for a suburban lifestyle based on desired quality of life: better housing, peace, space, safety, status, leisure opportunities. ■ A 'flight' from perceived inner-city problems: congestion, pollution, tension and violence, deteriorating physical environments and services. ■ Changes in family and household structures creates additional housing demand.
Political	■ Growth-oriented policies by central and local governments to attract investment, businesses and people. ■ Policies and programmes to improve the quality of living and working environments in core cities include relocation of economic activities and homes. ■ Governments respond to continued demands for more and better housing and improved transport infrastructure.
Environmental	■ Inadequate transport infrastructure means that inner districts of core cities, and especially the CBD, may no longer be the most accessible locations. ■ Ageing physical environments are decreasingly attractive to businesses and families, and are increasingly expensive to maintain. ■ Environmental regulations and standards are easier for developers to achieve on 'greenfield' sites than in 'brownfield' locations.

▲ **Figure 6.19** *A traffic jam at rush hour in Manila (the Philippines). As people commute further distances from home to work, gridlock has become a major feature of urban areas.*

Key Terms

Greenfield sites: sites developed on mainly agricultural land – i.e. sites that have not been previously built on.

Brownfield sites: Derelict sites within urban areas, which are redeveloped. This is also known as land recycling.

17 Group discussion: Is Britain experiencing a suburban explosion similar to that described for the USA in Figure 6.18? What features are similar? What features are unique to the USA?

18 Read the account by the managing director (right) carefully and give examples of each of the five sets of processes encouraging urban extension in Table 6.4 illustrated by his business decisions.

Key Terms

Strategic gaps used to avoid coalescence and protect te separate identity of settlements. They are often quite narrow and give people the feeling of having open land near to where they live.

Green wedges used to protect open land and shape urban growth, and to improve town–countryside links.

Rural buffers used to prevent towns from spreading to absorb surrounding villages, e.g. the circular belt around Swindon. They are smaller and less permanent than greenbelts.

We can separate the dispersion forces generating and accelerating the extension of cities into five broad sets of processes (Table 6.4). These processes work together in various ways. The managing director of a packaging company in Birmingham explains why they have moved to an industrial park near Redditch:

'We rely on road transport, and congestion in Birmingham was becoming a nightmare. Also, we wanted to expand, but did not have the space. So we moved to a new building on an industrial park out here. The local authority people were very helpful and have provided the infrastructure. Our development costs have been much lower than they would have been in Birmingham, and we are near to the motorways. There are plenty of recent housing estates and some lovely old villages out here, so we have no difficulty getting staff. Some of our existing staff moved out too. Most of them like it and it's much easier to get to work. Also, because of modern electronic communications systems, we have kept our headquarters in Birmingham where we are well-known.'

The following sections of this chapter make it clear that dispersion and decentralisation forces do not work in isolation, but alongside concentration and redistribution forces. We can illustrate this dynamic interaction by examining three elements of British planning policy which have been implemented over the past 50 years. First, **greenbelts** have been designated around a number of large cities and conurbations. These, in combination with planning controls on developments in rural areas, are intended to restrict urban sprawl, i.e. to constrain dispersion forces.

These Greenbelts are under constant pressure for development. For example, in 2001, planning permission was given for 1800 'detached executive homes' on 100ha of Glasgow's greenbelt. The main reasons given were the shortage of such houses in the area, the lack of suitable brownfield sites, the need to halt the drift of middle-class families away from the city, and the £1 million a year that the development would raise in council taxes. Nonetheless, the greenbelt idea survives and, in recent years, has been adapted by cities that use rural buffers to constrain growth and to prevent urban areas merging, for example the Southampton–Portsmouth urbanised area (Figure 6.20).

Second, urban **redevelopment** and **regeneration** programmes aim to reorganise the internal structure of cities while retaining relatively compact urban forms, i.e. sustaining concentration forces and reducing dispersion pressures. Third, the New Towns Policy was designed to satisfy the needs for urban growth while restricting impacts upon rural environments. These

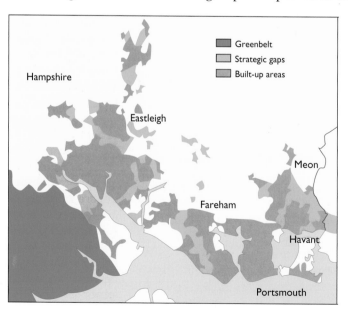

► **Figure 6.20** *Types of buffer zone*

Box 3

A selection of New Towns in the UK

■ Crawley
■ Cumbernauld
■ Cwmbran
■ Milton Keynes
■ Peterlee
■ Redditch
■ Runcorn
■ Skelmersdale
■ Welwyn Garden City

29 New Towns, developed over a thirty year period, vary widely (8 in Greater London, 13 in the rest of England, 6 in Scotland, 2 in Wales). In general, however, they are relatively compact urban forms, with post-industrial internal structures, built beyond greenbelts (see Box 3). They are attempts to apply an efficient balance of dispersion, concentration and redistribution forces (Figure 6.21).

In North America, available space, generally weak planning systems, vigorously growth-oriented policies in the public and private sectors, and the adoption of car-based lifestyles, combine to encourage suburban sprawl. Yet, there are also many examples of regeneration programmes in core cities.

► **Figure 6.21** *New Towns have been an important element of urban policy in many European countries. For example, in France, New Towns, such as La Defense, have been built around Paris in an attempt to control urban growth.*

6.6 The Central Business District (CBD)

Historic (pre-industrial) cities developed a core area in which religious, political and commercial buildings and residences of the wealthy were clustered, often set around open spaces, e.g. market places, piazzas, squares. Throughout the history of these cities, one of the distinctive features has been the intense nucleation of their core area (Figure 6.22). A centrally located nucleus later became an important element in the morphology of industrial cities, fed by the development of radial road and rail networks. The clusters of business towers in post-industrial cities illustrate the persistence of this urban component.

What is the CBD?

This core area of a city is generally known as the **Central Business District** (CBD) – a centrally located area within which a concentration of diverse business activities dominates the land-use pattern and townscape. Each city is unique, but common attributes of each CBD are:

■ High-density development, employment and activity.
■ High land values.
■ Concentration of a wide range of high-order tertiary and quaternary economic activities.
■ Progressive exclusion of housing and manufacturing (secondary) industry.

As city size increases, the CBD grows and becomes progressively separated into a set of land use sub-zones (e.g. retailing, finance, administration, entertainment).

Dynamism and change in the CBD is constant, through extension, replacement and regeneration.

▲ **Figure 6.22** *Prague is a very old town. The buildings in the core area of Prague are clustered together with an intricate road network that pre-dates the car.*

Figure 6.23 *The Leeds town hall and art gallery is a good example of civic pride.*

What sustains the CBD?

We can group the powerful forces of concentration which have produced the CBD into three broad sets of processes:

■ **Economic processes**: The CBD has been seen as the most accessible location, and therefore attractive to businesses which require access to each other and to large numbers of people (either staff or customers). Competition for these attractive locations forces land prices up, thus selectively excluding land uses (for instance housing) that are less able or willing to afford such costs (see 'the Bid-Rent Curve', Figure 6.13).
■ **Status, image and civic pride**: The image of living, working and having a business 'downtown' is often regarded as high status because you are able to afford it. Also, civic pride has resulted in funding from city councils and wealthy residents to develop public buildings such as galleries, libraries, museums and concert halls, in city centres (Figure 6.23).
■ **Planning policy**: The role of central and local governments in the planning and organisation of cities has strengthened over time. This includes the introduction of land-use zoning policies and development controls. Restriction of planning permissions for high-rise office blocks or large retail developments to central locations sustains a CBD while encouraging 'out-of-town' developments helps to cause CBD decline.

Decline in the CBD

During the second half of the twentieth century, many CBDs experienced a period of serious decline. The decline was noticeable first in North American cities such as Detroit and Chicago, before it spread to Europe, e.g. Lille, Liverpool) and even to Australia, e.g. Melbourne (Figure 6.25).

This change of fortune and the problems it creates have been explained by the complex interaction of a number of forces and processes which encourage movement from the CBD into non-central locations. The most significant shifts have been the reduction of retail and entertainment functions, partially countered by increases in office development (Figure 6.24).

Figure 6.24 *Ten factors influencing CBD decline*

■ Some planning policies encourage urban expansion, e.g. allowing development in peripheral and rural locations.
■ City councils offer incentives to encourage inward investment and new businesses, e.g. building on greenfield sites.
■ Investors and businesses are attracted by peripheral sites which have new buildings, lower costs and easy access.
■ Companies relocate in cheaper locations, often in the suburbs near to customers and staff.
■ Congestion reduces accessibility of city centres.
■ Progressive suburbanisation increases the time/cost/distance factor for use of city centres.
■ The costs of development and maintenance are high in central locations.
■ City centres often have an ageing environment and infrastructure.
■ City centres are perceived as crowded, polluted and unsafe.
■ Rises in car ownership have increased personal mobility.

Figure 6.25 *The derelict site of the centrally-located port of Melbourne (foreground) is being redeveloped as a sports and recreation complex. New port facilities have a peripheral location.*

19 Read the factors in Figure 6.24 carefully.

a List each one under the headings: economic, social, political or environmental factors. (You may find you can enter a factor into more than one list.)

b Use the factors in Figure 6.24 to suggest two reasons why there has been a shift in balance between retail and office business functions in many CBDs.

c Briefly discuss the idea that the items in Figure 6.24 can be divided into 'push' factors, i.e. processes that have pushed land uses from a CBD, and 'pull' factors, i.e. processes that draw land uses to non-central locations.

d Americans have called the decline in the retailing functions of traditional city centres, 'the doughnut syndrome'. Explain why this is an appropriate image for what has happened to many CBDs.

The CBD fights back

We should remember that certain business activities still show strong concentrating tendencies and seek central locations, especially in very large metropolises. A combination of the need for interaction and the desire for image and status retains the headquarter offices of transnational corporations and major companies in prestigious, ever-taller office towers in cities such as London, New York, Tokyo and Beijing. Modern electronic communication systems allow 'back office' functions of large corporations to be dispersed, thus saving money, but their headquarters remain centrally located (see chapter 8).

Since the early 1970s, central governments and local government authorities, increasingly in partnership with the private sector, have introduced vigorous regeneration programmes that are transforming many central areas (Figure 6.26). The main problems to be overcome in the regeneration and revitalisation process have been: the deteriorating physical environment and out-of-date infrastructure; the high costs involved; the scale of projects needed; the negative perceptions of individuals, investors, developers and businesses; the competitive alternatives and attractions offered by non-central locations and other cities.

As a result, regeneration policies and projects have focused upon:

■ Upgrading the physical environment with landscaped open spaces and modern buildings.

■ Improved access and internal circulation with new public transport systems and pedestrianisation.

■ Government financial and planning incentives to reduce costs for investors and developers, such as tax exemptions and eased planning consent for large office projects.

■ Partnership agreements between the public and private sector to share costs and make schemes economically sustainable. For example, the building of high quality but expensive, cultural complexes and conference centres.

■ Diversification of functions and land uses such as the development of integrated retail, business and entertainment projects and the encouragement of residential developments and small businesses, to make the CBD a 'living heart' and not just a '9 to 5' district.

■ Promotional campaigns to create a positive image for the CBD as a modern downtown area, with workplaces and high quality leisure opportunities in close proximity.

■ Enlargement to assist diversification and to upgrade declining CBD fringe areas.

Thus, the aim is to make the central area a place you would want to live, work and play (see San Diego example).

Figure 6.26 Birmingham's CBD has been extended and developed with many more amenities.

CBD Extension with 4 main elements.

ICC, NIA, theatre and open piazza complex.
Brindleyplace. Mixed zone of pubs, cafés, restaurants plus high density dwellings and the National Sea Life Centre.
Upgraded canals
Linear leisure developments along Broad Street

▲ Proposed Midland Metro Station

Pedestrian corridor over ring road

- - - - Pedestrian route

Environmental upgrading within existing CBD

1 Centenary Square
2 The Brindleyplace Square
3 Chamberlain Square
4 Victoria Square
5 St Philip's Square
6 International Convention Centre and Symphony Hall
7 National Indoor Arena
8 Hyatt Hotel
9 Repertory Theatre
10 Town Hall
11 Museum and Art Gallery

Example: San Diego

Regenerating San Diego's CBD

San Diego in southern California, is a rapidly growing 'spread city' metropolis of at least 1.5 million people. It is a port, navy base, business city and major tourist destination.

By 1970, the CBD was in serious decline as homes, shops, businesses and jobs moved out to the suburbs. In 1975, the city council set up the Centre City Development Corporation (CCDC) which has produced a Master Planned Community (MPC) to regenerate the downtown area. This covers over 1500 acres between the attractive San Diego Bay and a fringing freeway. The plan divides the area into eight neighbourhoods and has seven major components, which work together to diversify the area, and create an attractive working, living and leisure environment:

1 **Horton Plaza (see Figure 6.28)**: A multi-storey, high-density shopping and entertainment complex, developed from 1975. It aims to draw people back from the suburban malls, and to attract CBD workers and tourists. In total, 75,000 people work downtown.

2 **The Gaslamp Quarter Historic District**: A seventeen-acre area along two streets where 95 historic buildings have been conserved. In 1999 there were more than 100 restaurants, cafés, clubs, shops, art galleries, craft workshops and offices.

3 **Convention Centre and hotels**: Since 1990, a large bayfront convention centre and several high-grade hotels have been completed. The sunny climate and other attractions like beaches and restaurants draw conventions and business tourism. In 1997, an international Rotary convention attracted over 20 000 delegates.

4 **Waterfront redevelopment**: A disused dock, warehouse and pier facilities have been renovated as a strip for recreation, food, shopping and entertainment activities. These include a pedestrian promenade, Seaport Village, a boutique/ restaurant complex and pleasure boat tours of the bay.

Figure 6.27 *Downtown San Diego*

20 Read through the San Diego example:

a Give two examples of components of the MPC which make living downtown attractive.

b What types of people are likely to be attracted to downtown living?

c How has the waterfront area of disused port and dock facilities been used in the regeneration strategy?

d Why are the developments like Horton Plaza at high densities?

e Assess this statement: 'An essential part of successful CBD regeneration is the diversification of land uses.'

f Explain briefly the advantages of an integrated plan for the regeneration of a CBD.

5 **Local government and professional quarter**: Council offices, courthouses and other professional services have been retained by renovating old buildings, completing new buildings and upgrading the environment by landscaping and pedestrianisation.

6 **Baseball stadium**: A new 30 000 seat stadium was completed by 2003. Although this will create traffic problems, it is also likely that the sports fans will spend time and money downtown.

7 **Residential developments**: By renovation of old buildings and new construction, several neighbourhoods have been 'gentrified'. In 1999, over 20 000 people lived downtown, mostly in apartments and condominiums at medium to high densities.

Figure 6.28 *Horton Plaza is a huge new shopping complex in downtown San Diego*

▼ **Figure 6.29** *Key questions in the study of housing*

Location

| Where is it? | Why was it built here? |

Character

| What is it like? | Why was it built like this? |

Supply

| Who provided it? | Why did they build it? |

Demand

| Who lives in it? | Why do they live here? |

► **Figure 6.30** *House-building fluctuations and predominant house types in Birmingham, 1856–94. (Source: Gerrard and Slater 1996)*

6.7 Housing and the growth of the suburbs

The homes we live in cover more of a city's area and are more widely distributed than any other type of land use. Where we live has a crucial influence upon our quality of life. As a result, the study of residential land use is an important part of urban geography, based on asking and seeking answers to the questions in Figure 6.29.

Residential land-use distributions are complex, and it is dangerous to generalise. For instance, consider the following statement or key idea: 'as distance from a city centre increases: housing age decreases; housing densities decrease; housing values increase'.

If you base your response on cities you know, you will probably say: 'OK, it is sometimes true, but not always'. Nonetheless, there are discernible patterns, based on age, scale and style of building (Figure 6.30). We can understand these patterns in British cities in terms of three broad phases of development, based on the changing interaction between the forces of concentration, dispersion and redistribution.

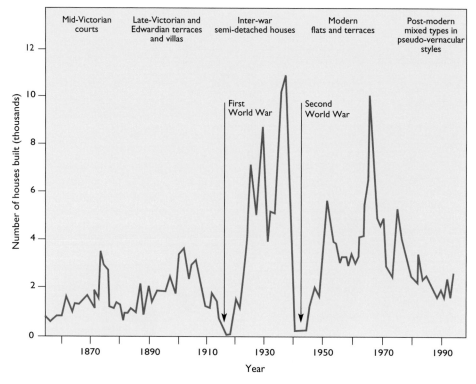

A concentration phase

Industrialisation and the factory system produced high-density, compact cities. The dominant residential outcome in England and Wales were districts of back-to-back and terraced housing (Figure 6.31a). In Scotland, much of Europe and in the USA, tenements (apartment blocks) were more common. What remain of those districts, built largely before 1914, are today part of the **Inner City** (see section 6.10). Many wealthier families too, lived close to city centres, in streets lined by elegant townhouses and detached homes. In some cities these survive today.

But emergent dispersion forces were also at work. Pockets of detached homes for the growing middle classes began to appear on urban fringes from the late nineteenth century. Also, growing concern for the living conditions of working populations saw the development of what we now call 'garden suburbs'. In Britain, this phase ended by 1920.

21 Look at the street maps in Figure 6.31.
a How big is the plot of a house in each example?
b Describe the different environments around the houses.

22 For a town or a city with which you are familiar, locate and name an area or street where each of the housing types in Figure 6.31 can be found.

▼ **Figure 6.31** *These three very different types of houses in Birmingham are found in different areas of the city*

a Late-Victorian, back-wing terraced houses

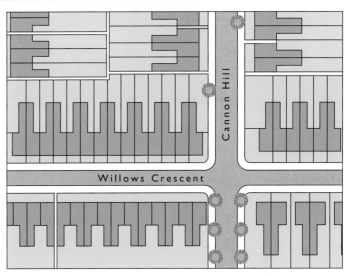

b Inter-war semi-detached houses

c Inter-war, neo-Georgian, cottage-style terraced houses

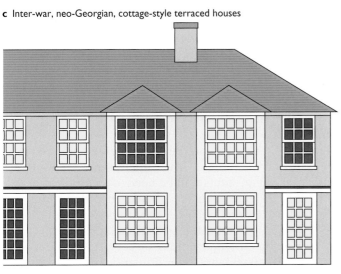

Figure 6.32 *Inter-war council housing in Birmingham. Houses like these were built all across the UK between 1919 and 1939.*

A dispersion phase

Today, many cities have a discontinuous zone of 'middle ring' suburbs, dominated by terraced and semi-detached houses, interspersed with smaller developments of detached houses. These are the main outcome of the suburban explosion that occurred during the 1919–39 period, i.e. between the two World Wars. Built at low densities, generally 8–12 dwellings per acre, these suburbs offered greatly improved living conditions for increasing numbers of families. However, this **decentralisation** caused extensive urban sprawl and a greater segregation of home and work, thereby creating the commuter lifestyle. A number of factors and processes combined to make this suburban revolution possible:

- Planning regulations were weak, and land was readily available.
- Wages and working conditions were steadily improving.
- Family expectations were rising and people were demanding a better quality of life.
- Central government policies focused increasingly on social welfare, including the right to better living conditions.
- Local authorities became stronger and more active.
- Buses, cars, and motorcycles were improving personal mobility.
- Mass-production techniques were introduced into the construction industry.

Most importantly, for the first time, the public sector became seriously involved in housing provision. 'Council estates', built by local authorities for affordable rental, accounted for increasing proportions of the total house building (Figure 6.32). Between 1919 and 1939, 1.3 million dwellings were built by local authorities in the UK, compared with 3 million built by the private sector.

An interactive phase

Over the last 50 years, the main issues which have dominated urban housing have been related to the shifting tensions between concentration, dispersion and redistribution forces. In general terms, suburban living is popular with large numbers of families (demand) while developers and builders are keen to provide homes people like, i.e. low to medium density single-family housing in pleasant surroundings (supply). The result is dispersion. In contrast, government policy has introduced greenbelts and **New Towns** with an aim to control urban sprawl while improving quality of life and social equity. **Regeneration** in existing urban areas has also been encouraged, to prevent decline. Such policies apply concentration and redistribution forces to restrict dispersion forces.

'The force is with you'

There is evidence from countries throughout the world that continued urban population growth and the desire for a suburban lifestyle are the two most powerful forces in creating new, low-density, decentralised urban forms. There are the extreme examples from free-market, capitalist systems, such as the 'non-cities' of North America, e.g. Phoenix (Figure 6.17). Yet **suburbanisation** is accelerating in communist societies too, such as in China (see Beijing example).

In Britain, where there is a shifting balance between the free market and government control, pressures remain high. Government forecasts are that at least 4 million additional dwelling units will be needed in England and Wales between 1996 and 2015. Private developers prefer greenfield sites, but in 1999, the government set a target of 50 per cent to be built in brownfield locations. The issue is complicated by regional differences: demand is greatest in the already crowded south-east region whereas the supply of brownfield sites is greatest in northern England.

Key Terms

Regeneration: The economic social and environmental improvement of a declining, ageing district.

23 Use the data for Beijing (Figure 6.33) to illustrate the process of residential decentralisation and suburbanisation.

Example: Suburbanisation in Beijing

The suburbanisation process at work in Beijing

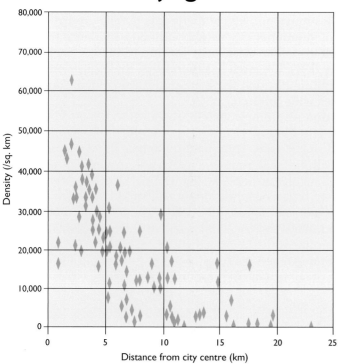

▲ **Figure 6.33a** *Distance–density relationships, 1992*

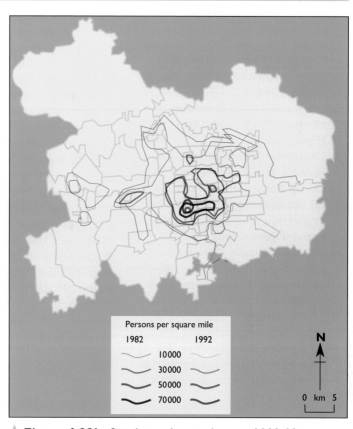

▲ **Figure 6.33b** *Population density changes, 1982-92*

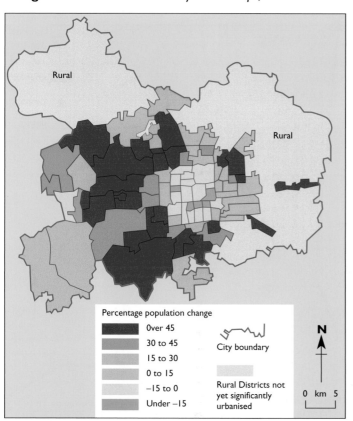

▲ **Figure 6.33c** *Population change, 1982–92*

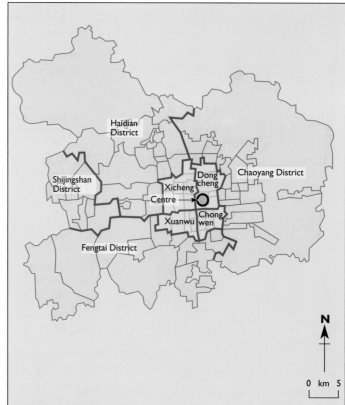

▲ **Figure 6.33d** *The main city districts*

24 Kitchener–Waterloo, near Toronto (Figure 6.34) is a rapidly expanding dispersed city. The two maps give some indication of its socio-economic structure. From the maps:

a Describe the distribution of (i) employment and (ii) amenities in Figure 6.34a.

b What evidence is there that the city has a multinuclear structure? (Figure 6.34b)

c Describe the distribution of districts with (i) above-average LQ for household incomes and (ii) above-average LQ for proportion of children living at home (Figure 6.34b).

d What evidence is there for suggesting that there is a relationship between the distribution of better-off families with children and the quality of urban environment?

e To what extent does Kitchener–Waterloo support the suggestion that suburbanisation is an élitist process?

6.8 Is suburbanisation an élitist process?

One claim frequently made is that it is mainly the 'better off' groups who benefit from suburbanisation and the emergence of decentralised cities. This hypothesis has been used to help explain the residential segregation of the different socio-economic groups identified in many cities. As more affluent, mobile families move out, less affluent, less mobile groups become increasingly concentrated in inner-city neighbourhoods. Evidence from Kitchener–Waterloo Metropolitan Area, near Toronto (Canada), supports this hypothesis (Figure 6.34). Research has found that the groups most likely to sustain the suburbanisation wave are high-income households and middle-income families with children.

This **social segregation** resulting from suburbanisation has also been associated with ethnic and cultural issues – for example in the 'white flight' process first identified in US cities. This suggests that as groups from different cultural and ethnic backgrounds, e.g. African–Americans or Hispanics, move into a neighbourhood, so the original white residents begin to move out. Over time, socially, economically and culturally segregated ghettos evolve – predominantly non-white neighbourhoods in inner-city districts, and mainly white, suburban, areas. For example, Washington DC, the capital of the USA, is the core city for the Greater Washington Metropolitan Area. A majority of the population of the core city is non-White; the ring of suburban cities around have mainly white populations.

▼ **Figure 6.34a** *Kitchener–Waterloo: Location of employment and amenities.*

▼ **Figure 6.34b** *Kitchener–Waterloo: Household income levels and the number of children living at home, measured by the Location Quotient (LQ).*

300

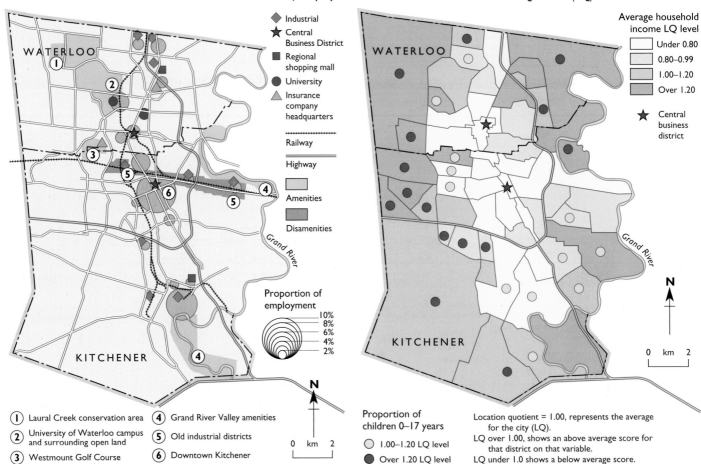

210

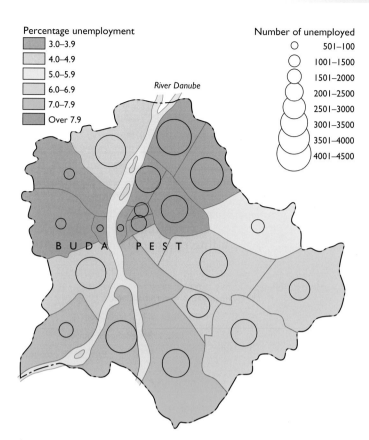

Percentage unemployment
- 3.0–3.9
- 4.0–4.9
- 5.0–5.9
- 6.0–6.9
- 7.0–7.9
- Over 7.9

Number of unemployed
- 501–100
- 1001–1500
- 1501–2000
- 2001–2500
- 2501–3000
- 3001–3500
- 3501–4000
- 4001–4500

River Danube

BUDA PEST

▲ **Figure 6.35a** *The number of unemployed and the rate of unemployment in Budapest*

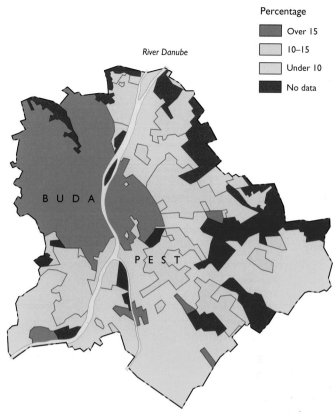

Percentage
- Over 15
- 10–15
- Under 10
- No data

River Danube

BUDA

PEST

▲ **Figure 6.35b** *The percentage of graduates living in districts of Budapest*

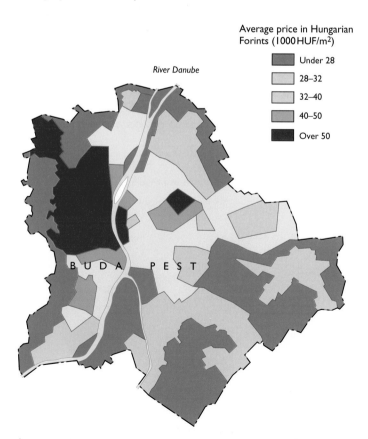

Average price in Hungarian Forints (1000 HUF/m²)
- Under 28
- 28–32
- 32–40
- 40–50
- Over 50

River Danube

BUDA PEST

▲ **Figure 6.35c** *Average price (in Hungarian Forints) of floorspace in apartments in districts of Budapest*

Social areas in cities

It is important to remember that suburbanisation is only one factor influencing the emergence of distinctive social areas in cities. Socio-economic factors, such as education, income and affordability of accommodation, influence locational choices and decisions – for example, cheaper housing stock is available in many older, inner-city areas. The example of Budapest illustrates this 'sorting' effect (Figure 6.35).

Cultural factors too, influence the social concentration and segregation processes. For instance, many immigrants settle near relatives, friends and others from similar cultural backgrounds. Neighbourhoods develop with services and support systems which sustain the attraction. The strong communities of people in cities such as London, Leicester, Birmingham and Bradford, whose origins are in the Indian sub-continent, are excellent examples.

One interesting question is: 'Will these concentrations of ethnic minority communities be sustained over time?' The North American experience has been that recent immigrant groups cluster together. Then, with the possible exception of the African–American community, they have become assimilated into mainstream society and gradually disperse to take on suburban lifestyles. There is some evidence that this is occurring in Britain, especially with more affluent households.

6.9 Housing policies and provision in the UK

Throughout the twentieth century, governments have battled with a complex dilemma: people vary in their housing needs and in their ability to afford accommodation, yet everyone should have the opportunity to live in decent conditions. The primary solution to this dilemma has been for the central and local governments to provide dwellings at subsidised costs for less affluent households (Figure 6.37). Since the 1920s, this social housing policy has produced 'council estates' in every UK city (see section 6.7). In addition, governments have encouraged private-sector provision by promoting home ownership as an alternative to renting, and by supporting the building industry in many ways, including passing planning legislation to make land available.

▼ **Figure 6.36** *Dwellings completed in the UK, 1945–96. In 1980, public sector housing was split into LA housing and HA housing.*

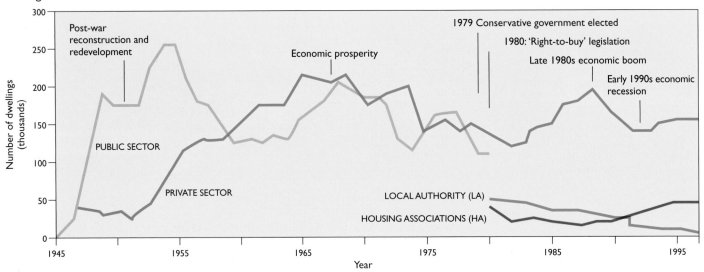

◄ **Figure 6.37** *High-rise blocks of council flats, like these in Hackney, London, were once thought of as being a good way of housing a large number of people in crowded urban areas in the 1950s and 1960s.*

▼ **Figure 6.38** *Housing tenure in the UK, 2000*

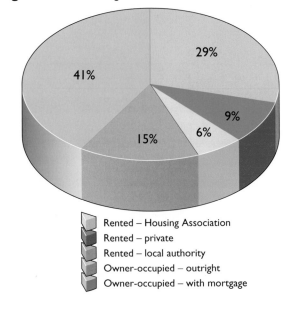

29%
9%
6%
15%
41%

- Rented – Housing Association
- Rented – private
- Rented – local authority
- Owner-occupied – outright
- Owner-occupied – with mortgage

25 Group discussion: Over the past twenty years, homelessness has increased greatly. Review carefully the information on housing policy and provision. Do the policies help to explain the rise in homelessness? (Think about the reasons for homelessness and who are most likely to become homeless.) What shifts in policy and provision could help to reduce homelessness?

317

26 From the information on Figure 6.36, divide the period since 1945 into three, and describe briefly the characteristics of each period.

The tenure revolution

As Figure 6.36 shows, the balance in housing supply between private and public sectors varies over time as government policies and economic conditions fluctuate. After 1945, priority was given to shortages caused by war damage and government authorities' schemes dominated. By the mid-1950s, rising prosperity encouraged the rapid growth of private house-building, especially by suburban expansion. Through the 1960s, a combination of large-scale inner city comprehensive redevelopment schemes (Figure 6.37), peripheral council estates and New Towns sustained high totals for public sector housing.

In 1980, the Conservative government introduced three important policies:

▪ Government spending on housing was restricted and the numbers of council dwellings being built fell severely.
▪ A Right-to-Buy (RTB) policy allowed council tenants to buy their homes (Box 4). This has continued and in 2000, RTB sales were still at least 60 000 a year.
▪ Housing Associations (HAs) funded through a government agency, the Housing Corporation, were introduced to replace local authorities as the main providers of social housing. In addition, existing council housing stock can be transferred to HAs if the tenants agree. For instance, Birmingham has the largest stock of council housing in England, but in 2001 the tenants voted against the transfer to a HA.

The combination of government policies, increased affluence and family expectations has caused a revolution in the way we occupy our homes (Figure 6.38). Between 1981 and 2001, the number of owner-occupied dwellings increased by over 40 per cent while rented dwellings fell by 15 per cent.

Regional variations in demand and supply

Between 1951 and 2001, the number of dwellings in the UK increased from 14 million to 25 million. Yet housing demand continues to rise. The number of people officially listed as 'homeless' almost doubled from 76 000 in 1990 to over 140 000 in 2000, despite a national total of 770 000 empty dwellings.

▼ **Figure 6.39** *Urban–rural tensions: UK government proposals, 2002*

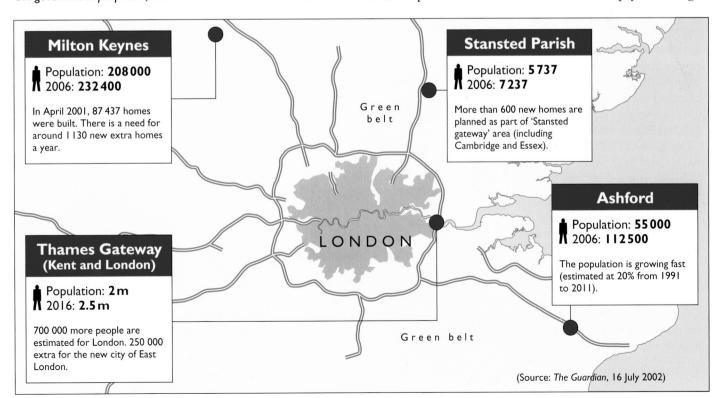

Milton Keynes

Population: **208 000**
2006: **232 400**

In April 2001, 87 437 homes were built. There is a need for around 1 130 new extra homes a year.

Stansted Parish

Population: **5 737**
2006: **7 237**

More than 600 new homes are planned as part of 'Stansted gateway' area (including Cambridge and Essex).

Green belt

Ashford

Population: **55 000**
2006: **112 500**

The population is growing fast (estimated at 20% from 1991 to 2011).

Thames Gateway (Kent and London)

Population: **2 m**
2016: **2.5 m**

700 000 more people are estimated for London. 250 000 extra for the new city of East London.

LONDON

Green belt

(Source: *The Guardian*, 16 July 2002)

26 Each of the five options for the provision of additional housing in the Cambridge sub-region has both advantages and disadvantages. For each suggest:

a two advantages and two disadvantages

b one group who might support and who might oppose the option.

27 From Box 4:

a What is meant by the 'Right-to-buy' policy?

b Outline the benefits of this policy?

c Why has it been more popular in some districts than in others?

d Why do critics claim that this has increased social segregation in cities?

Government forecasts indicate the need for 4 million more homes by 2015. To meet this demand, the government sets a target for each local authority who then must show how they intend to provide sites to meet this target for new dwellings. In the Cambridge sub-region, the government expects a provision of 2800 new homes a year for ten years. In 2002, the county council was considering five options:

■ build within Cambridge city
■ extend Cambridge city by building on the fringes
■ build within local market towns
■ extend local market towns
■ build a New Town of 6000 homes to 'swallow up' the villages of Oakington and Longstanton.

In addition, the University wants to develop 130ha within the city's greenbelt to provide housing for staff and students.

Demand varies regionally, with London and the South-East region experiencing greatest housing shortages. As a result, targets are highest in these regions. For example, in 2001 Surrey County Council were given a target of 47 200 although the current county plan made provision for only 35 000 new dwellings. In 2002 the government set a target for London and the South-East of 43 000 dwellings a year until 2016, with four growth areas identified (Figure 6.39).

In contrast, the North, North-East and North-West regions have surplus capacities for certain types of home. For example, in March 2002, the North-West region had 280 000 dwelling units either empty or in serious need of renovation. In the North-East region, 65 per cent of Newcastle's total is targeted for brownfield sites. Urban growth, therefore, is clearly linked to economic and environmental conditions.

300

Box 4

The 'Right-to-buy' policy

The powerful influence of government policy is vividly illustrated by the **Right-to-buy** legislation introduced by the Conservative Government in 1980. This gave tenants of government authorities the right to buy their homes at below market value, dependent on the number of years they had lived there. By 1997, over 2 million dwellings had been purchased, and less affluent families have the opportunity to own their own homes while costs have been reduced for local councils and for tax payers.

The most popular areas of Glasgow are Anniesland, Mosspark and Bailleston, which are mainly houses with some small blocks of flats. Springburn and Toryglen are lower in poularity. They are inner-city areas, redeveloped between 1968 and 1980, and have a large number of high-rise flats. Castlemilk, Drumchapel and Easterhouse are unpopular areas with peripheral estates built in the 1950s and many high-rise tenement flats.

Thus, the Right-to-buy sales have left local authorities with

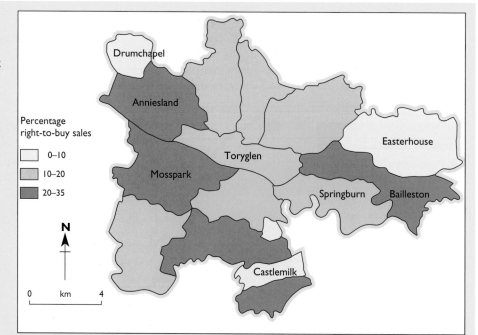

▲ **Figure 6.40** *Housing popularity and 'Right-to-Buy' sales in Glasgow, 1998*

the predominantly older, less popular estates and high-rise flats, with high maintenance costs. Progressively, households who can move out, will. This reinforces the segregation and isolation of certain groups within 'ghettos' or 'sink estates'. As a result, there are few alternatives for locating people with the most urgent need for 'social housing'.

6.10 Inner-city issues

The inner city may be defined as the zone of mixed, relatively high-density land uses surrounding the central urban core. Characteristics commonly identified in the inner city include:

- A declining economic base as businesses close or move out – partly as a result of the decline in manufacturing and the growth of service industries (see chapter 8).
- Outdated transport/access infrastructure and ageing industrial buildings, interspersed with modern premises.
- Older commercial and industrial premises available at relatively low rents, convenient for 'start-up' businesses.
- Above-average unemployment rates.
- Close to the labour-intensive CBD, but with a labour force often lacking in skills required for CBD business jobs.
- A complex mix of well-established neighbourhood communities and areas of rapid social change.
- A broad diversity of dwelling types, generally at high densities, and varying in age and quality.
- Above-average clustering of minority populations and groups classed as 'vulnerable'; a population with below-average earnings.
- Pockets of 'gentrification' and wealthy enclaves.
- Below-average scores on a variety of 'quality of life' indicators such as health or the availability of open space/parkland.

These characteristics indicate a number of ongoing problems. A series of government policies and programmes over the past 50 years have concentrated on four main aspects of economic and social regeneration: retaining and attracting economic activities; upgrading the physical environment, especially housing; improving the educational, work-related and social skills of residents; and strengthening a sense of 'community'. The two goals therefore, have been first, to improve the quality of life for residents and second, to create attractive environments for private-sector investors and businesses.

Many regeneration schemes are large-scale and involve a mix of land uses developed at high densities. As a result, they are expensive and carried out by private sector developers, sometimes in partnership with government agencies (Figure 6.42). For example, the New East Manchester urban regeneration company plans to renovate 7000 homes, build 12 500 new dwellings and create 15 000 jobs.

▼ **Figure 6.41** *The building of high-rise blocks of flats was seen as the way forward.*

'IT SEEMED LIKE A GOOD IDEA AT THE TIME'

A city councillor remembers:

'We were full of enthusiasm. We wanted to give people new homes and better lives. The old streets of crowded houses and factories seemed so out-of-date. We were excited by the neighbourhood designs that the architects and planners were showing us. One reason for building tall flats was to keep densities high. This would give us more space for schools, clinics, open space, shops, businesses, roads and so on. Even so, we could never fit in more than 60 per cent of the people from the old neighbourhoods. That's why we had to build similar estates on the outskirts of town – places like Castle Vale and Chelmsley Wood in Birmingham. We wanted to reorganise the land uses, so we separated homes from workplaces, and people from traffic. It all seemed such a good idea at the time.'

Canalside proposal put forward for site

Birmingham-based Midland and City Developments has put forward plans for 146 apartments and 1500sqm of retail and restaurant space on a canalside site between the city centre and the Jewellery Quarter.

The proposals form a key part of the regeneration of the canalside district, linking the city centre core with the Jewellery Quarter, a growing 'urban village'. The Jewellery Quarter, a longstanding centre for jewellery making, has been growing as a residential precinct in recent years.

Midland and City director Steven Byrne said that the apartments would be priced to make city living affordable for first-time and second-time home buyers.

The site lies between Fleet Street and a series of locks on the Birmingham and Fazeley Canal. A new canalside public open space would be provided, faced by one of three restaurants and adjoining a new pedestrian bridge to link Fleet Street to the opposite bank.

The site also lies opposite the new £30 million Bass headquarters and hotel and across the canal from the former science museum site in the Jewellery Quarter.

► **Figure 6.42** *Canalside proposal (Source: Planning, 6 April 2001)*

28 Why did renovation replace redevelopment in inner-city residential areas during the 1970s?

Regeneration of residential environments

The Inner City townscape today is the outcome of three main phases of activity: redevelopment, renovation and regeneration.

Redevelopment

Comprehensive redevelopment meant demolishing extensive sites and rebuilding new urban environments. For example, Birmingham built five Comprehensive Redevelopment Areas (CDAs) within 2 km of the city centre, rehousing over 50 000 people (Figure 6.41).

Renovation

By 1970, several factors were combining to shift policy towards rehabilitating existing neighbourhoods: comprehensive redevelopment was extremely expensive, took a long time to complete and disrupted communities, while many people liked their old homes, and the housing stock provided affordable accommodation for incomers. Rehabilitation of existing streets became a popular, quick and relatively cheap approach (Figure 6.43). In cities with high house prices, some areas became fashionable and have been 'gentrified' – bought and improved by young professionals, often as 'first homes'.

Replacement

By the 1990s, a number of serious problems were evident on estates built during the 1950–70 redevelopment phase. This was especially true of high-rise, high-density developments which were increasingly unpopular, were increasingly expensive to maintain and had been affected by the 'Right-to-buy' policy (see section 6.9). In response, government authorities have begun programmes of demolition and replacement, i.e. another phase of redevelopment. This is affecting not only inner-city estates, but also peripheral developments of the same period (Figure 6.44). Much of this activity has involved Housing Associations.

In parallel with policies to improve the physical environment, government authorities have attempted to involve local people in managing their own neighbourhoods. For example, resident committees take responsibility for the management and maintenance of their estates, streets and blocks of flats, thereby giving a sense of 'ownership'. In areas of privately-owned housing, 'community action' or 'neighbourhood' groups are encouraged.

▶ **Figure 6.43** *An 'enveloping' scheme in inner Birmingham. From the 1970s, government programmes have provided funds to help local councils to renovate older properties. For example, the Inner City Partnership Programme (ICPP) was introduced in 1978 and included 'enveloping' schemes, where the exterior fabric of whole streets was renovated. Further funds were available to assist home-owners to make internal improvements. Since the early 1990s, renovation has been an element of several broader programmes such as the City Challenge and the Single Regeneration Budget (SRB).*

1993

1998

◄ ▼ **Figure 6.44** *During the 1990s local authorities began to demolish some of the worst high-rise developments. This site at Bartley Green, an outer suburb of Birmingham, shows demolition and the replacement houses built on the same site in 1998.*

Suburban blues

It is not only in inner-city areas that problems have emerged. Many 'middle ring' suburbs, built between the 1920s and 1960s are now 'ageing', raising both physical and social issues. For example, the houses need to be upgraded and there is often inadequate provision for the movement and parking of cars. Residential developments tend to pass through a life cycle as the original residents move out or grow old, children grow up and leave, and so on. Thus, over time, the combination of physical decline, people's expectations and social change creates a less popular image of the area (Figure 6.45).

29 Look at the extracts in Figure 6.45.

a Construct a labelled diagram that summarises the cycle of suburban decline in the first extract.

b List the issues and problems in Braunstone. Does the proposed policy focus on the issues?

Figure 6.45 *Changes in suburbia reflect the decline of inner-city areas*

THE DEATH OF SUBURBIA

Ten years ago, when Ernest and Pat Vickers moved from inner London to Hayes on the western outskirts of the city, it was a step up the social ladder: their 1930s house was eminently 'respectable', neighbours were friendly, the community united. They intended to stay forever.

'Hayes was lovely when we arrived. Then people who had no respect for property started moving in. They fought with each other and threw furniture into the street. A driver destroyed three cars parked nearby. Nearly every house got burgled.'

Hayes grew in the 1920s as London's riposte to Hertfordshire's revolutionary garden suburbs. But now it manifests many of the social malaises once regarded by Middle England as safely confined to the inner cities. The garden gnomes have long been stolen; instead there are rows of boarded-up shops, run-down libraries and broken streetlights. Complaints about discarded syringes led to the closure of a town centre public toilet. Low-rent charity and bargain shops have replaced butchers and bakers.

Decline is almost always triggered by the closure of shops, even one single shop. A retail superstore is built in one suburb and this causes the closure of the hardware store on the high street of the other. Shoppers go to the superstore for their paint and wallpaper but stop at neighbouring shops for other products.

Quickly, local shops on the first high street go out of business, the high street dies and the community starts to die with it. The post office closes, the cinema closes and people start to leave the area because there are no shops. Property values stagnate. Developers buy up local houses and turn them into flats. Now the inner-city poor, desperate to escape, have their chance: they could not afford to buy a house in the suburbs, but they can afford to rent a flat.

(Source: The Sunday Times, 21 March 1999)

Reviving a Leicester Estate

BRAUNSTONE residents are being urged to play their part in a campaign to land a £50 million, ten-year aid package for the area.

The Braunstone Partnership wants to get the community involved in its plans to transform the estate using government cash.

On Tuesday, the council-led Leicester Partnership backed the Braunstone Partnership's bid for a £50 million slice of the Government's £800 million New Deal for Communities initiative.

Braunstone Partnership chairman, Bernard Greaves, said his organisation was already trying to recruit community support after winning £1.4 million from the Government's Single Regeneration Budget earlier this year.

The Partnership is carrying out household surveys to find out if Braunstone residents would be willing to form a community trust to run regeneration schemes.

He said: 'We already have community organisations involved in our partnership, such as the Braunstone Employment Project, Turning Point Women's Centre, working men's clubs, Youth Workers' Forum, Adventure playground, South Braunstone Action Group, Braunstone Motor project and North Braunstone Training and Resource Centre.

Thomas Smith (80), retired, Braunstone

'We could definitely do with a bowls club, which they could incorporate with a leisure club for the young people.'

Joyce Smith (75), retired, Braunstone

'The houses could all do with being modernised because they are all pre-war and most of them don't have central heating. I think there needs to be more police.'

Ravinder Gill (44), businessman, Braunstone

'The biggest need on the estate is something for the kids because they are running around the place being a nuisance. They need a place to go – like a youth club. Also, the houses need to be insulated and brought up to date.'

(Source: The Leicester Mercury, 29 October 1998)

Table 6.5 *Largest cities in LEDCs in rank order (million people)*

Argentina	
Buenos Aires	11.3
Concordia	1.2
India	
Calcutta	17.0
Mumbai (Bombay)	16.0
Delhi	8.4
Mexico	
Greater Mexico City	25.0
Monterrey	3.0
Thailand	
Bangkok	6.0
No other city over	0.5

30 Define the term 'primate city' and determine which of the countries in Table 6.4 have a primate city structure.

Figure 6.46 *Bangkok (Thailand) is a good example of a primate city. It is far bigger and wealthier than other cities in Thailand.*

6.11 LEDC cities

The current urbanisation in LEDCs has two distinguishing features: first, the rapidity of urban growth and second, in any particular country, the dominance of one, in very large countries perhaps two, huge metropolises (see section 6.12). Where a single metropolis dominates the urban system, a country is said to have a **primate city** structure, e.g. Bangkok in Thailand (Table 6.5 and Figure 6.46). One outcome of this powerful attraction of a single metropolis is that some of the world's largest urban agglomerations are emerging in LEDCs and NICs (see section 6.12).

Processes and issues

Because of the contrasts in their history (i.e. colonialism), economy (i.e. stage in industrialisation) society, religion, and political systems, cities in Asia, Africa and Latin America exhibit significant differences. However, we can identify certain shared characteristics of these cities:

- The rapid growth is sustained by a combination of large-scale rural–urban migration and high natural increase rates among urban populations. Fundamental changes in rural economies lead to overpopulation, under-employment and poverty (see chapter 7). However, up to 50 per cent of the growth is due to high fertility rates among the urban population.
- Urban population growth rates outstrip economic growth rates and the industrialisation process – numbers of people grow more quickly than number of jobs. This is very different to processes in MEDCs, where urbanisation was paralleled by rapid industrialisation.
- Very wide disparities in wealth and quality of life are evident. A small, wealthy élite dominates a majority who have low educational and skill levels, and hence, live in serious poverty.
- Most have weak or inconsistent government control of urban growth. Many governments lack the skills, money or power to control growth effectively.

Figure 6.47 *This is both a self-built home and a workshop for a Thai family on a main road on the edge of Bangkok.*

Figure 6.48 *Mumbai (India) has some areas of wealthy high-rise apartments, which are in great contrast to the slums on the edges of the city.*

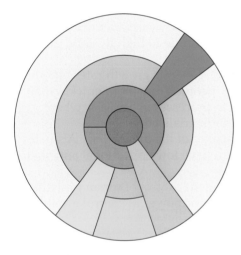

- CBD

- Expensive high-rise modern apartments with exclusive shops and services.

- Middle-class residential areas of older properties; mixture of apartments, homes and neighbourhood services.

- Modern industrial sector along main transport corridor.

- Mixed sector of industry, with high-density, slums along a transport route.

- Periferia: poor quality, permanent housing of high densities, but with some basic amenities.

- High-class suburbs for managerial and professional classes, and containing business and commercial nuclei.

- Favelas: spontaneous shanty and squatter townships. Very high densities. Frequently unofficial and residents have few rights.

- Low-cost government housing schemes: permanent small dwellings at high densities, with basic amenities.

Figure 6.49 *A simple model which can be applied to Latin American cities*

These characteristics help us to understand some of the images and problems of LEDC cities (Figure 6.47). For example, in Bombay (officially re-named as Mumbai in 1995) 50 per cent of its 15 million people live in slums, squatter settlements or on the streets. The average population density is 580 persons/sq.km, but in the Dharavi district over 200 000 people live in less than 1 sq.km. In Dharavi, up to 200 families share one stand-pipe, their only source of water, yet nearby, there are clusters of expensive, high-rise apartment blocks (Figure 6.48).

In many cities, the inadequacy of the transportation infrastructure and lack of environmental controls results in semi-permanent 'gridlock' (Bangkok and Calcutta) and serious air pollution (Mexico City) where air quality level falls below recommended 'healthy' standards on at least 150 days each year.

Newly arrived migrants and poorer sections of the population crowd on to low-value sites: inaccessible locations, poorly drained or steep land, strips along rail and road routes. Often they have no legal right to the land or to their homes. They create huge, high-density 'self-built' settlements with few facilities, known in different countries as favelas (Brazil), barrios (Colombia), bustees (India), bidonvilles (North Africa), or gecekondu (Turkey).

Because they lack legal status, these 'spontaneous' or 'informal' communities may be forced to move by government clearance programmes or formal housing projects. The displaced people then seek other low-cost sites, usually around the fringes of cities. It is important not to perceive these cities in terms of solely negative images. Fewer than one in two adults may have formal employment, but families show great resourcefulness, not only in building their own homes, but in gaining income (Figure 6.47).

In some countries in Africa, such as Kenya, governments, supported by international aid organisations, have developed **sites-and-services** schemes, in which the land and basic services of water, sewage and perhaps electricity, are provided, and the residents then build their own homes.

Urban forms in LEDCs

Figure 6.49 is a simple model that has been produced for Latin American cities. We can begin to analyse it by asking three questions:

- What is the distribution of land uses?
- In what ways is the morphology similar to and different from MEDC cities?
- What are the reasons for the similarities and differences?

One similarity with MEDC cities is the presence of a CBD core. Many LEDC cities have long histories that pre-date the current growth phase. For instance, the Indian sub-continent has a long urban tradition upon which the British colonial administrative and commercial system was built. The traditional and colonial periods can each be identified in the structure of the CBD and inner areas of many cities like Allahabad, Delhi and Mumbai (Figure 6.50a). In south-east Asia, cities such as Hong Kong, Macau, Shanghai and Singapore grew as colonial and commercial ports.

Industry too, tends to be located in sectors, and there are wealthy residential enclaves, both common features of MEDC cities. However, closer examination of the model reveals some fundamental contrasts with MEDC cities. Perhaps the most significant is the distribution of residential areas: in LEDC cities, the poorer population live on the peripheries; the more affluent live in inner districts and in some cases along a sectoral 'spine' (Figure 6.51). These patterns may be modified by political and cultural forces, such as the apartheid policy that helped to shape South African cities (Figure 6.50b).

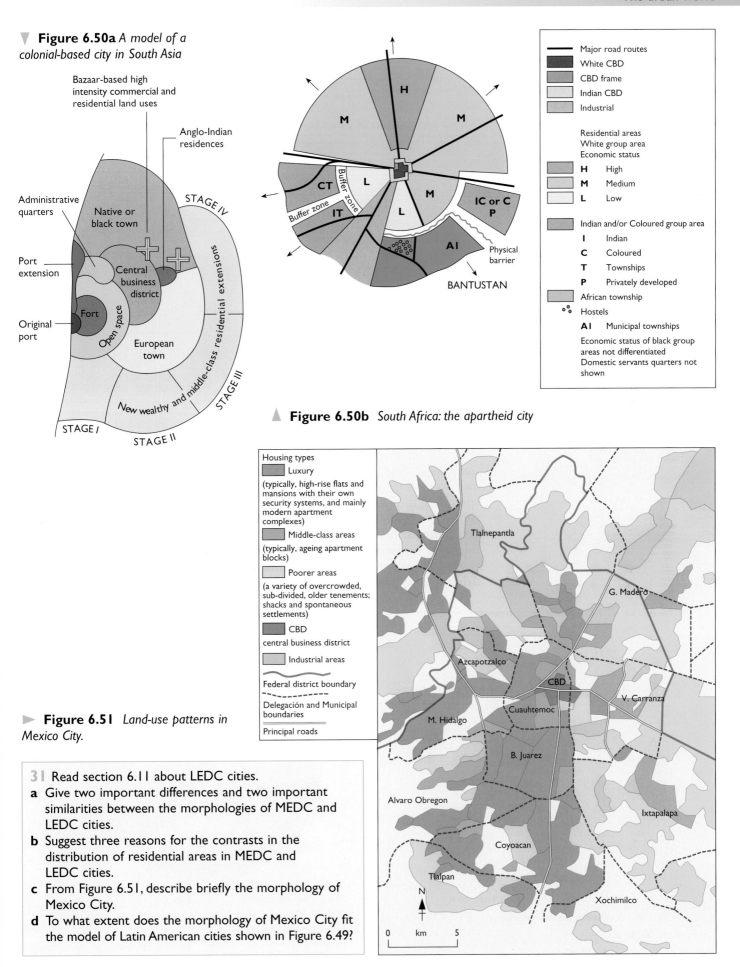

▼ **Figure 6.50a** *A model of a colonial-based city in South Asia*

Bazaar-based high intensity commercial and residential land uses

Anglo-Indian residences

Administrative quarters

Native or black town

STAGE IV

Port extension

Central business district

Original port

Fort

Open space

European town

New wealthy and middle-class residential extensions

STAGE I

STAGE II

STAGE III

Major road routes

White CBD

CBD frame

Indian CBD

Industrial

Residential areas
White group area
Economic status

H   High

M   Medium

L   Low

Indian and/or Coloured group area

I   Indian

C   Coloured

T   Townships

P   Privately developed

African township

Hostels

AI   Municipal townships

Economic status of black group areas not differentiated
Domestic servants quarters not shown

H

M

M

CT

Buffer zone

IT

L

L

M

IC or C
P

AI

Physical barrier

BANTUSTAN

▲ **Figure 6.50b** *South Africa: the apartheid city*

Housing types

Luxury

(typically, high-rise flats and mansions with their own security systems, and mainly modern apartment complexes)

Middle-class areas

(typically, ageing apartment blocks)

Poorer areas

(a variety of overcrowded, sub-divided, older tenements; shacks and spontaneous settlements)

CBD

central business district

Industrial areas

Federal district boundary

Delegación and Municipal boundaries

Principal roads

Tlalnepantla

G. Madero

Azcapotzalco

CBD

V. Carranza

M. Hidalgo

Cuauhtemoc

B. Juarez

Alvaro Obregon

Ixtapalapa

Coyoacan

Tlalpan

Xochimilco

N

0   km   5

► **Figure 6.51** *Land-use patterns in Mexico City.*

31 Read section 6.11 about LEDC cities.

a Give two important differences and two important similarities between the morphologies of MEDC and LEDC cities.

b Suggest three reasons for the contrasts in the distribution of residential areas in MEDC and LEDC cities.

c From Figure 6.51, describe briefly the morphology of Mexico City.

d To what extent does the morphology of Mexico City fit the model of Latin American cities shown in Figure 6.49?

6.12 Megacities

Millionaire cities, which have a population greater than 1 million, have become commonplace, but the **megacity** is a phenomenon of the late twentieth century. A megacity is generally defined as an urban agglomeration with a population of at least 10 million. In 1985, there were 11 megacities. By 2000 there were at least 20 of these urban giants (Figure 6.52).

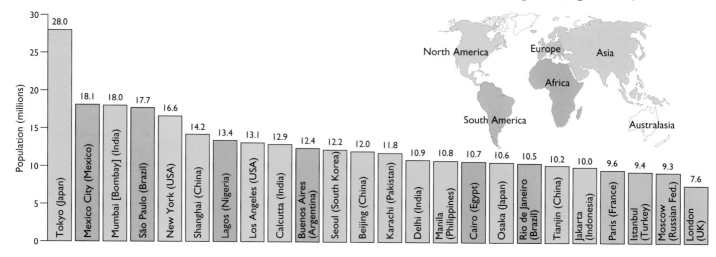

In LEDCs there are the extreme outcomes of uncontrolled urbanisation, where rural–urban migration and rapid population growth dangerously outrun economic growth, e.g. São Paulo (Brazil), Mexico City (Mexico), Calcutta (India). These urban areas tend to retain an essentially monocentric (mononuclear) structure: a core city enveloped by ever-spreading industrial and high-density residential sprawl, and increasingly choked by traffic 'gridlock' and pollution (Figure 6.53a). Bangkok (Thailand), is an excellent example of a city with these attributes. Its uncontrolled growth will see it attain megacity size early in this century.

Megacities in MEDCs are genuine 'global cities' in that the primary driving force behind their giantism is the globalisation of economic activity.

Figure 6.53a *The development of monocentric form in cities*   **Figure 6.53b** *The development of polycentric form in cities*

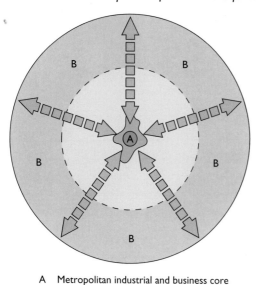

A   Metropolitan industrial and business core

B   Residential and services suburbs

Massive radial commuter flow

A   Metropolitan industrial and business core

D   Semi-independent outer peripheral zone

E   Edge city

Reduced radial commuter flow

New peripheral transportation corridor, taking large commuter flows

Box 5 Edge cities

These are economically and socially independent urban settlements with a suburban or peripheral location to large metropolises. They have some common characteristics:

- They have at least 5 million sq.ft of office space.
- They have at least 60 000 sq.ft of retail space.
- They have more jobs than bedrooms, i.e. they are not residential suburbs.
- They are perceived as a single place by local people.
- They are nothing like a city was 30 years ago.
- They are linked to the broader metropolitan system by a transport and communication network.

In 1997, there were 150 edge cities in the USA. These fell into three categories:

- **Uptowns**: built upon pre-existing settlements.
- **Boomers**: formed from sprawling suburban growth.
- **Greenfields**: developed beyond the pre-existing urban periphery.

Population, 1998

Tokyo Metropolis (central city):
12 million
Tokyo Megalopolis Region: 32 million
Tokyo Capital Region: 35 million

The plan for the Tokyo Megalopolis Region is to decentralise some of the highly concentrated business, commercial and industrial functions from the central city to a series of semi-autonomous nuclei in the suburban cities zone, i.e. to develop a polycentric structure. In the broader Capital Region a set of satellite towns, which are a form of 'edge cities' will be created over the next twenty years. The aim is to reduce the number of workers in the central city by 10 per cent by the year 2015, while coping with an overall regional urban population growth of 1 million.

New York (the USA) and Tokyo (Japan) are good examples. Through a varying combination of sub-regional planning and private investment, these are evolving with polycentric [polynuclear] structure (Figure 6.45b).

The combination of economic efficiency, lifestyle preferences and environmental quality controls interact to overcome the inadequacies of a monocentric structure. For example, the semi-continuous 'edge cities' (Box 5), linked by peripheral roads and mass transit systems, reduce the intensity of radial commuting flows to and from the core city (Tokyo, Figure 6.54).

A number of large metropolises are likely to become megacities within the next twenty years. Examples show that economic, social and environmental problems tend to intensify as urban size increases. As a result, governments are developing sub-regional plans to accommodate the expected growth, so that benefits will outweigh problems (see Shanghai example). Hong Kong has six edge cities, or planned New Towns, built from the 1970s, to accommodate 2.5 million people. They are linked to each other and to the Downtown business core by an efficient rail system (Figure 6.55).

▼ **Figure 6.54** *The city and wider region of Tokyo (Source: Tokyo Metropolitan Government 1996)*

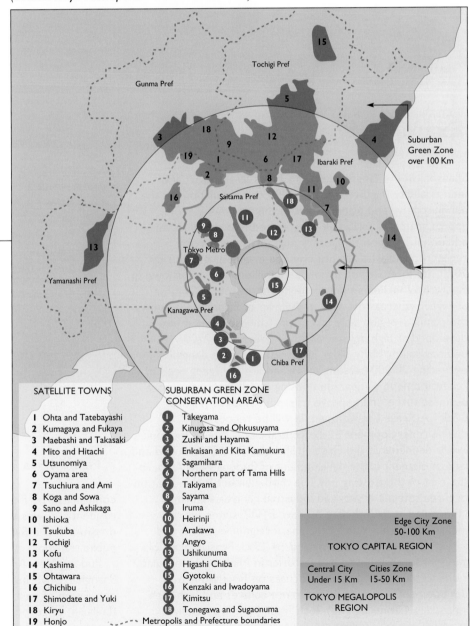

SATELLITE TOWNS	SUBURBAN GREEN ZONE CONSERVATION AREAS
1 Ohta and Tatebayashi	① Takeyama
2 Kumagaya and Fukaya	② Kinugasa and Ohkusuyama
3 Maebashi and Takasaki	③ Zushi and Hayama
4 Mito and Hitachi	④ Enkaisan and Kita Kamukura
5 Utsunomiya	⑤ Sagamihara
6 Oyama area	⑥ Northern part of Tama Hills
7 Tsuchiura and Ami	⑦ Takiyama
8 Koga and Sowa	⑧ Sayama
9 Sano and Ashikaga	⑨ Iruma
10 Ishioka	⑩ Heirinji
11 Tsukuba	⑪ Arakawa
12 Tochigi	⑫ Angyo
13 Kofu	⑬ Ushikunuma
14 Kashima	⑭ Higashi Chiba
15 Ohtawara	⑮ Gyotoku
16 Chichibu	⑯ Kenzaki and Iwadoyama
17 Shimodate and Yuki	⑰ Kimitsu
18 Kiryu	⑱ Tonegawa and Sugaonuma
19 Honjo	----- Metropolis and Prefecture boundaries

Edge City Zone 50-100 Km
TOKYO CAPITAL REGION
Central City Under 15 Km   Cities Zone 15-50 Km
TOKYO MEGALOPOLIS REGION

◀ **Figure 6.55** *Sha Tin New Town, Hong Kong, is linked to the business core by an efficient transport system.*

32 List the main features of the Greater Shanghai Plan.

33 Explain how the plan for Shanghai illustrates the polycentric form of modern megacities.

Example: Shanghai

Shanghai, in the Yangtse delta, is China's largest port and industrial city (Figure 6.56a). In 1995, the population for the administrative and planning unit of the city was approximately 7 million. (It is estimated that there are an additional 3 million temporary migrants who are not registered and who are not counted in official figures.) By 2020, the forecast population is 14 million. The area shown in Figure 6.52 includes the urbanising sub-region of Greater Shanghai, whereas the area in the map of Figure 6.56a shows the planning region only. Like many Chinese cities, it has developed at high densities, with an average of 2200 persons/sq.km. Many older, inner districts, have densities exceeding 20 000 persons/sq.km and are in urgent need of redevelopment, to upgrade housing standards, create more business space, and improve transport systems (Figure 6.56b). At least 1 million people will be rehoused.

The redevelopment of the central city is one of the four major elements in the Greater Shanghai Plan. The second and largest element is the development of Pudong New Area, to the east of the existing city. The third element is the creation of three satellite towns and industrial centres: Minhang, Hongquiao and Caohejing. All have ETDZ status, which means that, unlike much of China, foreign investment and foreign businesses are permitted. By 1995, Pepsi, Coca Cola and Johnson & Johnson had factories in Minhang. Hongquiao, close to the existing airport, focuses on hotels, office developments, trade and conference centres and high-tech industries. Caohejing High-Tech Trade Park is a rapidly growing area of electronics, aerospace and bio-engineering

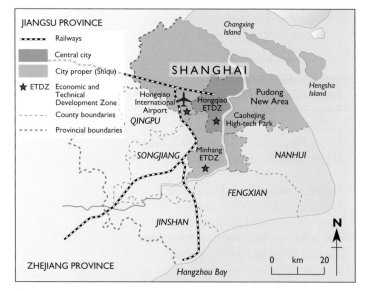

▲ **Figure 6.56a** *The city of Shanghai and the Pudong New Area*

companies. The fourth main element is a series of largely residential towns to be developed at medium to high densities in the five peripheral 'suburban counties', with public transport systems to the main employment centres.

Pudong New Area is adding a massive component to Shanghai. Covering an area of 522 sq.km, it will have a population of well over 2 million, and includes a large CBD (Lujiazui) of corporate headquarters, financial services, shopping malls, leisure and entertainment (Figure 6.56c).

In addition there will be three industrial and commercial zones, plus a huge new port in Waigaoquiao. These will also have ETDZ status, allowing foreign investment, businesses and international trade. A vital element is the new transport infrastructure linking Pudong to the rest of Shanghai, and the new airport which was opened in 2000.

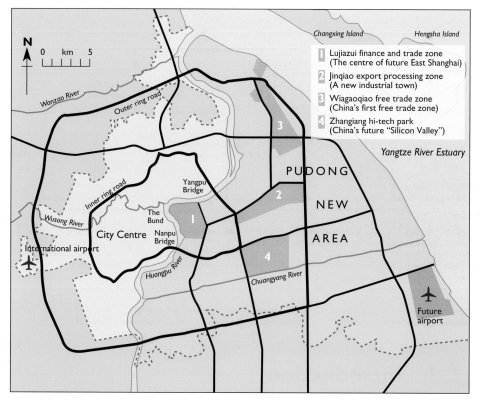

◀ **Figure 6.56b** *The Pudong New Area and its designated industrial zones*

▼ **Figure 6.56c** *The Pudong New Area is the new centre of corporate headquarters for global companies and financial institutions.*

Summary

Today more than 50 per cent of the world's population lives in cities.

In many countries, an increasing proportion of the total population is concentrated in a small number of very large urban agglomerations, including megacities of over 10 million people.

In MEDCs, the concentrated, high-density industrial city is being replaced by a dispersed, low-density post-industrial form.

Cities evolve by the interaction of the forces and processes of concentration, dispersion and redistribution.

The suburban explosion of dispersed cities is based upon increasingly mobile lifestyles supported by modern communication systems.

Suburban lifestyles are increasingly popular but may cause social segregation, economic costs and environmental problems.

Local and national governments have introduced vigorous regeneration programmes into older districts such as the CBD and inner city areas.

Significant contrasts exist between the economic, social and morphological characteristics of MEDC, NICs and LEDC cities.

In many LEDCs, cities are growing more rapidly than economic activity, resulting in serious economic, social and environmental problems.

Rural environments

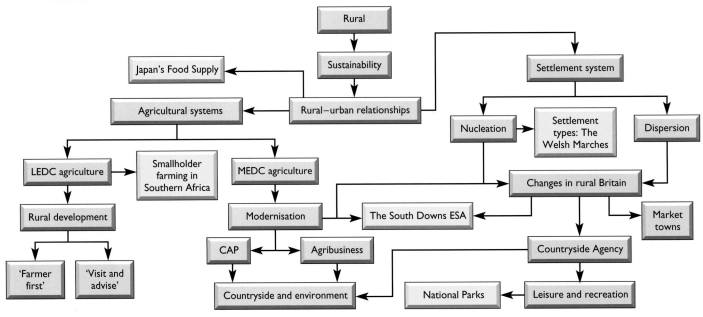

7.1 Introduction

Despite global urbanisation (chapter 6), approximately 50 per cent of the world's population still live in the rural environments that cover a majority of the world's land. Thus, the future of planet Earth will be strongly influenced by what happens in these rural environments. In both MEDCs and LEDCs, rural economies, societies and environments are experiencing rapid change, such as new types of farming and settlements, improved transport and communications technology and increasing demands for recreation and tourism. However, there is growing concern about the nature of many of these changes. These issues are often discussed in terms of their **sustainability** (see Box 1).

In the context of the UK countryside, the official government policy on sustainability is expressed by the Department of the Environment, Transport and the Regions as: 'Handing down to successive generations not only man-made wealth such as buildings, roads and railways, but also natural wealth, such as clean and adequate water supplies, well cared for farmland, landscape, a wealth of wildlife and ample forest' (DETR, 1998). These definitions make it clear that both change and sustainability have three broad, interrelated dimensions: economic, social and environmental (Figure 7.1).

This chapter focuses on the interactions between these dimensions. Issues arising from these interactions often produce tensions because of the different timescales involved:

- Economic processes and decisions often involve relatively short timescales – for example a development company buys farmland and builds houses with the intention of making a profit within a few years.
- Social change occurs within medium time-frames – the character of a village community, for instance, needs to change gradually if conflicts between resident groups are to be avoided.

Box 1 Some definitions of sustainability

- 'Development that meets the needs of the present without compromising the ability of future generations to meet their own needs.'
(Brundtland, 1987).
- 'Sustainable development seeks to improve the quality of human life without undermining the quality of our natural environment'
(English Nature).
- 'Sustainability involves development which meets economic and social needs at the same time as safeguarding and improving the environment'
(Countryside Commission, 1998).

Economic: Are the changes taking place in a rural economy likely to sustain that local economy over time?

Social: Are the social changes taking place likely to strengthen (i.e. sustain) or weaken the local community over time?

Environmental: Are the environmental changes taking place likely to conserve (i.e. sustain) or reduce the character, quality and, hence, the productivity, of the local environment over time?

▲ **Figure 7.1** *The dimensions of sustainability*

1 Read the definitions of 'sustainability' carefully.
a What is the single most important theme running through the definitions?
b Select one definition that you think gives the clearest summary of 'sustainability'.

2 **With a partner:** Look carefully at Figure 7.1. Give some examples of changes that could assist or threaten each of the economic, social and environmental dimensions of sustainability in that environment.

Key Terms

Rural: An area where the main economic activities and settlement patterns are related to agriculture.
••••••••••••••••••••••••••••••••••

■ Environmental decisions need to be based on an understanding of long-term processes, such as that the draining of a wetland may improve agricultural productivity, but may also destroy the habitat of valued species for ever.

Furthermore, the interactions between these three dimensions of sustainability work within a political context, such as the values, priorities powers and policies of the main 'gatekeepers' and 'stakeholders'. As this chapter will illustrate, the timescale perspectives of the interest groups and decision-makers have a crucial influence upon change and sustainability.

7.2 Changing rural–urban relationships

Traditionally, a **rural** area has been identifiable from three socio-economic characteristics:
■ Agriculture being the main economic activity and dominant land use.
■ Having a settlement pattern closely related to the type of agriculture.
■ Having a majority of the population is associated, directly or indirectly, with the agricultural system.

These characteristics suggest a clear distinction between 'urban' and 'rural', and remain useful definitions in many LEDCs. However, in MEDCs, urban–rural relationships have become more complex. For instance, a 1997 report by the European Commission states: 'Rural areas account for over 80 per cent of the territory of the EU and are home to more than one quarter of its population.' Yet, in 1999, agriculture employed only 5 per cent of the working population and yielded less than 2 per cent of the GDP of the EU. In the UK, almost 20 per cent of the population live in non-urban administrative districts, but only 2 per cent of the working population have jobs in farming. In England, the Countryside Agency defines rural areas as 'settlements with a population under 10 000'. This definition includes 9.3 million people, living in 16 700 rural towns, villages and hamlets. Notice the inclusion of 'rural towns', although 75 per cent of rural dwellers actually live in settlements with fewer than 500 people.

3 Read the section 7.2 on rural–urban relationships.

a Make a list of characteristics which you would use in defining a 'rural area' in the UK today. (It may help if you think of differences from what you consider to be 'urban'.)

b Why do you think the Countryside Agency includes 'rural towns' in its definition of rural areas?

c Give two examples which support the claim that rural–urban distinctions are becoming increasingly blurred.

▼ **Figure 7.3** Contrasting trends that lead to tensions

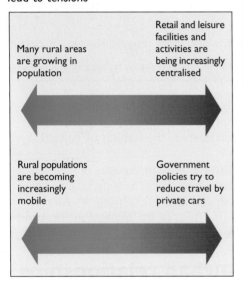

Key Terms

Agriculture: Farming (cultivating land, rearing livestock or growing crops).

Commercial agriculture: A farming system which produces products for sale.

Subsistence agriculture: A farming system which produces products for home use.

The explanation lies in the changing forms of cities and preferred lifestyles (see chapter 6). Boundaries between urban and rural environments are becoming increasingly blurred (Figure 7.3). Agriculture may remain the dominant land use in rural areas, but the majority of rural residents are likely to be economically and socially 'urban' (e.g. commuters and their families). For example, 39 per cent of people living in English rural areas commute to work. For this reason, the idea of an urban-rural continuum has been proposed as a realistic model (Figure 7.2).

▼ **Figure 7.2** The urban–rural environmental continuum

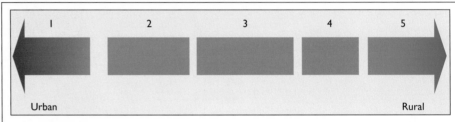

1 Urban built environment

2 Urban–rural fringe: May include greenbelt. Diverse land uses including agriculture modified by urban proximity (e.g. market gardening; Pick-Your-Own; declining arable land; pony stabling; garden centres). Settlements dominated by incomers – economically 'urban'. Country Parks and other leisure land uses.

3 Farming and commuter zone: Settlements increasingly modified by commuter families; strong development pressures. Farming remains productive but considerable leisure pressure from urban populations. Farming adjustments (e.g. 'working farms', Pay-As-You-Play golf courses). Affected by counter-urbanisation trend. Farm building conversions popular, property and land prices rise.

4 Deep countryside: Beyond significant commuting distance, but settlements influenced by holiday home purchases and retirement incomers. Farming little affected by urban influences, but property prices rise as more affluent families seek 'the rural lifestyle'. Some out-migration of younger people. Some counter-urbanisation trend.

5 Remote rural environments: Social and economic problems: marginal farming and steady out-migration. Incomers buy up properties as holiday and retirement homes. Tourism increasingly important to sustain local communities and economies. Conservation policies to sustain semi-natural environments (e.g. National Parks, ESA designations); CAP support schemes.

7.3 Agricultural systems

The importance and nature of agriculture

In all countries, **agriculture** is the largest user of land and the main economic activity of rural areas. We can define agriculture as: 'the control and use of plants and animals for the production of food, fibre, and raw materials for industry' (Raw and Atkins, 1995.)

We rely on agriculture, therefore, to feed us and to supply raw materials for industry. Ensuring food supplies is an important aspect of government policy in all countries, and agricultural products are a significant element in world trade (see Japan example).

Crops and animal products that enter the trade system are commercial products, that is, products for sale. In MEDCs, **commercial agriculture** is dominant, but in many LEDCs, a strong element of **subsistence agriculture** is maintained, that is to say, crops are grown and animals are reared to support the family, not for sale. It is important to understand that, there are few regions in the world today, where some element of commercialism has not been introduced into subsistence farming systems (Figure 7.5).

EXAMPLE: Japan

Ensuring Japan's food supply

Japan has a population of 126 million, over 80 per cent of whom live in urban areas. Less than 25 per cent of its land area is suitable for agriculture. In 1995, the country grew only 42 per cent of its food needs, and was the world's second largest importer of agricultural products, by value. Government policy has focused on making Japan self-sufficient in the production of rice, the staple of the Japanese diet. Varying proportions of all other foods are imported (e.g. fruit 50 per cent; meat 45 per cent; fish and shellfish 45 per cent; dairy products 30 per cent). The main sources of these imports are the USA (30 per cent), China (10 per cent) and Australia (8 per cent).

► **Figure 7.4** *Intensive rice cultivation takes place on these terrace systems which have been built across the hillsides over the last 2000 years.*

4 Look carefully at Figure 7.5.
a Describe briefly the landscape and vegetation.
b What evidence is there to support the idea that LEDC agriculture is labour-intensive?

Furthermore, some extreme forms of commercial agriculture are found in LEDCs (e.g. plantations for coffee, rubber, bananas, sugar, etc.). Thus, although we must beware of making generalisations, a fundamental distinction can be made between farming systems in the LEDCs and MEDCs. For instance, LEDC agriculture is **labour-intensive**; MEDC agriculture is **capital-intensive**. In 1998, the global agricultural workforce was just over 1 billion; of these, 95 per cent were in LEDCs, yet Western Europe and North America, with 14 per cent of the world's agricultural area, produced 30 per cent of global agricultural output.

► **Figure 7.5** *On this farm in Belize, Central America, land is being cleared for orange trees. Small farmers often add a commercial crop to their subsistence base.*

<table>
<tr><td>

Key Terms

Shifting cultivation: A form of agriculture where a plot is tilled for several years, then left to rest while the farmer clears and tills another plot.

</td></tr>
</table>

Agriculture in LEDCs

Despite rapid urbanisation (see chapter 6), LEDCs remain predominantly rural societies, dependent upon agriculture (Table 7.1).

In terms of area and population, agriculture in LEDCs is dominated by peasant societies and economies (e.g. in countries such as Ethiopia, Tanzania and Mozambique, more than 70 per cent of the population is involved). These societies range from semi-nomadic animal herders, through shifting cultivators, to intense arable farmers (Figure 7.6). Despite this diversity, whenever you study a specific example, the following

▼ **Table 7.1** *Percentage of the labour force involved in agriculture in selected countries*

Groups	Population (billion)	Percentage of workers in agriculture	Examples of countries
Low-income countries	3.3	67	Haiti, Mozambique, Nepal
Lower-middle-income countries	1.0	34	Bolivia, Morocco, Turkey
Upper-middle-income countries	0.6	20	Brazil, Greece, Malaysia
High-income countries	0.9	5	Australia, Netherlands, the USA

▼ ► **Figure 7.6** *Goat herding in Tunisia (right) and shifting cultivation in Amazonia (below) are two examples of diverse peasant agricultural societies*

5 From Table 7.1:

a Construct graphs or choropleth maps which represent the patterns shown in Table 7.1 (Think carefully about the most suitable form of graph or map.)

b Summarise, in one sentence, each of the patterns shown.

6 From Figure 7.6:

a Describe briefly the modifications and impacts made on the natural environment by each of the systems.

b Use the examples to define sustainable agriculture.

300

Figure 7.7 *Digging an irrigation canal on a fruit and rice farm in the Mekong Delta, Vietnam*

questions will help you to understand the agricultural system:

■ What is the balance between subsistence and commercial elements? For example, are a few products (eggs, fruit) sold in a local market, or is there a regular product such as coffee, bananas, or cattle, that is sold through a formal regional or national marketing system such as a co-operative, or by contract with a transnational company?

■ How much land does a family have, and how do they use this land? Most peasant holdings are small (frequently less than 2ha is cropped in a given season), and complex **intercropping** on very small plots is common. In shifting-cultivation systems, plots are cleared, tilled for a few years and then left **fallow**, e.g. the Amazonia example in Figure 7.6.

■ What access and rights does a family have to land and water resources? In many traditional societies, land is communally owned – individuals may have certain rights to their cropland, while grazing land is communally used. In seasonally dry regions where irrigation is required, the control of, and access to, water is a crucial factor. In communist countries, little land may be in individual private ownership.

■ What is the source of labour and how much is needed? Peasant agriculture is labour-intensive, and the family provides most of the inputs (Figure 7.7). As families, techniques and farming types change, will this labour source be available and be needed?

■ How is the system adapted to the local environment and how do families respond to fluctuating conditions? Farmers and herders have a close understanding of the local environmental character. They adjust their cropping and grazing systems to fit this character (Figure 7.8) and adapt to year-on-year climatic or market fluctuations. Problems which occur in resettlement schemes may be due to the new settlers not having this close understanding of the environment – an example is the failure of some projects in the Amazonian rainforests of Brazil.

Figure 7.8 *Soil–agriculture relationships at Ukiriguru, Tanzania. Farmers adjust their cropping patterns according to changes in soil and water conditions across the slope.*

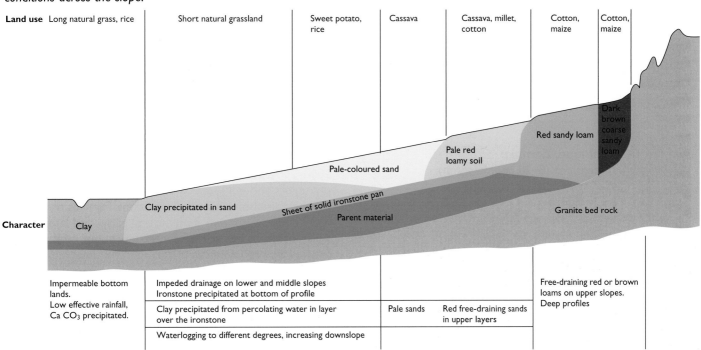

Land use	Long natural grass, rice	Short natural grassland	Sweet potato, rice	Cassava	Cassava, millet, cotton	Cotton, maize	Cotton, maize

Pale-coloured sand
Pale red loamy soil
Red sandy loam
Dark brown coarse sandy loam

Clay precipitated in sand
Sheet of solid ironstone pan
Parent material
Granite bed rock

Character — Clay

Impermeable bottom lands.
Low effective rainfall, Ca CO₃ precipitated.

Impeded drainage on lower and middle slopes
Ironstone precipitated at bottom of profile

Clay precipitated from percolating water in layer over the ironstone

Pale sands | Red free-draining sands in upper layers

Waterlogging to different degrees, increasing downslope

Free-draining red or brown loams on upper slopes.
Deep profiles

231

▼ **Figure 7.9** *Conditions and processes influencing sustainable agriculture (Source: Whiteside, 1998)*

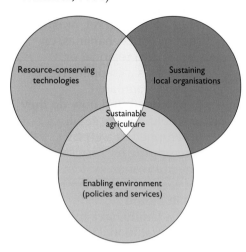

It is frequently claimed that traditional peasant systems are sustainable. However, it is clear when we apply the five questions above to individual situations that they are changing rapidly. A combination of continued population growth, rising expectations, continuing inequality of access to land and water resources and the increasing commercialisation of agriculture is threatening this sustainability. The seriousness of these threats becomes evident when we think carefully about the following definition of sustainable agriculture, which emphasises changing needs and expectations:

'Agriculture which meets today's livelihood needs without preventing the needs of neighbours or future generations from being met; this is achieved by the continuous efforts of men, women and children to adapt complex rural livelihoods to a changing environment, so as to protect and enhance the stocks of natural, physical, human and social 'capital' available to themselves and to future generations' (Whiteside, 1998).

The achievement of sustainability in these terms requires the successful interaction of several sets of processes (Figure 7.9). The difficulties involved are illustrated by the example of Southern Africa.

EXAMPLE: Southern Africa

Smallholder farming in Southern Africa

The ten countries of Southern Africa (Figure 7.10) vary in wealth and stage of development, but all are essentially rural, with high proportions of their population dependent upon agriculture (Table 7.2). The population of the region is growing at 3 per cent a year, and, although this growth rate is falling, the number of people is likely to double in the 1998–2018 period. Conditions and pressures vary across the region.

Agriculture remains dualistic, with a relatively small number of large commercial farms and a large number of smallholder (peasant) farms, having an emphasis on subsistence farming and trade locally. Typically, the commercial farms are white-owned, occupy the more favourable land, are mechanised and use modern techniques.

▼ **Table 7.2** *People, land and agriculture in Southern Africa*

Botswana, Namibia: Sparse population (2 people/sq.km);   dry environment.
Angola, Mozambique, Zambia: Low population densities (8–20 people/sq.km);   seasonal rainfall, long dry season.
South Africa, Zimbabwe: Moderate population densities (25–35 people/sq.km);   seasonal rainfall, long dry season;   very unequal land distribution leading to 'land hunger'.
Lesotho, Malawi, Swaziland: High population densities (45–120 people/sq.km);   seasonal rainfall, long dry season;   acute land pressures.

The smallholder farms in Southern Africa have been described as **Complex, Diverse and Risk-prone** (CDR):

- **Complexity**: farmers grow a large variety of crops using techniques such as intercropping (Figure 7.11) and planting in sequence by adapting to minor differences in the local environment. Maize is the main food crop. Products are sold according to immediate needs – for example, selling chickens in order to buy seed.

- **Diversity**: There is great inequality in income. In Zimbabwe, the top 10 per cent of households make 42 per cent of total income for the country while the bottom 25 per cent have only 4 per cent. Even within individual communities, the balance of crops, livestock and off-farm employment may vary widely, as does the expertise and the commitment of farmers.

- **Risk**: Holdings are usually small (0.5–2.0ha using hoe cultivation; 1–10ha if animals or a hired tractor are used for ploughing and tilling). Thus, families are at risk from weather (serious droughts like that in 1995–97, 2000–2002, and the floods in Mozambique in 2000), war (in Angola, and in Mozambique, for example), pests, robbery, disease, and price fluctuations. Families try to reduce risk by growing a mixture of early- and late-maturing crops because of uncertainty about rainfall, or by keeping animals in different places.

Farmers understand their local environments and are also opportunistic, in that they adapt quickly to changing conditions and seasonal fluctuations. None the less, the CDR nature of farming means that rural poverty is acute

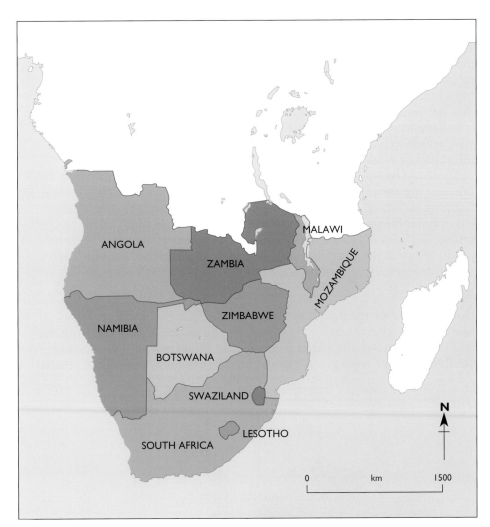

◄ **Figure 7.10** *The ten countries of Southern Africa*

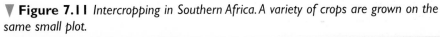

7 What benefits does intercropping have, and in what ways does it illustrate the labour-intensive nature of peasant farming?

8 Write a brief article for a newspaper or magazine, summarising why the sustainability of agriculture in Southern Africa is threatened. (Choose a 'punchy' headline; use a software package to design your article, if one is available.)

▼ **Figure 7.11** *Intercropping in Southern Africa. A variety of crops are grown on the same small plot.*

throughout the region. Continuing population growth and restrictions on access to additional land threaten long-term sustainability. Farmers inevitably make decisions based on short-term needs. For instance, in southern Malawi, population pressures and land shortages are particularly acute. Farmers react by reducing the fallow (rest) periods on their land, and by switching to maize from the traditional sorghum and millet. Maize yields can be higher, but it is less resistant to drought. Many smallholders are forced to grow almost continuous maize. Reduced fallow periods mean less animal grazing and reduced organic dung inputs to the soils. Only a small number of the better-off farmers can afford to buy fertiliser. Others are trapped in a downward spiral of reducing fertility and falling yields.

Seeking sustainability

The typical response to population growth and rising expectations is to increase production by **intensification** and **diversification**: grow more crops; rear more animals; sell more produce. If production grows more rapidly than population, then quality of life should improve – providing that the benefits extend to all members of a community. A second response is to increase income from off-farm employment: family members take paid jobs, which may involve short-term or long-term migration such as the migration of workers to the mines in Zambia and South Africa.

There are two ways for a rural community to increase output: first, by intensifying the use of existing land and raising its **productivity**; second, by bringing new land into use. Both approaches increase demands upon environmental resources, and the crucial question we must ask of any project is: 'Is it environmentally sustainable?' That is, will the character and productivity of the resource base be sustained over time? (Figure 7.12)

Two examples from the tropical rainforests of Amazonia illustrate this issue:

Brazil

During the 1980s the Brazilian government supported large-scale resettlement schemes in the Amazonian state of Rondonia, following the opening of **penetration roads** (Figure 7.13). Hundreds of thousands of landless families were given plots of land in the rainforests. They cleared

▲ **Figure 7.12** *In seasonally dry regions, one way to improve yields is to increase water supply. When water is drawn from a groundwater store, is the aquifer being replenished by percolation, or is it 'fossil' water, stored from a more humid climatic period? If it is 'fossil' water then it is not 'sustainable' over time.*

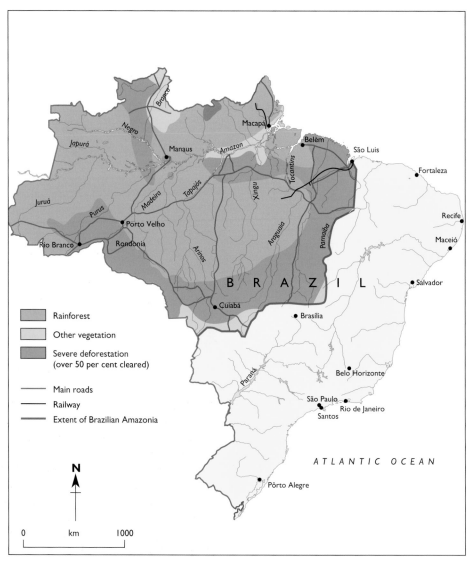

► **Figure 7.13** *The network of roads and railways in the Amazonian rainforest of Brazil shows the extent of the resettlement schemes supported by the Brazilian government.*

Figure 7.14 *Much of the land cleared in the Amazon rainforest for farming has been abandoned ten years after.*

9 Read the text on Rondonia (Brazil) and Ecuador.
a Give two significant differences between the farming techniques in Rondonia and Ecuador.
b Assess the examples from Rondonia and Ecuador in terms of the three dimensions of sustainability.

plots, but within five years around one-third had abandoned them, leaving a severely degraded environment (Figure 7.14). Reasons for the high failure rate are complex: no assessment of soil character was made by the government, the plots were too small to allow shifting cultivation (about 5ha), many settlers came from north-east Brazil and had little experience of the agricultural techniques required in a tropical rainforest or the types of seeds and crops which would survive. As a result, crops failed and the settlers moved on, to clear yet more forest. Estimates suggest that 15 per cent of Rondonia's rainforests were cleared between 1980 and 1995.

Ecuador

Over the past 25 years, increasing population pressures in the highland regions of Ecuador have caused increasing migration into the sparsely populated rainforests in the east of the country. Here, the average size of a farm holding is 40ha. The Ecuador government requires settler families to keep at least 50 per cent of their land under forest. Most of the migrants are experienced farmers and the failure rate is only 10 per cent. Successful farmers tend to pass through three phases:

1 **A coping phase**: using 'slash-and-mulch' techniques, the family clears plots and grows subsistence crops, using existing knowledge.
2 **An experimental phase**: new methods and crops are tried, and cash crops are introduced, especially coffee. This involves extending the cleared area and buying simple equipment like chain saws and ploughs
3 **A selection and adoption phase**: The family use their experience to choose the best combination of crops and animals. A crucial shift is the increase in cattle numbers, as a commercial element in the farm economy.

This appears, therefore, to be a success story. But in the medium-term, continued economic growth from increasing commercial coffee and cattle production may bring longer-term environmental threats from progressive forest clearance and pressure on fallow land.

Approaches to rural development

In many LEDCs, the process of change to agricultural systems has been assisted or even introduced by scientific, technical and financial inputs from MEDCs. This assistance ranges from large-scale programmes run by the United Nations Food and Agriculture Organisation (FAO) and rich foundations (e.g. the Ford Foundation) to local projects organised by voluntary agencies such as church-based charities. Over the past 40 years the emphasis has shifted from a **top-down** approach to a **bottom-up** approach. Three examples illustrate this shift:

■ **The technology–transfer approach**: This has been based on the belief that MEDC scientific methods are best – for example, the mechanisation and layout of farms, breeding programmes for animals and crops, the use of inorganic fertilisers and sprays, and so on. It was the dominant approach during the 1960s and 1970s, and it still affects agricultural systems throughout LEDCs. The most famous example is the so-called 'Green Revolution', which has been described by Raw and Atkins (1995) as 'the most significant technological advance in food production in the last 35 years' (Box 2). The most recent trend has been to extend the approach to other staple crops such as millet and sorghum, using genetic modification (GM) techniques.

■ **The 'visit and advise' approach**: In the 1970s the emphasis shifted towards working with communities to develop a more sensitive understanding of local environments, communities and systems. In this approach, teams of experts – agriculturists, botanists, soil scientists, hydrologists, economists, sociologists – study an area and its people. They carry out an audit of what is there and what is happening. From this audit they make recommendations and give advice to either the government or to the local community. The advice can involve both simple modifications to practices and fundamental shifts in the agricultural system (Figure 7.15a). The advice given may include the introduction of new technology and high-yielding cereal varieties.

■ **The 'farmer first' approach**: This involves 'bottom-up' strategies where local communities have 'ownership' of development projects. Once again, external advisers help the decision-making, suggesting alternatives, advantages, problems and sources of funding. In addition, they may provide training in appropriate organisational and management skills. This community-based approach has become increasingly popular since the mid-1980s. One of its main characteristics is that it attempts to balance economic benefits with social cohesion and environmental conservation – in other words, it takes into account the three main dimensions of sustainability. (Figure 7.15b)

Box 2

Technology transfer – the Green Revolution

■ The aim has been to improve crop productivity in LEDCs through the breeding of new high-yielding varieties (HYVs) of the staple cereals, wheat and rice. First introduced in the 1960s, the yields of HYVs are more than double those of traditional varieties (TVs). By 2000, HYVs produced two-thirds of Asia's rice output. HYV plants are stronger and so resist storm impacts, and have a shorter growing period, allowing up to three crops of rice a year where water is available. However, HYVs are susceptible to diseases and pests, so they require inputs of chemical sprays. Increased cropping intensity means more fertiliser and greater irrigation. One result of these additional demands has been that better-off farmers in communities have benefited most.

► **Figure 7.15a** *'Visit and advise':*
Nepalese hill farming

In the Likhu Khola drainage basin of
Nepal, in the foothills of the Himalayas
(altitude 500–1300 m), communities have
developed complex terrace systems. On
the lower slopes, khet terraces are
irrigated for intensive rice production;
across upper slopes, rain-fed bari terraces
grow mainly millet and maize. In recent
years, slope failures, gullying and river
flooding have become more frequent,
destroying parts of the terrace systems. In
1993, a team of British geographers,
funded by the Overseas Development
Agency (ODA), carried out research in the
area. They found that population pressures
were causing farmers to extend new bari
terraces across higher, steeper slopes,
involving deforestation and an increase in
exposed surface area. It was this exposure
and the resulting rapid runoff during the
heavy monsoon rains that were the
primary causes of the accelerated erosion.
From their findings, the researchers have
been able to make recommendations to
the local communities.

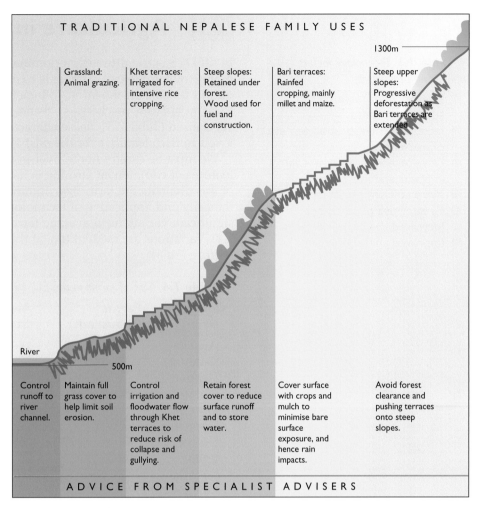

► **Figure 7.15b** *'Farmer first':The Community Baboon
Sanctuary, Belize.*

As the population grows and the agricultural area expands, the
tropical rainforest of Belize, is being cleared. In one district, a
group of eight villages have set up a nature reserve for rare Black
Howler monkeys (known locally as 'baboons'). Following advice
and support funding from the World Wildlife Fund for Nature
(WWF), a management committee of local people organises the
conservation of rainforest corridors along watercourses as a
habitat for the monkeys. Villagers have been persuaded by the
idea that 'wildlife pays, so wildlife stays' – **ecotourism** pays for
guided tours to view the monkeys and to stay in farmers' homes.
Young local men earn money as guides. In 2000 there were
7000 visitors. The management committee makes decisions about
the distribution and use of profits. This is a good example of
'bottom-up', community-based development where farming,
conservation and economic development co-exist.

10 There are three approaches to rural development.
a Write one-sentence definitions for each of the three
approaches to rural development and assess each of
the approaches in terms of the three dimensions of
sustainability.
b To what extent is diversification an element of the
rural development process?

7.4 Agriculture in MEDCs

In MEDCS, agriculture is predominantly commercial in nature. The primary motive of commercial farmers is to maximise profits, with most of the output for sale off the farm. To benefit from economies of scale, farms tend to be large-scale, mechanised, technologically advanced, and relatively specialised (Table 7.3). These characteristics have enabled productivity levels to rise steadily over the past 40 years.

We must be careful not to over-generalise, because one of the key features of commercial farming is its diversity. Three important dimensions of this diversity are farm size, dominant-enterprise type, and farming intensity and application of technology. At a national scale, there are significant variations in average farm size across the European Union (Table 7.4). Variations are evident too, at the regional scale, such as in France (Figure 7.16a).

▼ **Table 7.3** *Technology factors influencing productivity*

- Investment in research and advisory services
- Increased sophistication and use of machinery
- Development of disease control in plants and livestock
- More efficient use of fertilisers
- Greater use of pesticides
- Increased use of hybrid seeds
- Improved animal breeding.

314

▼ **Table 7.4** *Size of farms in the EU, 1995. (Source: Eurostat, 1998)*

Country	Farms (thousands)	Average size (ha)	Percentage under 2ha	Percentage over 20ha
Austria	255	14	13	30
Belgium	80	18	20	29
Denmark	78	36	2	59
Finland	185	15	11	21
France	890	34	13	50
Germany	630	29	16	34
Greece	900	6	42	4
Ireland	160	28	2	48
Italy	2400	8	52	6
Luxembourg	4	34	13	56
Netherlands	120	16	16	31
Portugal	570	8	54	5
Spain	1400	18	33	15
Sweden	90	38	–	45
UK	235	70	5	60
Total	**8000**	**18**	**34**	**19**

Figure 7.16 *Regional variations in French agriculture (Source INSEE, 1998)*

300

▼ **a** *average farm size in France*

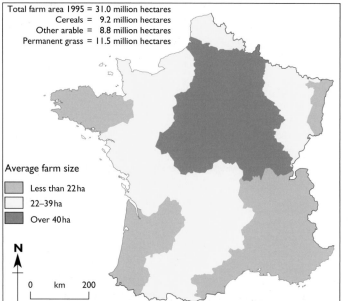

Total farm area 1995 = 31.0 million hectares
Cereals = 9.2 million hectares
Other arable = 8.8 million hectares
Permanent grass = 11.5 million hectares

Average farm size
- Less than 22 ha
- 22–39 ha
- Over 40 ha

▼ **b** *Percentage of arable land in France*

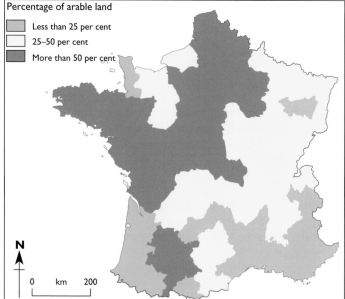

Percentage of arable land
- Less than 25 per cent
- 25–50 per cent
- More than 50 per cent

11 From Figure 7.16a and Figure 7.16b, write a brief description of the regional variations in French agriculture. (Use an atlas to help you with regional names, etc.)

12 From Table 7.4:

a Rank the countries of the EU in terms of average farm size (column 3).

b Divide the 15 EU countries up into three classes according to farm size distribution (columns 4 and 5): mainly large farms: diversified farm sizes; mainly small farms.

c In two or three sentences, summarise the structure of EU farming in 1995 (use the 'Total' row of data).

d What is distinctive about the UK farm structure?

Distinguishing farm types by their dominant enterprise (their main products) also produces well-defined patterns at scales from countries to individual farms. The simplest classification is between arable crops and animal production, usually measured by product value (Figure 7.17). Even within one dominant enterprise type, there is considerable diversity in terms of farming intensity and application of technology, e.g. animal rearing varies from low-tech, extensive 'range' grazing to high-tech, intensive feedlot production.

▼ **Figure 7.17** *US farm types by product value in the top five agricultural states (Source: US Annual Abstract of Statistics, 1997)*

314

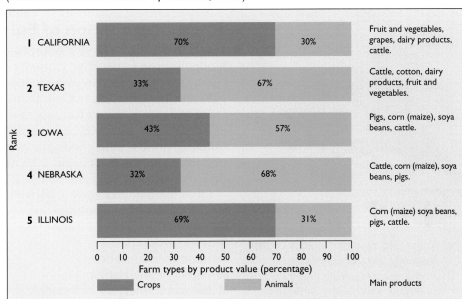

▼ **Figure 7.18** *Changes in major crops, Saskatchewan, 1981–99*

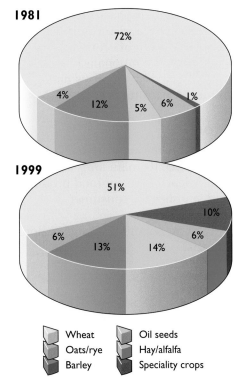

(Source: from Seaborne, A.*Geography*, 86[2],2001)

Factors influencing agricultural systems

It is clear that social, economic, political and technological factors exert strong influences upon farming practices – size of holding, tenure type, skills and motives of farmers, market trends, government policy, availability and application of technology. Yet all these must be fitted within an environmental context. Even rich, technologically advanced countries such as the USA remain vulnerable to fluctuations in environmental conditions. The impact can be seen in North America. In the USA, agricultural land has fallen from a peak of 1160 million acres in 1950 to 930 million acres in 2000. Arable land fell by 10 per cent during this period although production rose. Losses were greatest in the eastern states. In contrast, cropland has been expanding in the Mississippi valley, Florida, the Central Valley of California and the mountain North-West as a result of improved water management and irrigation techniques. Even productive, specialised regions such as the Canadian Prairies have changed. For example, in Saskatchewan the fall in world wheat prices is the main reason for the fall in planted area, from 8.6 million hectares in 1981 to 7.5 million hectares in 1999. One response has been towards diversification, for example, to oil seeds and speciality crops, although grains still dominate (Figure 7.18).

Despite technological advances, environmental characteristics remain important. Physical character, climate and weather, soil type, water availability, ecological elements (pests, predators) combine to influence farmers' options and decisions. For example if we examine the pattern of dominant enterprises in England and Wales (Figure 7.19) there is a general relationship with physical and climatic conditions. Grass/livestock farming dominates in the wetter, often higher, cooler regions of the west and north;

▼ **Figure 7.19** *Main farm enterprises in England and Wales (Source: Raw and Atkins, 1995)*

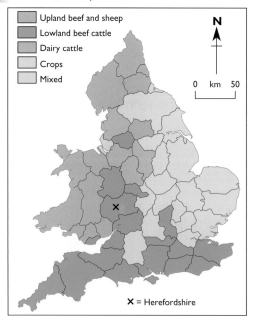

X = Herefordshire

---

13 Use Figure 7.19 and an atlas.

a Name two counties in each of the dominant enterprise classes.

b By the use of specific county examples, illustrate significance of altitude and climatic characteristics as two of the factors influencing agriculture.

---

crop-based systems dominate in the drier, sunnier, generally lower regions of the east. In the UK, uplands above 350 m are generally unsuitable for cultivation: precipitation is greater than 1500 mm a year and the growing season is less than 200 days. Figure 7.24b (see page 244) shows this altitudinal land use gradation in Wales.

Remember however, that this is a very general and simplified relationship. There is considerable variation within a particular region. For example, Herefordshire (x on Figure 7.19) is classified as an area of lowland beef and sheep, but has over one-third of its farmed area under crops (although a proportion is for animal feed). There are also specialised orchards and hop farms.

The modernisation of European agriculture

Two parallel trends have modernised European farming systems: First, rationalisation forces have caused a progressive loss of small peasant farms and a steady increase in average farm size. For example, in France between 1950 and 1990, holdings of less than 20ha declined from 2 million to 750 000.

Second, productivity forces, generated by advances in science and technology, have caused progressive rises in crop and livestock yields. For example, annual milk yields per dairy cow in the EU increased by over two-thirds between 1960 and 1995 and, overall, EU farm output more than doubled between 1955 and 1995.

From the data in Figure 7.20, we can summarise the main outcomes of this modernisation revolution as:
- larger farms/fewer farms
- fewer workers
- more mechanisation
- higher capitalisation
- increased intensity of inputs (e.g. fertiliser, drainage, irrigation)
- increased intensity of production and outputs
- greater specialisation
- increased productivity
- new agricultural landscapes.

As we consider this list, three important points need emphasising. First, changes vary widely across the EU, and a diversity of agricultural systems and landscapes remain. Second, the changes have brought economic, social and environmental costs as well as benefits. Third, the rationalisation and productivity forces have been energised and directed by vigorous national government and EU policies (e.g. the Common Agricultural Policy (CAP)).

The CAP of the European Union

The Common Agricultural Policy (CAP) provides the support and control framework for rural economies throughout the EU. Its main aims are:
- To increase productivity and attain self-sufficiency in major food products.
- To assure price stability and regular food supplies for consumers.
- To ensure economic viability for farmers.
- To sustain rural communities, especially in more remote, marginal regions.
- To protect EU farmers from cheaper imports.

Strategies

- Price support (which in 1995 was approximately 90 per cent of total expenditure); subsidies and grants
- identification of 'less favoured' regions which qualify for additional support
- production control through **quotas**
- shifting subsidy levels; management directives and guidelines (including environmental and hygiene regulations).

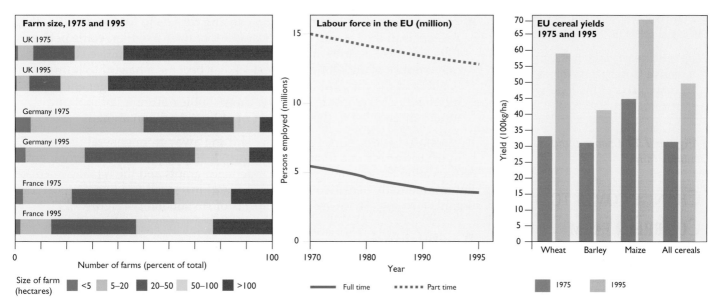

▲ **Figure 7.20** *Changes in European agriculture, 1975–95*

14 From the data in Figure 7.20, give three examples which illustrate the modernisation of agriculture in Europe during the 1975–95 period. (Look at the graphs carefully, to extract precise figures for your examples.)

15 From the Common Agricultural Policy (CAP):

a For each dimension of sustainability – economic, social, environmental – give one example in the CAP that has targeted that dimension

b What evidence is there that the CAP has achieved economic sustainability goals, but at the expense of threats to environmental sustainability?

Outcomes

The CAP has been so successful in raising productivity and outputs that, by the mid-1980s, there was over-production of several major products (e.g. cereals, potatoes, dairy products, beef). In response, the EU bought and stored the surplus output (e.g. the highly publicised butter and grain 'mountains'); imposed quotas (e.g. on milk production) and established the Set-Aside programme where farmers are paid to take land out of production (e.g. until 2002, farmers with more than 40ha arable could set aside 15 per cent for non-production).

Criticisms of the CAP

■ **Cost:** until the mid 1990s, at least 60 per cent of the total EU budget was spent on the CAP. In 1995, English farms received an average of £14 000 in support payments. By 2001 the cost of the CAP had been reduced to around 50 per cent of the total budget. In 2002, UK farmers received at least £5 billion from the CAP.

■ **Support of inefficient and marginal farms:** small farms are still numerous in several EU countries and are the basis of rural society (e.g. Greece, Italy (Table 7.4)). They cannot be economically viable, but attract high support payments in order to sustain rural communities. UK hill-farming districts, dependent upon extensive sheep grazing, are also classed as economically 'marginal'. They have been heavily supported, partly to conserve attractive landscapes such as in National Parks and **Areas of Outstanding Natural Beauty** (AONBs) .

■ **Larger, wealthier farmers benefit most from the support system:** for example, the so-called 'grain barons' of eastern England, with large cereal hectarages, can afford high fertiliser, pesticide and mechanical inputs to achieve high yields, claim full support prices as well as **Set-Aside** payments. In 2001 subsidies for wheat were £247 per hectare. More than 20 large arable farmers in East Anglia received over £500 000 a year, and six agribusinesses more than £1 million.

■ **Overemphasis on productivity and profits at the expense of environmental quality:** the removal of more than 40 per cent of the UK's hedgerows and the increased use of agrochemicals and autumn ploughing and planting, have reduced the habitats for a number of bird species.

Key Terms

Agribusiness: Large scale, capital-intensive farming, sometimes linked to other parts of the food system.

▼ **Figure 7.21** *Change comes to a Cumbrian farm*

CHANGE COMES TO A CUMBRIAN FARM

MIKE CARTNER HAS A 200-acre dairy farm in Cumbria, west of Carlisle. He says of the CAP, 'It puts false value on the price of the final product, encourages farmers to overstock, thus damaging the environment. ... But the more the subsidy, the more animals you need to make ends meet because, with oversupply, prices are depressed.'

In response, in 1998 he converted to organic farming. The idea was to farm less intensively because milk from organic-fed cows commands higher prices.

'I was getting on a treadmill and would have had to double the size of my herd to get the necessary value for the business. The animals are better on an organic farm, and you don't need to put nitrogen on the fields to make them grow.'

He received a grant of £28 500 to help with the organic conversion, and receives special 'environmental management' grants to turn some of his fields back to hay meadows, and restore hedgerows to encourage wildlife. He lost his herd in the 2001 foot-and-mouth outbreak, but compensation helped him to restock. In 2002 he was producing 1000 litres a day from his 72 cow herd, and was paid 19p/litre.

(Adapted from The Guardian, 11 July 2002)

■ **Encouragement of intensification causing threats to human health**: the BSE crisis in the UK caused the slaughter of over 2 million cattle between 1996 and 1999; in 1999, a government Food Safety Committee claimed that overuse of antibiotics in livestock has become a hazard to human health and the continuing debate concerning the effects of genetically modified (GM) crops. In 2001, the catastrophic outbreaks of foot-and-mouth disease led to the slaughter of at least 4 million animals.

In response to criticisms, policies are shifting away from general price support towards more focused targeting, such as support for environmental conservation through hedgerow-maintenance grants and the establishment of Environmentally Sensitive Areas (ESAs) and increased funding support for organic farming. These are other aspects of sustainability (Figure 7.21).

Agribusiness

Agribusiness can be defined as 'large-scale capital-intensive farming which is sometimes linked to other parts of the food system'. This definition includes the large cereal farmers of eastern England. Another excellent example is Bernard Matthews, an East Anglian farmer. He developed an intensive turkey-rearing enterprise, then extended into the processing and marketing of his own brand through popular teevision commercials.

Companies and transnational corporations (TNCs), rather than individual landowners are increasingly involved in agribusiness. For example, in the USA, two-thirds of the fruit and vegetable output of California and the feedlot beef of Arizona and Texas is produced by companies and corporations (Figure 7.22). Agribusiness is widespread too, in the LEDCS. Two huge US-based TNCs control banana production in Central and South America. European TNCs such as Unilever control all stages of production from agrochemical inputs, through land ownership, to processing and marketing e.g. oil-palm plantations in Africa and south-east Asia. In total, Unilever owns plantations covering more than 100 000ha, and is the world's largest producer of margarine (Flora) and ice-cream (Walls).

► **Figure 7.22** *An intensive cattle feedlot in Arizona owned by a large meat production corporation*

Key Terms

Dispersed settlement: Individual homes scattered across an area.

Nucleated settlement: The clustering of homes into villages.

7.5 Rural settlement systems

Thinking about rural settlements

One thousand years ago, the multi-roomed circular structure at Bandelier National Monument, New Mexico (Figure 7.23) was home to at least 30 Anasazi families. Its location and form were determined by four sets of influences:

- **Environmental:** Situated on a flat canyon floor, with water supply, and trees to provide shade from summer heat and shelter from winter storms.
- **Social/cultural:** Social and religious life centred around extended family groups. The purpose of the central open space, enclosed by the buildings, was to provide the setting for this interaction. The clustering also offered protection.
- **Economic:** Irrigated agriculture could support a sizeable community within a small area, so that people could walk to and from the family plots.
- **Technological:** The ability to use stone, timber and to make mud bricks allowed the construction of the two-storey, multi-room dwellings.

A significant change in one or more of these factors could trigger changes in the settlement. In this case, the settlement was abandoned around 800 years ago, when the climate became drier, agricultural output declined, and the canyon could no longer support a village population.

We can understand rural settlements, therefore, in terms of the natural environments and human contexts in which they have evolved. Remember, the location, form and distribution we see today may be a result of past influences and conditions, not of those at work today. Unlike Bandelier, most rural settlements persist and adapt as conditions change.

▼ **Figure 7.23** *An Anasazi settlement at Bandelier National Monument, New Mexico, USA*

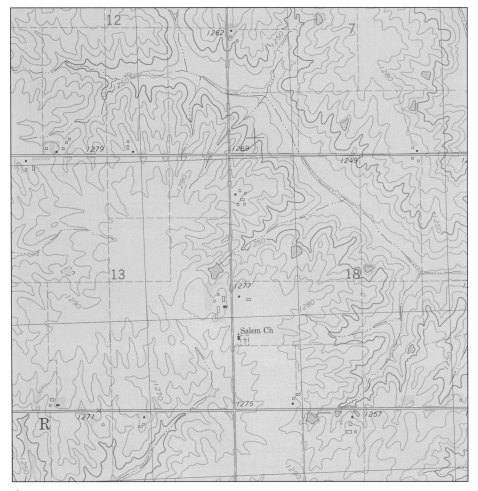

▲ **Figure 7.24a** *Dispersed farm settlement in the US Mid-west*

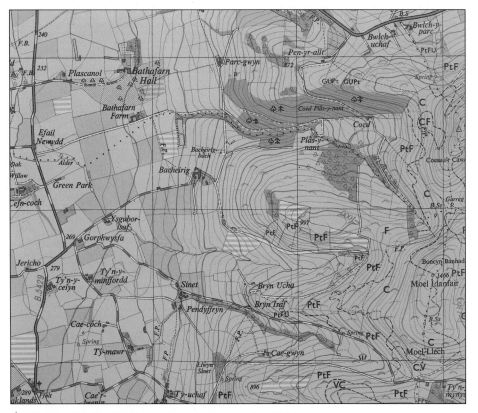

▲ **Figure 7.24b** *Dispersed farmsteads in upland Wales*

Types of rural settlement

The changing interaction of natural and human influences, frequently over long timescales, has created a wide diversity of settlement types and patterns. Yet two fundamental types can be identified: **dispersed** (Figure 7.24a and Figure 7.24b) and **nucleated** (Figure 7.24c and 7.24d).

The four map extracts, all at the 1:25000 scale, illustrate the diversity of rural settlement in terms of location and form. We can 'read' the maps and, with not too much difficulty, answer the 'Where?' and 'What?' questions. However, a more careful reading helps us to answer some of the 'Why?' questions. In each example, the answers lie in a distinctive interrelationship of natural and human environmental factors over time.

Dispersed settlement: the US Mid-west (Figure 7.24a)

In the mid-nineteenth century, US government surveys superimposed a grid system on this flat, fertile landscape, as immigrants moved west. Each square on the grid is a 640-acre 'section', which is the landholding unit to which a settler family could lay claim and build a home. This has become the productive Corn Belt. Over time, some sections have been divided into half and even quarter sections (although some have become part of larger units as a result of rationalisation).

Dispersed settlement: upland Wales (Figure 7.24b)

The topography of much of Wales consists of long valleys cut into a series of uplands and mountains. Over many centuries, farmers have adopted a livestock-rearing system which makes the best use of the valleys, the slopes and the uplands. The location of the farmsteads and the road/track network reflect this three-component land-use system.

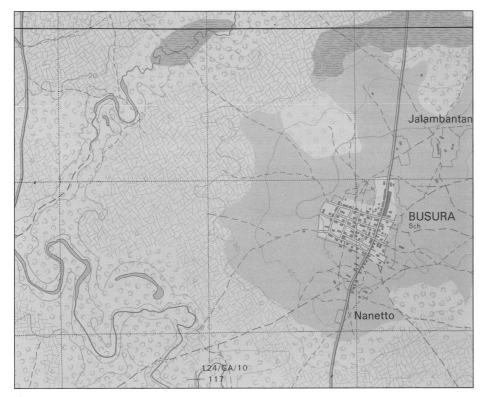

▲ **Figure 7.24c** *Clustered nucleation in The Gambia*

▲ **Figure 7.24d** *Linear nucleation in the Netherlands Polders. The buildings are shown in red.*

Nucleated settlement: The Gambia (Figure 7.24c)

The topography of much of The Gambia consists of low, broad ridges above shallow, seasonally floodable valleys. The climate is tropical with a wet-season and a dry-season. The natural vegetation is forest. The village location and the land-use pattern are closely adapted to the variations in terrain and drainage. The land is productive, and families can cluster in villages yet still have access to their small plots and the forest resources.

Nucleated settlement: the Netherland Polders (Figure 7.24d)

Polders have been created by progressive wetland reclamation and massive sea-dyke, land-extension schemes. Today they are among the most productive and intensively farmed landscapes in Europe, with high rural population densities. Many polders are below sea level and are protected by tall dykes. These are the highest elements in the landscape and exert a strong control on the settlement pattern and farm location.

16 For each settlement map:
a Describe the location and form of the settlements
b Outline the relationship between the settlement and the land-use pattern.
c Suggest the main factor influencing (i) the location and (ii) the form of the settlement.
d From the information given in the maps, discuss this statement: 'The location and form of rural settlements are determined by the human response to environmental resources.'

Key Terms

Settlement hierarchy: The classification of settlements into a set of levels or orders, according to their size and functional range.

Settlement systems

Individual settlements do not exist in isolation, but as elements in a functional system: farms and hamlets are linked to villages; villages use local towns for markets and services; larger towns have a sub-regional role, and so on. This system has been formalised into a **settlement hierarchy**, with rural settlements forming the lower levels or orders (Figure 7.25).

Settlements which provide services for a surrounding area are known as central places. They have **spheres of influence** related to the number, range and quality of services they offer, and to the competition from other central places. Hierarchies evolve and change over time, and good examples may be found in long-settled societies such as in the UK (see the Welsh Marches example).

► **Figure 7.25** *The principles of settlement hierarchies*

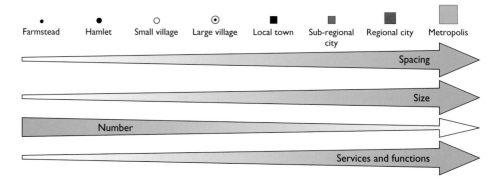

EXAMPLE: The Welsh Marches

The Welsh Marches

The Welsh Marches are the border zone between rolling mixed farming landscapes and more extensive livestock rearing over the Welsh uplands. A settlement pattern of nucleated villages has evolved, on average 3–5 km apart (e.g. Richard's Castle, Orleton, Luston). Farms were clustered in these villages, with their fields generally within 2 km – walking distance. As population increased and further land was enclosed and improved in the seventeenth and eighteenth centuries, a scattering of hamlets and dispersed farmsteads emerged. A set of local market towns, 5–10 km apart, serviced these lower-order settlements, the spacing being determined by the limits of a day's travel (Figure 7.26). Livestock and produce were sold and bought in these towns (e.g. Ludlow, Tenbury Wells, Leominster). The sub-regional centres of Hereford, Shrewsbury and Worcester, some 30–50 km distant, provide higher-order services and administration. This traditional hierarchical structure is becoming less well defined as mobility increases, lifestyles change, and service provision, businesses, and so on, become reorganised (Figure 7.27).

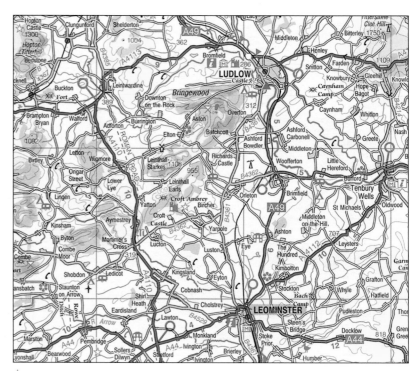

▲ **Figure 7.26** *The Welsh Marches area showing the local market towns of Ludlow, Leominster and Tenbury Wells*

Figure 7.27 *Changing shopping patterns and their effects on traditional market towns are reflected in this newspaper report (Source: Ludlow Journal, 27 August, 1999)*

VOW TO FIGHT A SUPERSTORE

Supermarket giants Sainsbury's or Tesco have been revealed as the possible mystery buyers of the cattle market in Tenbury Wells, a site which could be home to a massive new superstore. But a local action group opposed to the plans claims that the giant superstore would force local shops out of business and create traffic chaos in the town.

Developers Chelverton West Ltd, of Bristol, submitted an application for outline planning permission to build a supermarket covering 21,000 square metres.

Mr Ben Bydawell, the spokesman for the local action group opposed to the plans said: 'Tenbury is a small, unique market town, with its own range of specialist shops, and it likely that a lot of them will go out of business if these plans go ahead.'

He said that a traffic survey carried out by the developers showed that the new superstore would bring in an extra 3,000 cars, making two-way journeys, to the site each week.

'The local plan points out that because of the town's narrow roads, Tenbury has already reached the limit of its development potential.'

7.6 Changing rural Britain

Table 7.5 *Some problems in rural England in 2000*

In 2000, 1 in 10 people lived in a village, but:

- 74% had no post office
- 72% had no village shop
- 73% had no youth groups or clubs
- 29% had no bus service
- 92% had no doctor's surgery
- 48% had no community meeting place
- 91% had no community internet facility.

(Source: Rural Services in 2000, Countryside Focus, Dec 2001/Jan 2002)

Environments and landscapes tend to evolve via periods of dynamic equilibrium (periods of gradual adjustment), interrupted by surges of accelerated change. During the past 50 years, rural Britain has been experiencing one such surge, energised by the modernisation of agriculture, the decentralisation of cities, the redistribution of economic activity, and the emergence of more mobile lifestyles (see chapter 6). The twin environmental impacts have been the creation of new agricultural landscapes (Figure 7.28) and what has been called the 'suburbanisation' of the countryside (Figure 7.29). For instance, villages in more accessible, prosperous regions have experienced strong development pressures and growth, while communities in more remote regions have declined (Table 7.5 and the Orleton example, page 248).

Government forecasts show that England alone will need 4 million additional homes by 2016. At least 50 per cent of these are likely to be on greenfield sites. This will put increasing pressures on rural communities and environments, especially in prosperous regions such as the South-East (Figures 7.29; 7.30).

A related issue is affordability. In the South-East region, housing shortages force prices up. In more remote but attractive rural areas, prices rise as the limited housing stock is bought by affluent 'incomers', e.g. retirees. In its 2002 *State of the Countryside* report, the CA claimed that in seven out of eight English regions, homes in rural areas were less 'affordable' than in urban areas. Their index or measure of 'affordability' uses a combination of average earnings, average house prices and the standard rules for approving mortgages. In rural areas, 57 per cent of new house buyers have to commit more than one-half of their income; in urban areas the figure is only 32 per cent.

▼ **Figure 7.28** *Increased pressure on prosperous regions in the South-East (Source: The Guardian, 16 July 2002)*

LEAFY VILLAGE ENGULFED BY SUBURBIA

TEN YEARS AGO Sevington was an isolated hamlet deep in the Kent countryside, home to only 100 people.

Today the nearby town of Ashford has enveloped the village and new housing estates, busy roads, industrial estates and out-of-town shopping centres have sprung up.

Residents, rural campaigners and local politicians fear the government's moves to push through more housing developments in the South-East could cut further into areas like this.

There are already 750 new homes being built in Ashford borough every year, but the government could force the area to more than treble this figure.

Local politicians argue there are too few brownfield sites to cope with such an increase and say they will be forced to give planning permission for estates to be built on green fields like those still to be found south of Kesington.

But even putting aside the concerns that countryside will be spoiled by the new houses, there are concerns that not enough money is being spent on infrastructure and on affordable housing.

The homes will be there, but will health and social services cope?

And will there be enough schools and sufficient leisure facilities for a growing population?

Neville Green, chairman of the campaign group the Ashford Rural Trust, said that the new developments lacked a sense of identity.

'They are sprawling estates without a focal point, without a church or pub.'

(Source: The Guardian, 16 July 2002)

CHANGES IN PERMANENT GRASSLAND 1992–97

ENGLAND HAS LOST an area of grassland the size of Bedfordshire since 1992 – the equivalent of about 100 football pitches every day, according to the Council for the Protection of Rural England (CPRE) report, 'Meadow Madness'.

The losses, from the chalk grasslands of the South Downs to the flood-plain meadows of Herefordshire and the culm grasslands of Devon, threaten to destroy some of England's most important wildlife habitats.

According to the report, more than 122,000 hectares of grassland have disappeared in England since 1992, with the greatest losses in southern and western parts of the country.

Meadows are generally used for grazing livestock and producing hay. Many have never been disturbed by ploughing, and are less likely to be heavily fertilised than other grassland, making them an ideal habitat for wildlife.

In 1992, reforms to the Common Agricultural Policy reduced the incentives offered to farmers for expanding arable land, which temporarily halted the losses to England's meadows. But over the past four years, grasslands have again been disappearing at an alarming rate.

Intensive dairy farming, greater profits from growing cereals and other subsidised crops, and increased potato production, are the principal reasons for the continued loss of grasslands, says the report.

It also blames weaknesses in planning policy, which means that meadows have been lost to urban development, particularly in the South-east.

The CPRE recommends that extra funds should be given to farmers who protect grasslands, and urges an extension of planning controls to include the ploughing of pasture land.

▲ **Figure 7.29** *The changing face of agricultural landscapes*
(Source: The Guardian, 15 March 1999)

17 From Figure 7.29:
a List two reasons given for the continuing loss of grasslands.
b What role do grasslands play in sustaining biodiversity?
c What suggestions are made to ensure the conservation of grasslands?
d Who are the decision-makers and other interested parties who are likely to influence the future of grasslands?

18 From Figure 7.30:
a Name (i) the three counties with the greatest number of new homes planned (ii) three counties with planned increases of over 20 per cent, 1991–2016.
b The figures in the table and on the map are not necessarily the same. Why not?
c What evidence is there that the greatest pressures for housing development will be in relatively rural counties? (Remember, at least 50 per cent of the new home figures listed will be in rural areas.)

County	No. of houses planned by 2011
1 Avon	44 500
2 Bedfordshire	49 300
3 Berkshire	40 000
4 Buckinghamshire	64 000
5 Cambridgeshire	60 000
6 Cheshire	43 300
7 Cleveland	n/a
8 Cornwall	45 000
9 Cumbria	27 500
10 Derbyshire	60–88 000
11 Devon	99 000
12 Dorset	52 900
13 Durham	22 400
14 East Sussex	45 700
15 Essex	106 000
16 Gloucestershire	50 000
17 Hampshire	44 000
18 Hereford & Worcs	50 750
19 Hertfordshire	65 000
20 Humberside	awaits plan
21 Isle of Wight	10 000
22 Kent	116 000
23 Lancashire	67 400
24 Leicestershire	53 000
25 Lincolnshire	66 900
26 Norfolk	61 000
27 North Yorkshire	44 600
28 Northamptonshire	62 400
29 Northumberland	16 000
30 Nottinghamshire	69 000

County	No. of houses planned by 2011
31 Oxfordshire	30 500
32 Shropshire	36 000
33 Somerset	44 300
34 Staffordshire	51 800
35 Suffolk	38 800
36 Surrey	36 000
37 Warwickshire	31 100
38 West Sussex	42 200
39 Wiltshire	65 000

Conurbations – no figures available

40 Greater London
41 Greater Manchester
42 Merseyside
43 South Yorkshire
44 Tyne & Wear
45 West Midlands
46 West Yorkshire

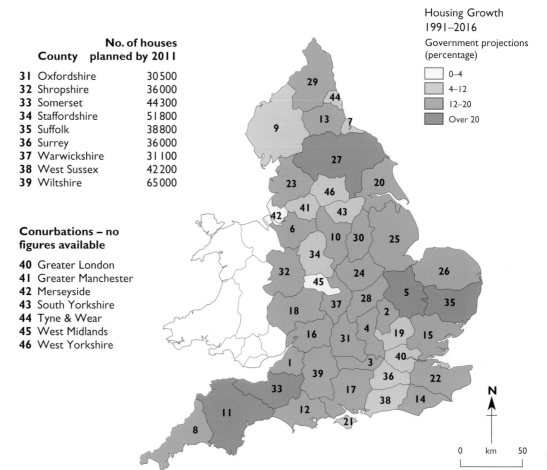

▲ **Figure 7.30** *A comparison of housing growth projections, 1991–2016 (Source: The Guardian, 14 November 1997*

Government and EU policies

Government and EU policies have exerted a strong influence upon these processes of change. Based on policy emphases, we can divide the post-1945 period into two broad phases:

1 1945–85: *A production and rationalisation phase*, dominated by two main areas of planning policy:

■ The revival and modernisation of the agricultural industry through intensification, mechanisation and specialisation

■ The redistribution of population and economic activity which has restructured rural–urban relationships.

2 1985–present: *A diversification and conservation phase*, with a shift in policy priorities towards conscious efforts to achieve economic, social, and environmental sustainability. The principal interrelated aims are:

■ To restructure the agricultural industry in response to the economic and environmental costs of intensive farming. For example, farmers now receive less in direct product-support payments and more under agri-environment schemes such as **Environmentally Sensitive Areas** (ESAs) (see the South Downs example).

■ To diversify farming and rural economies. For example, relaxing planning controls, and the availability of start-up support grants to allow farmers to diversify – by the introduction of leisure activities for instance.

■ To sustain **biodiversity** and control pollution. For example, stricter limits are set on nitrate and other inorganic fertiliser and pesticide application. This helps to reduce eutrophication and algal blooms in streams and lakes.

Key Terms

Environmentally Sensitive Area (ESA): An area designated by the UK government as possessing environmental conservation value, within which farmers are given grants to farm in traditional and sustainable ways.

249

- To conserve attractive qualities in rural environments and improve recreational access and opportunity – for example, grants to upgrade footpaths, and to conserve drystone walls and hedgerows.
- To control urban growth. For example, since 1997, stricter planning controls on large out-of-town commercial developments. yet expansion pressures are intensifying (see Chapter 6)

Nonetheless, forces from the previous phase persist. Thus, between 1985 and 1995, some 160 000 km of hedgerows were removed, and others became dilapidated. Agricultural jobs continued to decline – in England alone the workforce declined by 60 500 between 1987 and 1997.

EXAMPLE: Orleton, Herefordshire

The changing English village

The OS map extract below shows the village as it was in 1945, set into a fringe of cider apple orchards. In 1951, approximately 280 people lived in the village, by 1998 there were almost 600. In the County Development Plan of 1953, Orleton was defined as a 'capital' village, where growth was to be encouraged by infilling and extension. Today the village and its local area support a primary school, a doctor's surgery, a shop/post office, two pubs, a church and a chapel, a bakery, and a range of social groups such as a playgroup, 'Evergreens' for the elderly, and local football and cricket teams. No working farms remain within the village, and all of the orchards have disappeared (Figure 7.30).

▼ **Figure 7.31** *The changes in the village of Orleton from 1950*

306

a 1990s: 40 affordable starter homes to encourage local families to stay in the area.

b 1970s: 35 bungalows, largely occupied by older and retired households.

c 1950s: 25 council houses built for local people. Today most have been sold under the 'Right-to-buy' programme.

d 1960s: New primary school built.

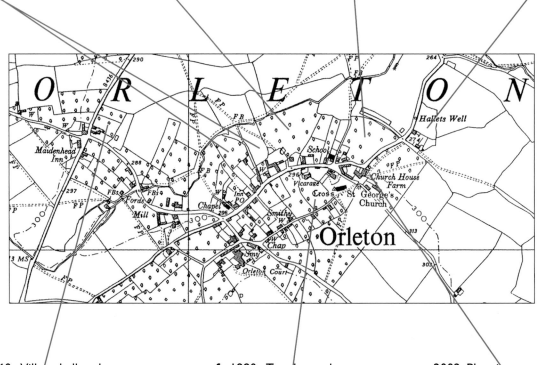

e 1960s: Village hall and recreation ground (village hall rebuilt in 1998).

f 1980s: Two larger houses and a new vicarage built on church land.

g 2002: Planning permission sought for conversion of farm buildings to dwellings.

Market towns and the changing rural system

Markets are an essential component of all rural economies and societies. Before motor vehicles increased mobility, a network of towns evolved, each one acting as a service centre for a rural district. (See the Welsh Marches example, page 246) Many of these market towns are losing this crucial function in MEDCs today. In the UK and other EU countries, livestock trading is increasingly concentrated in a small number of widely spaced, large markets. For example, in the central section of the Welsh Marches, Ludlow's modern, out-of-town livestock market has taken trade from other traditional markets such as Cleobury Mortimer, Leominster and Tenbury. Retail and other services too are becoming increasingly concentrated in and around larger settlements. (See Chapter 8) A survey in Cleobury Mortimer showed that residents did 70 per cent of their 'main weekly' shopping in supermarkets in Kidderminster or Ludlow. Similar trends can be seen in North America, Australia and New Zealand (Figure 7.32)

In response, in 2000 the UK government launched a £37 million Market Towns Regeneration Programme, linked to regional development policies. For example, the West Midlands Market Town Programme is a 'key priority' in the Regional Economic Strategy (RES), and received £2.5 million, 2000–2003. The aim is to stimulate and support initiatives that bring inward investment, diversify the economies and give distinctive 'images' to the towns (Table 7.6). Regional strategies also identify Rural Regeneration Zones (RRZ) that include market towns. Recommended actions focus on sustainable development, and that the towns do not become commuter 'outliers' for cities (Box 3).

▼ **Table 7.6** *Market towns: Examples of regeneration initiatives*

- **Town management and shop care schemes**, e.g. Whittlesley appointed a town centre manager and developed an environmental improvement plan, including the use of empty shop windows to display community information
- **Customer Loyalty Schemes**, e.g. Leominster ran a scheme where people were awarded discounts for using local shops
- **Niche marketing**, e.g. Hay-on-Wye has established itself as a specialist centre for second-hand books and now runs a well-known Literary Festival based around this specialism; Ludlow markets the quality of its restaurants
- **Town festivals**, e.g. Upton-on-Severn holds an annual Jazz festival; Ludlow and Leominster have successful annual festivals including art, music and theatre
- **Tourism promotion**, e.g. Thornbury has tourism information available in the town and the large Cribbs Causeway Mall
- **Themed markets**, e.g. Wareham has held a French market in co-operation with its 'twin town' in Normandy.
- **Local produce**, e.g. Ludlow promotes the marketing of local food and craft products.

(Source: Barrett et al, Geography, 86[2], 2001)

▼ **Figure 7.32** *The broad main street of Coleraine, central Victoria, Australia. This small market town is surrounded by a sheep and cattle rearing district. It is struggling to survive as agricultural business and other services move to the sub-regional capital of Hamilton.*

19 Give two reasons why many small market towns are struggling to retain their economic functions

20 Why is it important for people living in rural areas that market towns continue to flourish?

21 Why does the UK government believe that market towns are important? (Think what could happen if many market towns did continue to decline.)

22 In what ways do the proposals for regeneration illustrate the idea that sustainability must involve economic, social and environmental elements?

Box 3 Some 'actions' recommended for Bewdley in the West Midlands RRZ

■ Develop a community plan
■ Develop links with the West Midlands conurbation through tourism promotion and marketing of local food and produce
■ Improve access to work and services in the town, thereby reducing the need for travel
■ Improve sustainable transport modes such as walking and cycling
■ Conserve and enhance the built environment, especially buildings and spaces of historic interest
■ Improve support of cultural and leisure services including libraries, museums and leisure centres; the introduction of festivals
■ Initiatives to encourage young people including improved access to leisure opportunities
■ Develop the potential contribution and involvement of the 50+ age group

(Source: Caffyn, University of Birmingham, 2002)

EXAMPLE: The South Downs ESA

Since 1945 over three-quarters of the chalk grasslands of the South Downs have been lost to ploughing and intensified grazing practices. Because of their location, the Downs are also under intense presure for development and recreation.

'The South Downs ESA has been designated because of its high scenic value characterised by open rolling downland and because of its important wildlife habitats, particularly the wild flowers and insects associated with the traditional chalk grassland. The river valleys which cross the Downs are an integral part of the landscape, and the pastures provide breeding sites for birds such as lapwing and snipe' (MAFF, 1994). The primary aim is to conserve the remaining permanent grasslands and return arable land to permanent pasture where practicable. The scheme is voluntary for farmers.

▼ **Figure 7.33** The area of the South Downs stretching right across the map, makes up the ESA area.

▼ **Figure 7.34** *The main themes of the Countryside Commission and the Countryside Agency policies*

CoCo, 1996: A Living Countryside,
- Encouraging local pride.
- Promoting sustainable leisure activities.
- Achieving long-term benefits from farms and woodland.
- Planning for sustainable development.
- Providing better information.
- Care for scenic gems.
- New look for the urban fringe.

CA, 1999: Tomorrow's Countryside – 2020 vision
- Diverse character and outstanding beauty.
- Prosperous and inclusive communities.
- Economic opportunity and enterprise.
- Sustainable agriculture.
- Transport that serves people and doesn't destroy the environment.
- Recreational access for local people and visitors.

Key Terms

Stakeholder: An individual, group or organisation with a direct interest in a particular resource or issue.

Gatekeeper: An individual, group or organisation with some degree of responsibility for, and authority to, make decisions concerning a resource or issue.

7.7 Stakeholders and gatekeepers in rural policy change

Approximately 85 per cent of the UK land area is under some form of agriculture. This land is owned and managed by fewer than 200 000 families – for example, in England in 1998 there were around 145 000 farm holdings, and 3 per cent of these, each over 300 ha, covered 27 per cent of all farmland. This concentrates decision-making powers over the countryside within a very small proportion of the total population. However, all decision-making occurs within a framework of national and EU planning, and within agricultural and environmental policies. Furthermore, there is the influence of international market forces and Agenda 21 programmes from the 1992 Rio de Janeiro environmental conference.

Evidence suggests that a high proportion of the population have an interest in the countryside and may be classed as 'stakeholders'. For instance, in 1996, 1.3 billion day-visits were made to the UK countryside, and surveys show that more than 60 per cent of the population make some recreational use of the countryside. A majority of these visitors live in cities, but we must remember that around 18 per cent of people do live in rural areas. One of the main features of recent policies is to increase the role of local populations, and to some extent urban visitors, in the decision-making process. That is, to give a sense of 'ownership' and 'empowerment' to a wide range of stakeholder groups, from national organisations such as the Ramblers' Association, the National Trust, the RSPB, to parish councils and local communities.

The Countryside Agency

In 1968 the government set up the Countryside Commission (CoCo). Until 1999, CoCo, alongside the **Ministry of Agriculture, Fisheries and Food** (MAFF), the Nature Conservancy (later, English Nature and the Scottish and Welsh equivalents), and Rural Development Commissions (RDCs), were the main government agencies influencing rural policies and management. In 1999, CoCo was replaced by the **Countryside Agency** (CA) and RDCs were merged into new **Regional Development Agencies** (RDAs), whose role includes rural economic development. In 2002, MAFF was replaced by DEFRA.

The CA, like its predecessor, is an enabling and facilitating agency. It publishes its vision and preferred goals for the countryside and its management (Figure 7.34). It then promotes, encourages and supports other institutions, organisations and groups who actually carry out the programmes and projects. The main strategy of CoCo/CA is to promote sets of initiatives, focusing upon specific aspects of countryside resource conservation and management (Box 4). Three distinctive features are (i) the encouragement and use of partnerships between the public, private and voluntary sectors; (ii) the emphasis on community-based, 'bottom-up' approaches; and (iii) an integrated approach to countryside conservation and management.

Farming and the countryside

From what we have learned in this chapter so far, we can make a clear, strong statement – *a sustainable countryside cannot be achieved without a sustainable agricultural industry.*

The complexity of balancing the three main dimensions of sustainability (Figure 7.1) in a period of rapid change is well illustrated by the issues facing British farmers (Figure 7.35a and Figure 7.35b).

Two important responses to the growing crisis in UK agriculture have been diversification and the adoption of more environmentally friendly methods. Signs of this restructuring include increases in the hectarage of new crops such as rape seed and flax; the expansion of Pick-Your-Own fruit and vegetable enterprises; direct marketing through Farmers' Markets; the encouragement of organic farming using Farmers' Markets as an outlet; and alternative land uses and enterprises (like planting woodland, and building developments for leisure). Nonetheless, UK farming remains in essentially the pattern of dominant enterprises previously shown on Figure 7.19. Remember, too, that policies aimed at the diversification of rural economies may strengthen the argument for high-impact developments such as quarrying, housing, industry, roads, and tourism complexes.

Box 4 A selection of rural initiatives from CoCo and the CA

■ **Countryside Stewardship (from 1991)**: Farmers and landowners are offered grants for conserving and restoring important landscapes, wildlife and historic features. In 1996, for example, farmers and the parish council in Wressle (Yorkshire) began a project to restore 8 km of hedgerows, to create margins around fields and along streams for wildlife, to restore ponds and plant trees. By 1998, CoCo and MAFF had distributed £21 million to local projects like this.

■ **Heritage Coasts (from 1973)**: The aim is to protect and manage the most attractive, unspoilt stretches of coastline in England and Wales. By 1998, there were 50 Heritage Coasts along approximately 1400 km of coastline such as the Scilly Isles, Embleton Bay (Northumberland), Purbeck Cliffs (Dorset).

■ **Community Forests (from 1987)**: The aim is 'to improve the countryside around towns and cities and create a pleasant environment on people's doorsteps' (Countryside, 1996). By 1997 there were 12 Community Forests in England, with over 3,500 ha of new plantings and 8500 ha of improved woodland from joint CoCo/Forestry Commission projects. The work is done mainly by local communities and the British Trust of Conservation Volunteers (BTCV). In 1996–7, £5.5 million came from the EU and the National Lottery, and £2.6 million from private funding. Examples are: Mersey Forest (St Helen's), Marston Vale (Bedfordshire), Great North Forest (Durham) and South Yorkshire Forest (Sheffield).

■ **Village Design Statements (VDSs) (from 1996)**: Guidelines and advice are available on how local people can prepare and present a Village Design Statement and influence the working of the local planning system. This is a 'bottom-up' approach to involve residents in deciding what type of built environment they prefer. In 1998, 167 villages were working on a VDSs, such as Cartmel (Cumbria), Elstead (Surrey) and Cottenham (Cambridgeshire).

■ **Milestones Project (from 1987)**: The major CoCo programme whose aim was to have all 240 000 km of the national footpath system working by the new millennium.

■ **Parish Paths Partnership Scheme (PPPS) (from 1992)**: An important element in the Milestones Project. Offers grants to parish councils and local groups for surveying and improving footpaths. By 1998, over 1500 parishes were managing 35 000 km of rights of way; £5 million had come from CoCo and increasing amounts from the National Lottery.

■ **Land Management Initiatives (from 1999)**: A series of pilot projects covering arable, lowland pastoral, upland and wetland farming systems aimed at demonstrating sustainable land management. These projects are run jointly by the CA and the MAFF.

■ **Rural Action (1992–99)**: Funded jointly by CoCo, English Nature and the RDC. The aim has been to help people to protect and improve their local countryside, e.g. nature reserves, tree planting, woodland management, local history projects, access improvements. Grants totalled £43.5 million in 40 counties.

■ **Greenways Initiative (from 1998)**: The aim is to create: 'a network of largely car-free, off-road routes connecting people to facilities and open spaces in and around towns, cities and to the countryside. For shared use by people of all abilities on foot, bike or horse-back.' (Countryside, 1998). By 1999, counties had identified 17 000 km of potential greenways, and the CA hopes to co-ordinate these into a national network. The CA runs seven pilot projects, provides advice and lobbies government for increased funding.

► **Figure 7.35** *Two contrasting farms – do either meet the three dimensions of sustainability?*

a *The large-scale farmers of eastern England, sometimes called 'grain barons', have been economically successful. Despite fluctuations in market conditions, reductions in CAP support payments, and tougher environmental regulations, these farmers are likely to remain economically sustainable. However, environmental sustainability – such as biodiversity, soil and water quality, and scenic character are threatened by the creation of huge fields ('prairies'), and the intensive use of inorganic fertilisers and pesticides. Social sustainability – for example retention of viable communities and services – is threatened by the low labour demands of agribusiness.*

◄ **b** *The reduction of CAP support and the collapse of sheep prices during the late 1990s are threatening the economic sustainability of Welsh hill farms. Social sustainability, too, is in doubt, as the number of farmers decreases, their average age increases and government policies include 'retirement' incentives. However, these farmers are stewards of some of Britain's most valued landscapes. Environmental sustainability and landscape conservation depend upon continued land management by farmers.*

311

23 Make a list of the 'stakeholders and gate keepers' involved in countryside issues. Discuss the assertion that the main obstacle to achieving sustainable management of the countryside is the large number of 'stakeholders' involved.

24 Use the two examples of Figure 7.35 to illustrate why it is so difficult to achieve sustainable management of rural environments.

317

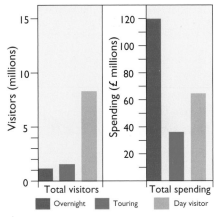

▲ **Figure 7.36** *Worcestershire – the economic impact of tourism, 1995 (Source: Heart of England Tourist Board, 1996)*

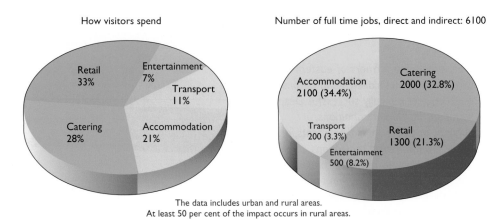

How visitors spend

Retail 33%
Entertainment 7%
Transport 11%
Catering 28%
Accommodation 21%

Number of full time jobs, direct and indirect: 6100

Accommodation 2100 (34.4%)
Catering 2000 (32.8%)
Transport 200 (3.3%)
Retail 1300 (21.3%)
Entertainment 500 (8.2%)

The data includes urban and rural areas.
At least 50 per cent of the impact occurs in rural areas.

---

25 From Figure 7.36:

a How many visitors did Worcestershire receive in 1995?

b Using an estimate that 50 per cent of total visitor spending occurs in rural areas, estimate the amount spent in rural Worcestershire.

c Summarise the economic impact of tourism on rural Worcestershire. (Note the different types of spending and jobs.)

26 **With a partner:** Think carefully about the meaning of the Sandford and Edwards principles:

a Explain how their purpose is to assist sustainability;

b Suggest two reasons why they may be difficult to put into practice.

▼ **Figure 7.37** *Are gorge-walking and rock-climbing appropriate inside National parks?*

Leisure, conservation and the countryside

In 1991, the president of the NFU said, 'Leisure is the crop of the 1990s'. He meant, of course, that the growth of demand for rural recreation and tourism provides considerable potential for profitable investment by farmers, communities and businesses – in other words, an increased supply of opportunities (Figure 7.36). For example, during the 2001 foot-and-mouth outbreak in the UK, rural communities and businesses lost more money from the decline in recreation and tourism than from the disasters in the farming community. This is an example of diversification assisting economic sustainability. Furthermore, because visitor satisfaction is related to the conservation of environmental quality, there is strong motivation for environmental sustainability.

There are inevitable tensions, however. Think of the implications of these three statistics: 80 per cent of the UK population lives in urban areas; 60 per cent of the population say they like to use the countryside for recreation; less than 2 per cent of the population owns 80 per cent of the countryside. For many years, demands for access to rural recreational resources have exceeded supply, and recreation pressures have caused environmental concerns, such as footpath erosion, and disturbance to farm animals.

Seeking solutions to the tensions and conflicts caused by this competition for resources has been an element of government policy for over 50 years. One popular approach has been to give special status to selected areas. For instance, the twin goals for National Parks set out in the National Parks and Access to the Countryside Act of 1949, were to 'preserve and enhance' Park environments and to 'provide for the public enjoyment' of the Parks. One criticism has been that the Act identified the conservation–recreation tension but did not clarify how it should be resolved: the managers of the Peak District and Lake District National Parks must each produce plans to cope with over 15 million visitor-days a year – and still conserve the precious environments! Two recommendations in the reports of Royal Commissions on National Parks which have been adopted as Park policy illustrate the continuing attempts to sustain Park goals:

■ **The Sandford Principle**: That where development and conservation goals are in conflict, conservation shall normally be given priority (from the Sandford Report, 1974).

■ **The Edwards Principle**: That recreational activities should be 'appropriate' to the goals of National Parks, that is, the provision of opportunities for quiet recreation (the Edwards Report, 1991). This reflects evidence that pressures on the Parks are not caused only by growing visitor numbers. Over the past 20 years a widening variety of action sports have become increasingly popular (Figure 7.37).

EXAMPLE: Pressures and responses in National Parks

The 11 National Parks of England and Wales cover approximately 10 per cent of the surface area. In 1999, the New Forest and the South Downs were approved as National Parks, and proposals were made for the designation of two National Parks in Scotland (Loch Lomond and the Trossachs; the Cairngorms). One important understanding is that two-thirds of the total designated area is in private ownership. In no park do the park managers own more than 10 per cent of the area. This influences the decision-making process.

Pressures: The number of visitor-days is increasing all the time – a visitor-day is defined one person visiting for one day – if you make a two-day visit, then you are counted as two visitor-days (see Table 7.5).

Loss of tranquillity: Tranquillity is a valued feature of rural environments. For a national survey in 1995, the **Council for the Protection of Rural England** (CPRE) defined a 'tranquil area' as 'sufficiently far away from visual or noise intrusion of development or traffic to be considered unspoilt by urban influences'. This means at least 3 km from the busiest roads, 2 km from other major roads and 3 km from large city built-up areas. Two maps produced from this survey show the erosion of tranquillity in Cumbria, which includes the Lake District National Park. Loss of tranquillity is especially important in National Parks, where quiet enjoyment is a key element in management policy (Figure 7.38).

A response: One strategy used by National Park managers is to set out a zoning system within a Park (Figure 7.33). The objective is to meet both conservation and recreation objectives by protecting the most valued and sensitive areas, and by providing for a wide range of preferred visitor experiences, such as opportunities for quiet, low-density recreational use in more remote, less accessible and semi-natural areas. In contrast, the mass, car-bound visitors and caravanners can be concentrated within designated development zones.

▼ **Table 7.6** *Visitor-days in UK National Parks, 1995*

National Parks	Visitor-days (million)
Lake District	>15
Peak District	>15
Yorkshire Dales	8.5
North Yorkshire Moors	8
Snowdonia	7
Pembrokeshire Coast	5
The Broads	5
Dartmoor	4
Brecon Beacons	4
Exmoor	1.5
Northumberland	1.5.

▼ **Figure 7.38** *Zoning in the Peak District National Park*

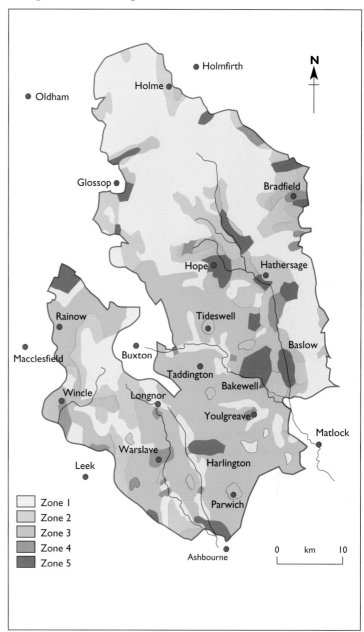

Zone 1
Zone 2
Zone 3
Zone 4
Zone 5

Explanation of zones
Zone 1: Wild areas with a general absence of human influence; low density, quiet recreation.
Zone 2: Remoter areas of farmland and woodland with poor access; low-density recreation opportunities.
Zone 3: More generally accessible areas; moderate visitor densities.
Zone 4: Specific localities suited to moderate levels of recreational use.
Zone 5: Accessible areas of 'robust' landscape, able to take high visitor densities and facilities.

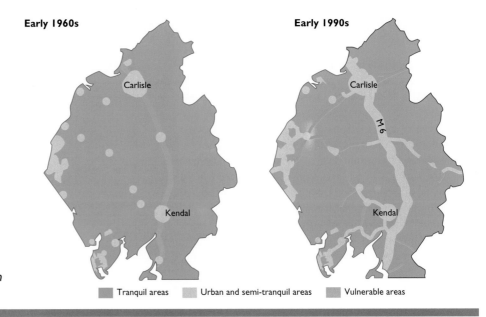

Early 1960s   **Early 1990s**

Carlisle   Carlisle

Kendal   Kendal

M 6

■ Tranquil areas   ■ Urban and semi-tranquil areas   ■ Vulnerable areas

► **Figure 7.39** *The loss of tranquility in Cumbria (After: CPRE, 1995)*

---

27 From Figure 7.39:

a Describe briefly the changes shown. (Use an atlas or OS maps to give greater locational accuracy).

b To what extent do the maps suggest that the National Park goal of sustaining quiet and tranquil environments is threatened? (Note particularly, which parts of the Park are under threat, and why.)

Key Terms

Access: The terms on which people can enter and use an area or resource.
....

---

28 Discuss briefly:

a What is meant by a zoning policy for recreational opportunity

b How it allows both concentration and dispersion of recreational experience and impact

c What are its advantages and disadvantages.

By the mid-1990s, over one-quarter of England and Wales had some form of protected status – National Park, Area of Outstanding Natural Beauty (AONB), Country Park, Community Forest, Nature Reserve, Heritage Coast, Site of Special Scientific Interest (SSSI). Each has its own distinctive balance of priorities between development, conservation and recreation, and different powers and funds to enforce and implement these priorities.

Access, Ownership and Policy

To enjoy the countryside we need attractive destinations that are accessible, that is to say, can be reached in an acceptable time. Equally important, however, is whether we have rights of **access**, and on what terms, to countryside resources. Remember, the majority of UK land and water resources are in private ownership. This is why organisations such as the Ramblers' Association have campaigned so vigorously to retain the public rights-of-way network (such as footpaths and bridleways), and for '**Right to Roam**' laws introduced in 2001 to improve access rights over open landscapes. This explains too, the CoCo Milestones and Parish Paths Partnership schemes (see Box 4), and the Sustrans strategy for a National Cycle Network (10 000 km). Recreational demand is most intense around towns and cities, i.e. in locations accessible from the homes of 80 per cent of the population. Thus, many policies aim to increase recreational opportunities (improve access) in accessible locations. For example, local authorities, supported by CoCo and the National Trust, have established Country Parks within 25 km of urban areas to provide free access for informal recreation. The result has been to bring the supply of resources closer to the origin of demand. Country Parks are also intended to take some of the pressure off National Parks.

7.8 So – what sort of countryside do we want?

Each year, thousands of families move from cities to rural homes, and millions of people take drives or walks into the countryside. All have a vision of what they expect or perceive the countryside to be. Such often romanticised images are often built on nostalgia, a belief in 'the rural idyll' and 'the good life' to be found. Many issues discussed in this chapter arise in part from these deeply held, often romanticised images – for example, development pressures in villages, recreational impacts, opposition to the 'new agricultural landscapes' of modern farming systems and the erosion of tranquillity (Figure 7.39).

From this perspective, the British countryside is a social construct: a diversity of environments containing sets of attractive and valued attributes. The problem is, of course, that different groups have different images and perceptions, and hence what they want the countryside to be. For instance, since 1998 a series of mass rallies in London have been organised by interest groups and stakeholders who feel strongly that the countryside, and the way of life they love, is under threat (Figure 7.40). To others, this protest is all about the 'not-in-my-back-yard' 'NIMBY' syndrome: an élitist attempt to secure a countryside in their own preferred image. So – whose 'image' should be used to determine sustainable rural management policies?

◀ **Figure 7.40** *The Countryside Alliance rally held in London, 2002*

Summary

Sustainable development in rural environments must include economic, social and environmental dimensions.

Despite rural development processes, there are still significant differences between LEDC and MEDC farming systems.

Many LEDCs remain mainly rural societies and find it difficult to balance short-term economic needs with longer-term social and environmental sustainability.

Approaches to rural development are increasingly community-based.

In MEDCs, there are growing concerns over social and environmental impacts of modernised commercial farming and agribusiness.

Rural settlements have a wide range of nucleated and dispersed forms.

British rural settlements and environments are changing in character and function in response to accelerating shifts in economic systems and preferred lifestyles.

Since 1985, UK government policy for rural areas has been increasingly based upon the twin themes of diversification and conservation.

The Countryside Agency is the main government organisation concerned with countryside management policies in the UK.

Leisure and recreation are increasingly important in influencing the economic, social and environmental character of the countryside.

Economic activities

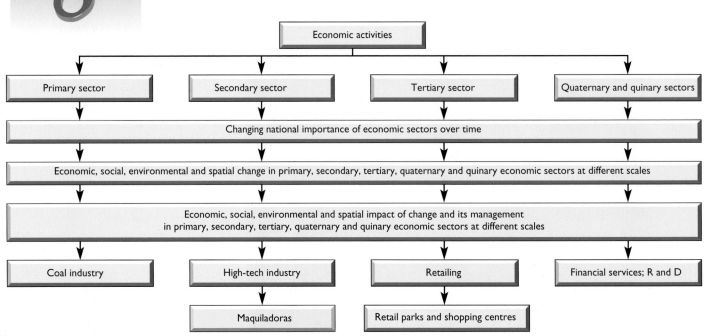

8.1 Introduction

Economic activities which create jobs and wealth are the life blood of human society. They include the production of food and raw materials, manufacturing and the provision of services. These activities are dynamic and geographically varied.

The global economy and the economies of many nation states have experienced considerable changes in the last 30 years or so. Rapid urbanisation in **LEDCs** underlies the huge employment shift from agriculture to industry and services. The economic transformation of countries such as Taiwan and South Korea, from pre-industrial to post-industrial status, has occurred with breathtaking speed. Similar changes are taking place in China, which within 20 years will be the world's biggest economy.

The impact of change is geographically uneven and occurs at different scales. While some parts of the world benefit, others lose out. The global share of manufacturing will continue to shift from **MEDCs** to LEDCs. Some places will benefit from new investment and opportunities; others which are unable to adapt, will decline.

These changes are a response to a number of powerful forces. The forces include technological change, the operation of supply and demand, globalisation, and the increasing interdependence between countries. Moreover, the pace of change, fuelled by innovative technologies such as the internet, can only increase in future.

8.2 Types of economic activity

There are five sectors of economic activity: primary, secondary, tertiary, quaternary and quinary.

316
318

I The Standard Industrial Classification (SIC) 1992, provides detailed information on employment in the UK (Figure 8.1). Using the employment data in Figure 8.1 for March 2002:

a Calculate the proportion of the UK workforce in primary, secondary (manufacturing and construction), tertiary, quaternary and quinary activities.

b Comment on any difficulties you had in allocating employment to each of the five main economic activities.

United Kingdom				March 2002 (thousands)	
Sections and sub-sections		Male		Female		All
		Full-time	Part-time	Full-time	Part-time	Total
A–Q	**ALL SECTIONS**	**10 786.1**	**1 749.9**	**6 405.5**	**5 053.5**	**24 994.9**
A	Agriculture, hunting and forestry	140.4	22.1	38.8	30.1	231.4
B	Fishing	6.1	0.7	0.7	.0	8.5
C	Mining and quarrying	63.1	0.7	7.6	2.2	73.6
	CA Energy-producing materials	37.7	0.5	5.0	1.2	44.4
	CB Other materials	25.4	0.2	2.6	1.0	29.2
D	Manufacturing	2 563.2	71.9	756.3	213.4	3 604.8
	DA Manufacture of food products; beverages and tobacco	288.8	13.8	123.1	42.7	468.3
	DB Manufacture of textiles and textile products	106.5	5.2	68.4	20.3	200.4
	DC Manufacture of leather products incl. footwear	11.4	0.4	6.1	1.3	19.2
	DD Manufacture of wood and wood products	54.2	1.9	13.7	7.5	77.4
	DE Manufacture of pulp, paper and paper products; publishing and printing	263.9	11.2	123.0	38.4	436.7
	DF Manufacture of coke, refined petroleum products and nuclear fuel	25.3	0.1	5.0	0.8	31.2
	DG Manufacture of chemicals, chemical products and man-made fibres	154.7	2.6	58.8	9.6	225.7
	DH Manufacture of rubber and plastic products	164.1	4.2	37.2	12.1	17.6
	DI Manufacture of other non-metallic mineral products	102.1	1.2	20.6	4.0	127.9
	DJ Manufacture of basic metals and fabricated metal products	379.8	8.2	63.5	20.0	471.4
	DK Manufacture of machinery and eqpt.	267.6	3.2	49.6	13.8	334.2
	DL Manufacture of electrical and optical equipment	291.6	6.4	101.0	18.8	417.8
	DM Manufacture of transport equipment	317.7	3.9	40.9	6.3	368.8
	DN Manufacturing n.e.c	135.5	9.5	45.5	17.7	208.2
E	Electricity, gas and water supply	70.4	0.9	25.1	4.2	100.7
F	Construction	974.4	224.0	103.8	75.2	1 175.8
G–Q	**SERVICE INDUSTRIES**	**6 968.4**	**1 631.3**	**5 473.1**	**5 727.3**	**19 800.0**
G	Wholesale and retail trade; repair of motor vehicles, motorcycles and personal and household goods	1 665.0	399.3	882.6	1 403.9	4 350.9
	Motor vehicles	419.2	2	6.0	75.933.9	555.0
	Others, wholesale	749.0	41.9	243.5	100.6	1 135.0
	Retail trade, except motor vehicles	503.0	281.0	550.1	1 061.8	2 395.9
H	Hotels and restaurants	280.6	207.5	303.8	533.7	1 325.6
I	Transport, storage and communication	978.0	72.2	305.5	106.4	1 462.1
	Land transport	393.9	24.6	57.1	22.3	498.0
	Water transport	15.9	0.8	5.1	0.9	22.6
	Air transport	42.9	1.1	30.6	7.9	82.5
	Supporting transport activities	201.4	13.4	122.7	36.9	374.4
	Post and telecommunications	323.9	32.3	89.9	38.4	484.6
J	Financial intermediation	444.5	13.8	436.2	141.4	1 035.9
K	Real estate, renting and business activities	1 351.6	300.2	867.9	894.7	3 414.5
	Real estate activities	127.3	22.4	87.1	56.4	293.1
	Renting of equipment and household goods	83.7	13.1	29.3	14.6	140.7
	Computer and related activities	287.0	8.4	104.0	28.1	427.5
	Research and development	56.4	0.9	30.2	4.6	92.1
	Other business activities	797.3	255.4	617.2	791.1	2 461.1
L	Public administration and defence; compulsory social security	642.7	44.2	459.6	182.4	1 328.9
M	Education	362.0	203.2	602.6	777.7	1 945.6
N	Health and social work	392.5	126.5	1 029.2	996.0	2 544.2
O, P, Q	Other community, social and personal service activities	389.6	144.0	322.9	304.8	1 161.4
	Sewage and refuse disposal	60.1	1.9	9.9	4.9	76.8
	Membership organisations n.e.c	61.1	34.7	65.7	59.2	220.7
	Recreational, cultural and sporting	209.7	91.2	150.9	175.7	627.5
	Other service activities n.e.c.	58.8	16.2	96.4	65.0	236.4

▲ **Figure 8.1** *The Standard Industrial Classification*

Primary sector

Primary activities include agriculture, mining and quarrying, fisheries and forestry. These activities produce food, energy and raw materials. In terms of employment, agriculture is the leading activity in the primary sector. Agriculture employs nearly one in two of the world's workers; 95 per cent of the world's agricultural workforce are in LEDCs.

Secondary sector

Secondary activities cover the manufacturing sector. Manufacturing is a complex activity which by processing, fabricating and assembling, adds value to products. Some industries are involved almost exclusively in one of these operations. Sugar refining is a processing industry which manufactures sugar from sugar beet; shoe manufacturing uses leather to make footwear; and car makers assemble vehicles from components made by hundreds of different suppliers. However, some industries may involve more than one manufacturing activity. For example, the iron and steel industry both processes raw materials to produce steel and turns steel into finished products such as beams, rails and plate.

Services: the tertiary, quaternary and quinary sectors

Unlike the primary and secondary sectors, the service sector provides intangible products such as transport, health care and recreation. Today the service sector in MEDCs is so large and wide ranging that it is sub-divided into three separate sectors: tertiary, quaternary and quinary. However, the distinction between these three divisions of the service sector is by no means clear cut.

Tertiary sector

The tertiary sector includes service activities which are closely linked with industry such as transport, communication and utilities such as water, gas and electricity supply. Some classifications also place retailing in the tertiary sector.

Quaternary sector

Quaternary activities are producer-oriented services such as legal and financial services, accountancy and advertising. For the most part, these services act on behalf of other commercial organisations rather than for private individuals.

Quinary sector

Quinary activities are consumer-oriented services. They include health care, education, government, entertainment, tourism and recreation.

Economic development and sectoral change

As a country's economy develops and grows, both absolute and relative shifts occur in employment between economic sectors. The sector model of economic development (Figure 8.2) describes these changes.

Initially, primary activities dominate employment. This is known as the **pre-industrial stage**, with between one-half and three-quarters of all employment in agriculture. Many LEDCs in Africa and Asia are still in this stage. In the USA, until 1880, more than half of all employment was in agriculture.

With industrialisation, a shift in employment occurs from agriculture to manufacturing. This marks the beginning of the **industrial stage**. For

319

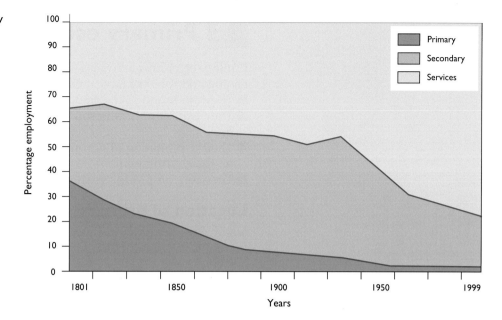

► **Figure 8.2** *Employment change by economic sector in the UK, 1801–1999*

example, manufacturing in the USA grew from 12 per cent of employment in 1820 to a peak of 35 per cent in 1947. However, while the relative importance of manufacturing declined after 1947, absolute employment increased until 1979 (23 per cent of employment). Today, the manufacturing sector accounts for just 12 per cent of employment in the USA.

The **sectoral shift** in jobs from agriculture to manufacturing occurs for two reasons. First, the growth of new employment opportunities in manufacturing in towns and cities, and second, improved technology causes a loss of jobs in agriculture, which thus becomes less labour-intensive. Running parallel with this sectoral shift of employment from agriculture to industry is the increasing concentration of population in towns and cities (See chapter 6).

Eventually, employment in manufacturing begins to decline. This happens as **automated production** replaces labour and as competition from more cost-effective foreign manufacturers displaces domestic production. In this **post-industrial stage**, service activities dominate the economy. Economic growth brings rising prosperity. Firms contract-out accounting, marketing, advertising and so on, to specialist service providers. Individuals, with their basic needs satisfied, spend more on holidays, cultural pursuits, education and health care. The effect of these changes is to inflate the service sector, which will typically account for three out of every four jobs in MEDCs such as the USA and the UK.

► **Table 8.1** *Employment by economic sector and per capita income, 1998*

2 From Table 8.1:
a Allocate the countries to one of the three stages of economic development.

b Comment on the relationship between wealth and the sectoral distribution of employment amongst the countries.

	Percentage primary	Percentage secondary	Percentage services	Income ($ per person per year)
Algeria	11	37	52	1580
Argentina	6	32	62	7460
Botswana	28	11	61	3300
Costa Rica	20	22	58	3810
France	4	25	71	24090
Kenya	81	7	12	350
Netherlands	4	23	73	24970
Russia	7	34	59	16600
Sierra Leone	70	14	16	130
Taiwan	3	33	64	17400

316

8.3 Primary economic activities

Primary economic activities such as agriculture, mining and quarrying, fishing and forestry have a number of key characteristics.
- They are found at the site of resources.
- They exploit natural resources such as climate, soil, plants, animals and fuel.
- Their exploitation of non-renewable resources often has disastrous effects on the environment.
- They rely on products whose supply and price are extremely variable.

Location of primary activities

Natural resources are usually both localised in their distribution and immobile. Agriculture's resources (climate, soil and land) are area-based and thus tied to specific locations. Mineral ores and fossil fuels are even more localised, and because they are finite, the site of production eventually has to move. The economic activities which exploit these resources are constrained to locate on ore fields, coalfields, oilfields and gas fields. Biotic resources such as forests also have fixed locations. Even fisheries, concentrated in shallow-water continental shelf areas such as the Grand Banks and the North Sea, are localised in occurrence.

Types of natural resource and sustainable production

The natural resources which sustain primary economic activities can be **renewable** or **non-renewable**. All biotic resources such as plants (or crops) and animals are renewable, but most energy resources (oil, gas, coal) are non-renewable. As long as production rates do not exceed rates of replacement we say that levels of production are **sustainable**. Current levels of production of timber and fish, on a global scale, are unsustainable. As a result, the total area of forest and world fish stocks dwindle each year. Alternative energy resources such as water, solar, wind and geothermal power are inexhaustible.

However, most of the world's energy comes from **fossil fuels** such as coal, oil and gas. These resources are finite and, except in very long geological timescales, non-renewable. Eventually, the exhaustion of accessible resources (or unacceptably high costs of production) leads to the closure of mining and other extractive industries. Examples include the abandonment of deep-mining in coalfields such as South Wales and Lancashire, and the exhaustion of the earliest gas fields in the southern North Sea.

▶ **Figure 8.3** *World tin prices, 1997–2000*

3 From Figure 8.3:
a Describe the variation in world tin prices between 1997 and 2000.
b The last tin mine in Cornwall closed in March 1998. From the evidence of Figure 8.3, suggest one possible reason for the closure.

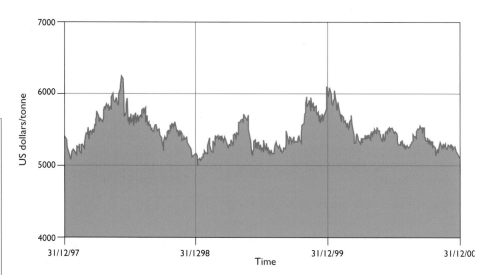

Some mining operations are extremely vulnerable to fluctuations in commodity prices (Figure 8.3). In 1985, the world price for tin collapsed from over £10,000 to £3,000 per tonne. Unable to compete with lower-cost and higher-grade mines operating in East Asia and South America, Cornish tin mines gradually closed down. South Crofty, Cornwall's last tin mine, closed in March 1998. Closure occurred despite ore reserves of 3.2 million tonnes, containing about 45,000 tonnes of tin (Figure 8.4).

▼ **Figure 8.4** *The local effect of the end of a mining industry (Source: The Guardian, 7 March 1998)*

Cornwall mourns the last tin mine

THOUSANDS of years of Cornish history drew to an emotional close yesterday when miners emerged into the daylight after completing the last drilling and blasting shift at Britain's only remaining tin mine.

As they left the gates of South Crofty for an uncertain future in one of the most economically disadvantaged parts of Britain, the grim-faced tinners were greeted by a vigil of hundreds of men and women, come to mourn the passing of an industry that once formed the bedrock of the local economy.

'It is not just the jobs that are going, it's a whole way of life that is being destroyed,' said Mark Kaczmarek, a skilled underground worker for more than 17 years.

Like many of the 200 miners who now find themselves without a job, Mr Kaczmarek says he has no idea what he will do for work in an area where male unemployment, at 12.4 per cent, is significantly higher than the national average.

The South Crofty mine at Pool, near Redruth, has been living under the threat of closure since last August when the mine's Canadian owners decided to cease production in the face of heavy losses caused by the strength of sterling and falling world prices.

Efforts to mount a rescue package finally collapsed last month when the government announced it was unable to support an application for £5.9 million regional Selective Assistance grant.

Ministers said that they were unable to commit public money to a project that was not viable, but pledged to support local efforts to alleviate the impact of closure on the Camborne–Redruth area.

The loss of 200 jobs at Crofty will strip £2.5 million out of the local economy annually. But forecasters fear a wider knock-on effect for local retailers and other service businesses, some of whom are not expected to survive.

'The age and skill structure of the South Crofty workforce will make it difficult for the majority of employees to find work in West Cornwall,' said a recent Cornwall County Council study of the likely impact of the mine's closure.

The report warned that specialist skills would be permanently lost and that jobs would be particularly hard to come by for older members of the Crofty workforce – many of whom have worked there for over 20 years.

Closure of the mine marked the demise of an industry that dates back to the recovery of alluvial tin from the region's moors and streams.

Hard-rock underground mining for tin and copper peaked in the mid-nineteenth century when around 50,000 people were employed in the county's 400 mines.

The discovery of cheaper overseas deposits brought a sharp decline in the industry's fortunes, however, and forced thousands of Cornish miners to emigrate to countries where their skills were in demand. By the end of the First World War only a handful of Cornish tin mines remained.

4 Read the newspaper article, Figure 8.4, on the closure of the South Crofty tin mine in Cornwall. Draw a spider diagram to show the social and economic impacts of closure of the South Crofty mine on the local area.

Resource exploitation and the environmental impact

Primary economic activities have a major impact on the environment at all scales. Among these primary activities agriculture has the greatest impact, for two reasons. First, agriculture occupies around 40 per cent of the world's land surface, and second most agricultural activities change natural environmental systems. Agriculture causes deforestation and replaces forests and woodlands with its own landscapes of open fields, hedges, ditches and so on (Figure 8.5). Associated with this change is a reduction in **biodiversity** and a loss of habitat (see chapter 4). In dry regions, **desertification** and **salinisation** stem from mismanagement of resources by farmers. Even in MEDCs, agriculture causes environmental problems.

► **Figure 8.5** *Large arable farms have destroyed hedgerows and natural ecosystems.*

Accelerated soil erosion on farmland by wind and water is widespread. In regions of intensive arable farming such as East Anglia, the pollution of water courses and groundwater by nitrate fertilisers threatens both wildlife and human health.

Mining and quarrying have obvious effects on the environment. Deep mining creates huge heaps of spoil, surface subsidence and shallow areas of standing water known as 'flashes'. Open-cast mining and quarrying produce large holes in the ground, destroying the original landscape and causing additional problems of noise, airborne dust, and water pollution. Meanwhile, hydro-electric power (HEP) often leads to the flooding of valleys; forestry plantations destroy wildlife habitats and clear felling of plantation trees causes soil erosion and loss of visual amenity.

Changes in the UK coal-mining industry

In 2001, the UK coal-mining industry produced just over 30 million tonnes of coal and had a workforce of 11 500. Sixty per cent of its production came from the industry's 17 collieries (**deep-mining**), the rest from **open-cast mining** as Figure 8.6 shows. The 1990s was a decade of steep decline for the coal industry. During this period 50 collieries were closed, and production was more than halved. Decline continued in the early twenty-first century, with mines at Selby, Pontefract) and Clipstone (Nottinghamshire) due to close by 2004. The causes of decline were: the substitution of gas for coal in electricity generation; imports of cheap foreign coal; and the government's energy policies.

The dash for gas

Eighty per cent of UK coal production goes to electricity generation, which makes the coal industry highly sensitive to changes in the electricity market. In 1990, when the electricity industry was privatised, coal accounted for two-thirds of the primary energy used in electricity generation. By 2001, this proportion had fallen to just one third.

The new privatised electricity companies preferred gas-fired power stations (Figure 8.8). Gas-fired stations had two main advantages over coal-fired stations. They were cheaper and faster to build and they were less polluting (see Box 1). The result was a 'dash for gas'. Between 1990 and 2000 nearly 25 GigaWatts of gas-fired power stations were commissioned. This new capacity displaced around 60 million tonnes of coal.

► **Figure 8.6** *UK coal production, 1994–2001*

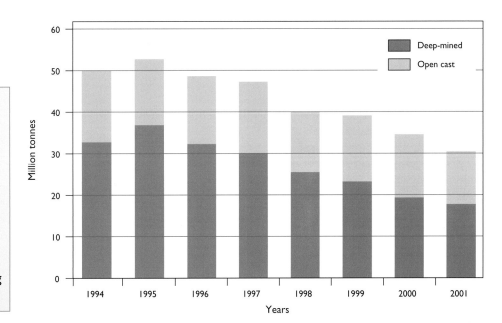

5 From Figure 8.6:
a Describe and explain the trends in coal production in the UK between 1994 and 2001.
b Discuss the economic and environmental issues associated with open-cast mining.
c **Group discussion**: debate the arguments for and against open-cast mining.
d Write down your own views on the open-cast mining issue, making a logical case for your beliefs.

◄ **Figure 8.7** *An open cast coal mine in Derbyshire. Open-cast mining has a huge environmental impact on the landscape.*

► **Figure 8.8** *The change to gas-fired power stations such as this one in the Tees estuary required a large initial capital investment. However, because gas-fired power stations are fairly flexible in their locational requirements, new jobs were created in areas with no previous tradition of electricity generation.*

6 Forecasts of the UK's energy economy in 1999 suggested that on present trends, three-quarters of the country's electricity would be generated from gas by 2020. However, UK gas reserves were equal to just nine years' supply at 1999 levels of consumption, and by 2003, the government said that the UK would become a net importer of gas. Explain how these forecasts support the case for a strong UK coal industry.

Imported coal

Some electricity producers operating coal-fired power stations began to import cheap coal from countries such as Colombia, the USA and South Africa. In 2001, these coal imports amounted to 23.5 million tonnes.

Government energy policies

The coal industry's contract to supply coal to the electricity producers ran out in 1998. This contract had given the coal industry a guaranteed market and helped to stabilise production at around 50 million tonnes a year. By 1998, electricity producers wanted to burn less coal, threatening the coal industry with further contraction. At this point, the government stepped in and called a temporary halt to building new gas-fired stations. However, the government's freeze on planning consent for new gas-fired power stations was lifted in November 2000. This triggered further decline in UK coal production.

The environmental impact of coal

The local impact: deep-mining

Deep mining in long-established coalfields such as South Yorkshire and South Wales has blighted large areas with spoil heaps and land subsidence. Land subsidence damages property and natural drainage. Shallow pools or

Box 1 Coal and global environmental issues

■ Coal combustion produces several air pollutants, including carbon dioxide (CO_2), sulphur dioxide (SO_2) and nitrous oxides (NOx). Coal has the highest carbon content of all the fossil fuels. CO_2 emissions per unit of energy obtained from coal are 80 per cent higher than from natural gas, and 20 per cent higher than from fuel oil. As CO_2 forms 99 per cent of the greenhouse emissions responsible for global warming, a switch from coal- to gas-fired power stations has clear environmental benefits.

■ International attempts to reach agreement on limiting greenhouse gas emissions began in 1992 at the Rio Earth Summit and the subsequent Climate Change Treaty. Five years later at Kyoto, 32 MEDCs agreed in principle to reduce their greenhouse gas emissions by 7 per cent (compared with 1990 levels) by 2008–12. Thanks to the 'dash for gas', the UK had achieved this target by 2000. However, in 2001 the USA, responsible for over one third of all greenhouse gas emissions, refused to ratify the Kyoto Protocol at the Hague conference.

■ Apart from global warming, coal burning is a major source of sulphur dioxide, which is linked to acid rain (Figure 8.9). Acid rain destroys aquatic life in acidified lakes and rivers, causes die-back in forests, and damages stone buildings through accelerated chemical weathering. Prevailing winds can carry pollutants hundreds of kilometres. Pollutants from the UK, for example, fall as acid rain in Scandinavia.

■ Electricity producers in the UK have responded to the environmental problems caused by SO_2 pollution. Two of the UK's largest coal-fired power stations – Drax and Radcliffe – have been fitted with equipment to reduce SO_2 and NOx emissions. Current research is focused on clean coal technologies which allow coal to be burnt more efficiently releasing less CO_2.

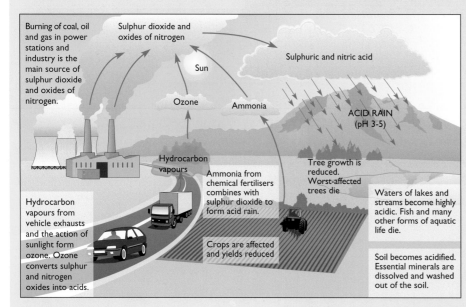

◀ **Figure 8.9** *Sources of atmospheric pollution and acid rain (Source: Raw, 1988)*

'flashes' occupy subsidence hollows which are widespread in the coalfields. In some parts of the UK, coal-fired power stations are found alongside collieries (Figure 8.10). This combination leads to the most severe environmental damage. In addition to colliery wastes and subsidence, ash from the power station furnaces creates its own spoil heaps; there is localised air pollution (especially SO_2) from power stations; and visual pollution caused by cooling towers, smokestacks and coal heaps.

The local impact: open-cast mining

Figure 8.6 shows that the massive fall in coal output in the 1990s came from the decline of deep mines, but that open-cast coal production remained more or less constant. Open-cast operations now account for over one-third of total coal output.

Although open-casting is cheaper than deep mining, it carries with it severe environmental disbenefits. Open cast-operations often occupy large sites (i.e. more than 500 hectares) and create holes up to 150 metres deep. The growing importance of open-casting has led to strong opposition from local pressure groups and environmental organisations such as Friends of the Earth. Opposition is particularly strong in former mining areas where deep mining has ended, but open-casting (which employs fewer workers) has expanded (Figure 8.11). Table 8.2 provides a summary of the arguments for and against open-cast mining.

The economic and social impact of change in mining communities

The collapse of the coal mining industry since the mid-1980s has devastated mining communities throughout the UK coalfields. For example, the

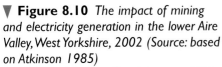

▼ **Figure 8.10** *The impact of mining and electricity generation in the lower Aire Valley, West Yorkshire, 2002 (Source: based on Atkinson 1985)*

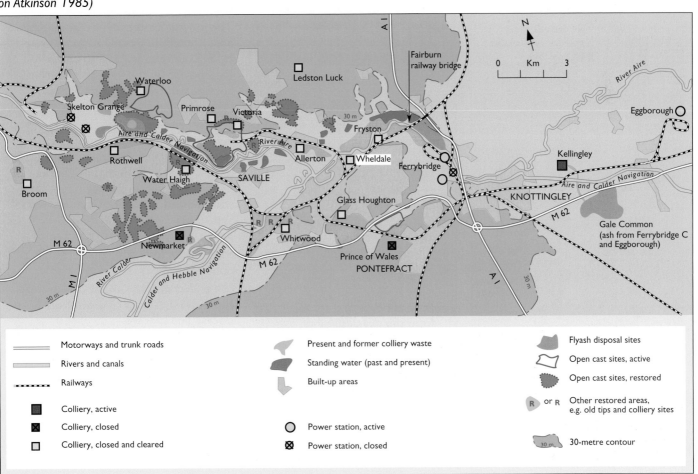

Motorways and trunk roads	Present and former colliery waste	Flyash disposal sites
Rivers and canals	Standing water (past and present)	Open cast sites, active
Railways	Built-up areas	Open cast sites, restored
Colliery, active		R or R Other restored areas, e.g. old tips and colliery sites
Colliery, closed	Power station, active	
Colliery, closed and cleared	Power station, closed	30 m 30-metre contour

► **Table 8.2** *The issue of open-cast mining*

Arguments for	Arguments against
■ Open-cast sites are eventually filled in and restored for farming.	■ Open-casting causes fundamental changes in the landscape. Wildlife habitats are destroyed; ancient patterns of fields and settlement disappear; natural drainage is disrupted; mature hedges and trees disappear.
■ Some open-cast workings are on land previously damaged by deep mining. Restoration will improve the quality of the land.	■ Noise and dust from the blasting and the use of heavy machinery causes inconvenience and distress to local communities.
■ Open-casting is temporary, lasting on average for just five or six years.	■ There is an increase in heavy lorry traffic in the vicinity of the open-cast site, often along narrow lanes.
■ Open-cast mining provides jobs and cheap coal for electricity generation.	■ The large earth embankments around open-cast sites, designed to reduce noise, are unsightly.
	■ Even with land restoration, the original landscape cannot be restored. It takes many years for farmland to recover its productivity and for trees and hedges to mature.

► **Table 8.3** *Social and economic problems of former mining communities*

▼ **Figure 8.11** *Coalfields and coal production sites in the UK, 2002*

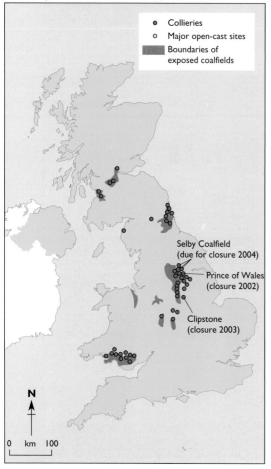

Social and economic problems	Example: South Yorkshire
■ High levels of unemployment.	■ Unemployment 50 per cent higher than the national average; long-term unemployment twice the national average.
■ Deindustrialisation – industry shed 170 000 jobs between 1978 and 1998, including 42 000 in steel and 30 000 in coalmining.	■ The lowest disposable income of any area of the UK. Income support claimed by 84 000 households with children. Deprivation levels in the four local authority districts of South Yorkshire increased significantly between 1991 and 1998.
■ High levels of poverty, crime and, in some instances, drug abuse.	■ A large proportion of the population living in substandard local authority and former Coal Board housing. Amount of derelict land increased five times faster than the national average between 1978 and 1998.
■ Poor housing and poor environment.	■ Poor levels of achievement in schools; adults often have poor qualifications or none at all.
■ Poor educational attainment and lack of skills.	■ Life expectancy below national average; above-average rates of death from heart disease and cancers.
■ Poor health.	■ Depopulation of 0.02 per cent between 1994 and 1998. Young people (often the most progressive) migrate, leaving a declining and ageing population.
■ Population decline resulting from the net outflow of people.	■ Many mining communities are in rural areas. Poverty means that many people often have no access to private transport. Public transport is inadequate.
■ Isolation.	

coalfield of North Derbyshire and North Nottinghamshire alone lost 55 000 jobs between 1981 and 1999. During the 1990s, some of the most severe poverty in the UK was found in coalmining communities (Figure 8.12). Table 8.3 describes some of the social and economic problems of these communities.

The **multiple deprivation** suffered by former mining communities (Table 8.3) results in the social exclusion of many people living in coalfields. Low incomes, unemployment, poor health and low educational attainment create barriers which deny people access to jobs and a meaningful place in society.

Tackling the social and economic problems of the coalfields

A wide range of initiatives have been put in place to tackle the social and economic problems of coalfield communities. Some initiatives come from the British government, and some from the EU (see Box 2 and Figure 8.13). In 1997, the British government set up a Coalfields Task Force to develop a programme for the regeneration of the coalfields. As a result, the government agreed to spend over £1 billion a year on **regeneration** of these areas. Priority is given to job creation, and improving housing and transport. Most funding is available from the government's Single Regeneration Budget. Because many coalfields are in assisted areas, firms can also get regional aid through grants which cover 40 per cent of the cost of new investment projects. EU **regional aid** is also available in a number of coalfield regions. Re-organisation of regional aid in 1999 meant that two British coalfield regions – South Yorkshire and the Welsh Valleys – qualified as Objective 1 regions (see Box 2). Objective 1 regions are the poorest in the EU and have a GDP per head of less than 75 per cent the EU average. In these regions, the EU's regional funds will pay up to 75 per cent of the costs of economic development projects. Many more coalfield regions qualify for generous grants as Objective 2 regions.

► **Figure 8.12** *As in this abandoned street in County Durham, the decline of the coal mining industry has taken its toll on the villages that depended on it.*

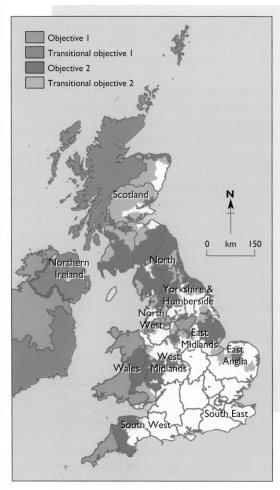

Figure 8.13 *Objective 1 and Objective 2 areas in the UK, 2000–2006*

Box 2 Objective 1 and Objective 2 areas in the UK

- Objective 1 areas include Merseyside, West Wales and the Valleys, South Yorkshire and Cornwall, and cover 8.5 per cent of the UK population (Figure 8.13). The EU describes these areas as 'substantially lagging behind the rest of Europe' – that is, with a GDP per head which is less than 75 per cent of the EU average. In future, Objective 1 money will be spent on projects aimed at improving competitiveness, promoting new businesses and innovation, and creating sustainable economic development. In the past, most money has gone to improving infrastructure by building roads and industrial estates. In 2001, 83 million people in the EU lived in regions with Objective 1 status.

- Objective 2 covers four types of area: industrial, urban, rural and those dependent on fisheries. Industrial areas eligible for Objective 2 assistance have unemployment above the national average, a higher percentage of jobs in the industrial sector than the EU average, and an absolute decline in industrial employment. Rural areas eligible for Objective 2 assistance have population densities less than 100 km², employment in agriculture more than twice the EU average, unemployment above the EU average, and declining populations. In 2001, 68 million people in the EU were covered by Objective 2. Objective 2 grants aim to regenerate economies by diversifying employment, developing new technologies, creating jobs, improving infrastructure and promoting small manufacturing enterprises and tourism. Between 2000 and 2006, Objective 2 areas in the UK will receive EU structural funds worth £2.6 billion.

8.4 Secondary economic activities

Industrial changes in MEDCs

Deindustrialisation

We saw in section 8.1 that the proportion of the workforce in manufacturing in MEDCs has fallen steeply over the last 25 years. Job losses in manufacturing in the UK and the EU occurred most rapidly between 1975 and 1985. This process of **deindustrialisation** hit hardest the older, 'smokestack' industries such as iron and steel, shipbuilding, chemicals, textiles and heavy engineering. However, while deindustrialisation caused a massive shake-out of employment in many manufacturing industries, it also led to significant improvements in productivity (Figure 8.16).

Figure 8.14 *The effects of deindustrialisation: employment changes by manufacturing sector in the EU, 1983–93*

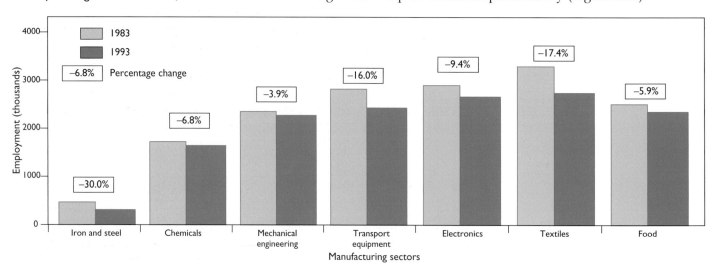

Key Terms

Deindustrialisation: The impact of industrial decline on a previously industrialised area.

The geographical impact of deindustrialisation varied according to a region's industrial structure. Regions dominated by traditional heavy industries such as north-east England, the Ruhr (Germany) and Nord-Pas-de-Calais (France), experienced huge job losses and economic decline. In contrast, regions with a diverse industrial structure such as south-east England, Bavaria and the Ile-de-France, faired much better. Their mix of light industries and growth industries such as high-tech made them far less vulnerable to deindustrialisation (see Box 3).

► **Figure 8.15** *Deindustrialisation in the inner city. Derelict Lister's woollen mill, Bradford, West Yorkshire*

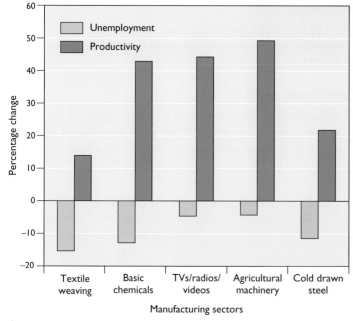

▲ **Figure 8.16** *Changes in employment and productivity in selected EU industries, 1993–99*

Box 3 Causes of deindustrialisation
A combination of factors brought about the rapid deindustrialisation in the UK and EU between 1975 and 1985.
- Global competition – many manufacturing industries such as textiles and iron & steel, based on low technology, grew rapidly in NICs and LEDCs. These countries held a comparative advantage of lower labour costs. Some new industries in NICs were also protected and subsidised by governments as part of an export-led industrialisation programme.
- Recession in MEDCs, following the oil price rises of 1973 and 1978, reduced the demand for many manufactured goods.

Many traditional manufacturing industries in MEDCs (e.g. iron & steel and motor vehicles in the UK) suffered from underinvestment and overmanning. Low productivity and high labour costs made these industries uncompetitive.

New industrial growth in the UK: central Scotland

The growth of new industries in the UK since the mid-1980s has offset the contraction in manufacturing caused by deindustrialisation. At the same time, **reindustrialisation** has transformed the industrial structure of regions such as central Scotland, South Wales and north-east England.

Key Terms

Foreign direct investment (FDI):
Investment in a country by a foreign company (usually a TNC). FDI in manufacturing usually comprises factories, plant and machinery.

Until the 1970s, central Scotland was a traditional industrial region, relying on smokestack industries such as shipbuilding, textiles and steel. Following deindustrialisation, manufacturing was reborn. Scotland attracted huge **foreign direct investment** (FDI) from US, EU and East Asian transnational corporations (TNCs). Much of this investment was in electronics, especially in semi-conductors and consumer equipment such as VCRs and PCs. Investment concentrated in the central belt between Glasgow, Edinburgh and Dundee – an area eventually dubbed 'Silicon Glen'. The recent emergence of research and development (R&D) to support the semi-conductor industry is a sign that 'Silicon Glen' has become a mature high-tech region. By the late 1990s, electronic products accounted for 40 per cent of Scotland's exports. Foreign-owned companies such as NEC, Canon and Mitsubishi had around 330 plants in Scotland which between them, employed 82,000 people.

The attractions of central Scotland for FDI include:
- A well-organised development agency – Locate in Scotland – which markets the region.
- The existing cluster of electronics companies in central Scotland which creates a group of suppliers and buyers of components.
- The availability and quality of the Scottish workforce.
- The availability of regional aid (grants) in assisted areas.

Globalisation and industrial growth in NICs and LEDCs

The industrialisation of countries such as South Korea, Taiwan, Thailand, China and Brazil in the last 20 years reflects the increasing globalisation of industry. **Globalisation** means that firms source materials, manufacture products and target markets on a worldwide scale. The globalisation of manufacturing is the outcome of :
- A lowering of transport costs.
- Improvements in communications (e.g. telecommunications, the Internet, IT, etc).
- The gradual removal of barriers to world trade.

The key players in the globalisation process are very large transnational corporations (TNCs).

Transnational corporations (TNCs)

TNCs are very large firms with production units in more than one country. Even so, few TNCs are truly global. A typical TNC such as General Motors, will produce two-thirds of its output and employ two-thirds of its total workforce in its home country (Table 8.4). A firm organised on a global scale

7 Make a search of newspapers and magazines for advertisements designed to attract investment to Scotland.

a List the locational advantages claimed by the advertisements for Scotland.

b Which advantages, if any, are given prominence?

c Comment on the effectiveness of the advertisements.

▶ **Table 8.4** *The world's ten top TNCs*

Company	Industry	Percentage workforce from outside home country
Shell	Energy	77.9
Ford	Vehicles	29.8
General Electric	Electronics	32.4
Exxon	Energy	53.7
General Motors	Vehicles	33.9
Volkswagen	Vehicles	44.4
IBM	Computers	50.1
Toyota	Vehicles	23.0
Nestlé	Food	97.0
Bayer	Chemicals	54.6

has a number of advantages. By serving a global market it can achieve economies of scale in production and marketing. It can also exploit areas of lower costs (e.g. lower wages, less restrictive safety and environmental laws). However, many firms simply become transnational by buying foreign firms. Others, such as car parts suppliers, may be forced to follow their main markets (i.e. car assemblers) which have located overseas.

Industrial growth in LEDCs

Industrial growth in the economically developing world is very uneven. In the last 25 years, rapid industrialisation has occurred in countries such as Brazil, China and India. Industrialisation has, meanwhile, made little impression on most African countries.

The critical factor influencing industrial growth in LEDCs is foreign direct investment (FDI). Most FDI is from TNCs based in North America, Europe, Japan and East Asia. Today, 40 per cent of all FDI goes to LEDCs. However, the bulk of this investment is concentrated in a handful of countries, with China being the major recipient in the late 1990s. Despite its rich natural resource base, Africa receives almost no FDI.

Investment by TNCs in LEDCs usually occurs for two reasons: to serve local markets and to exploit cheaper resources such as labour and materials.

FDI to enter local markets

TNCs may locate factories in LEDCs in order to sell the output in local or regional markets (Figure 8.17). Such locations will already have a relatively well-off population which can afford to buy cars, computers and other consumer goods. The south-east region of Brazil with major cities such as Sao Paulo, Rio de Janeiro and Belo Horizonte, is an attractive location for this reason. TNCs such as Volkswagen, Fiat and Ford have plants in this region which serve a market of more than 40 million people. The region also has a large, skilled workforce and a sophisticated **infrastructure**. This tends to suggest that FDI goes to countries and regions which have the best growth prospects. It also explains why most LEDCs have so far attracted little FDI and have achieved minimal industrial growth.

8 Read the text on TNCs:

a Suggest two possible reasons why most TNCs have the bulk of their workforce in their home country.

b Why do you think that oil companies often have a large proportion of their workforce outside their home country?

c Nestlé, a Swiss food manufacturer, has 97 per cent of its workforce outside Switzerland. Suggest one possible reason for this.

▶ **Figure 8.17** *This General Motors car plant in Brazil is a good example of foreign direct investment in an LEDC.*

275

EXAMPLE: Maquiladoras in Mexico

FDI to exploit low-cost labour

In Mexico, the border region with the USA has seen astonishing industrial growth in the last 25 years (Figure 8.18). It has attracted a flood of foreign investment from the USA, Japan and East Asia. Four thousand firms employing 1 million workers have located factories known as maquiladoras, in the border zone (Figure 8.19). About half the production from the maquiladoras comprises textiles and consumer electronics goods such as TVs and VCRs. Industrial output in the border zone is worth about $7 billion a year.

The USA–Mexico border zone has two attractions for industry. First, it is one of the few places where a MEDC (in this case the world's richest country) and a LEDC share a common border. Not surprisingly, labour costs are much lower in Mexico: wages in Tijuana (on the Mexican side of the border) are just one fifth of those in San Diego (USA) a

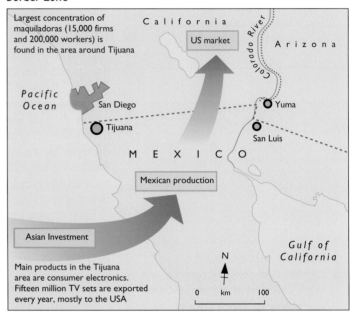

▼ **Figure 8.18** *Maquiladora production in the Mexico–USA border zone*

▲ **Figure 8.19** *Mexican maquiladoras: labour is cheaper in Mexico than the USA.*

few kilometres away. Second, a border location gives firms easy access to a huge American market. This reduces transport costs and enables firms to respond quickly to changes in market demand. Other factors have also played a part in the success of maquiladoras. Firms locating in the border zone do not have to pay government duties on imported materials. Because labour is also relatively cheap, components (e.g. for electronic equipment) can be sourced from other maquiladoras in the region at lower cost than in the USA.

A similar industrial zone has developed in the Pearl River delta in China, close to Hong Kong. Here, five million Chinese work in factories largely owned by Hong Kong and Taiwanese businesses. The principal attraction is the low labour costs.

Industrial change and globalisation

The globalisation of the world economy has brought both advantages and disadvantages. As TNCs have sought to lower their costs and exploit global markets, they have invested overseas. FDI has created jobs, promoted world trade, strengthened the trade balances of individual countries and introduced new skills. Thanks to globalisation, firms can operate more efficiently, accessing labour and materials where they are cheapest. But globalisation also has some disadvantages:

■ By increasing global interdependence, any economic downturn affecting a leading country or group of countries, is likely to have an impact worldwide. Thus, the financial crisis in East Asia in 1997 and 1998 led to the closure (and mothballing) of several South Korean factories in the UK with the consequent loss of jobs.

- Over-reliance on FDI creates branch-plant economies. Branch plants perform routine tasks, such as assembly, which require limited skills. They often rely on relatively cheap labour and have no involvement in research and development or decision-making. During a recession, branch plants are the first to close (Figure 8.20).
- Foreign TNCs (compared with firms based in a single country) are less committed to the workers and regions in which they locate. Being multinational it is fairly easy for TNCs to transfer production to other countries. Thus, branch-plants and branch-plant economies are vulnerable to closure and disinvestment.

EXAMPLE: Dyson – dilemmas for an expanding company

The Dyson dual cyclone vacuum cleaner was one of the innovative product successes of the 1990s. In 1995 Dyson, the inventor, set up business in a disused factory in Malmesbury, Wiltshire. This location decision was influenced by personal choice and accessibility (close to the M4 motorway). The factory grew to employ 1500 people, 43 per cent of whom lived within eight kilometres. The company actively supported local businesses and so became an important component of the sub-regional economy.

As sales grew the company wanted to open a second factory, to create a further 1200 jobs. Planning application was rejected by the North Wiltshire District Council because of concerns over traffic congestion in Malmesbury's historic town centre.

By 2001, Dyson faced a dilemma: demand exceeded the capacity of the Malmesbury factory, and more money needed to be invested in research and development (R & D) for new products in a highly competitive market, yet costs in the UK continued to rise. To resolve the problem, in 2002 a decision was taken to close the main production plant in Malmesbury and relocate in Malaysia where costs, especially labour costs, are much lower and capacity is greater. The R & D and, in the short term at least, the production of the latest contra-rotating washing machine, were to remain in Malmesbury, retaining 900 jobs. A knock-on effect on the local economy is inevitable, although the boom district of Swindon is within commuting distance.

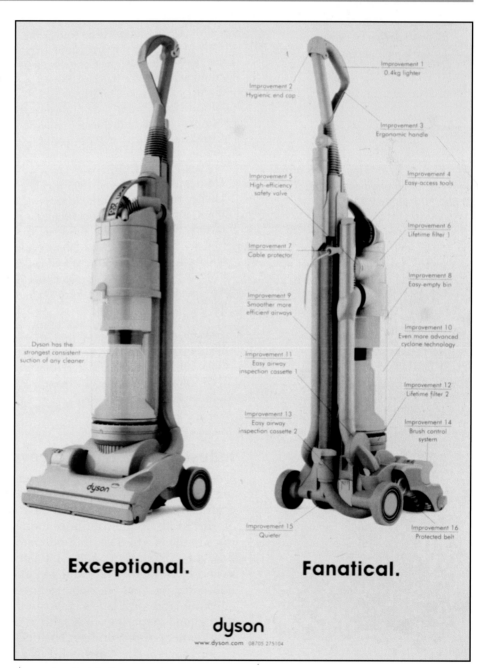

▲ **Figure 8.20** *The dual cyclone vacuum cleaner was Dyson's first success.*

9 Read the newspaper article (Figure 8.21) on the closure of the Homepride factory in Maryport, Cumbria, by the US-owned Campbell Soup Company.

a Explain why, despite its profitability, Campbell's decided to close the Maryport plant.

b According to Workington's MP, what is the main reason for the closure of the plant?

c Why is the closure of the plant a particular blow to Maryport?

d How does this example of plant closure illustrate the disbenefits to small towns of relying on TNCs for employment?

► **Figure 8.21** *The dangers of relying on TNCs for employment (Source: The Times, 30 January 1996)*

FOOD PLANT'S DEMISE WILL DENT TOWN'S HOME PRIDE

US CONGRESSMEN will this week become involved in a campaign to save a food manufacturing plant in the small struggling English town of Maryport.

Mr Dale Campbell–Savours, the Labour MP for the area, will urge Democrats to raise in Congress, the imminent transfer by the US-owned Campbell Soup Company of all production from its recently acquired Homepride factory in west Cumbria, which makes cooking sauces.

Campbell's, which paid Dalgety £58.6 million in August 1995 for the Homepride brand and the Maryport plant, announced in October that the factory was to close by the end of March. Employees said that on takeover, the company talked of investment, but it became clear that investment was to be in the brand, not in the plant.

The company insisted that it only discovered on acquisition that the plant's costs were twice as high as those of its other UK factories. Mr Campbell–Savours strongly disputed these calculations: 'I'm absolutely determined that Campbell's should not get away with nobbling one of our great brand names and then with closing down a highly profitable factory with a probable loss of 123 jobs – all in the name of disposing of a competitor's brand.'

The company said it must transfer production to other UK plants to increase competitiveness, safeguard UK jobs and promote the brand.

The Maryport plant's annual operating profit was £3.9 million on £36 million turnover. But Campbell's claimed its volumes were static and its share of the growing market declining. Substantial change would have been needed, it argued, even if one of the rival bidders had acquired it.

The company's responsibility to Maryport, Campbell's insisted, had been met by offering all employees transfers to plants in other parts of the UK. It had offered to assist those not wishing to move by providing redundancy pay-off above the statutory minimum.

Industrial change and government policies

Most MEDCs have pronounced regional differences in prosperity, unemployment and economic growth. In the UK, for most of the twentieth century, the North was consistently less prosperous than the South. This North–South divide continues today (Figure 8.22). In 2000, unemployment rates in the North-East (5.5–6.0 per cent) were double those in the South-East (1.6–2.8 per cent), and GDP per capita was 84 in the North (compared to an EU average of 100) and 148 in London (Figure 8.22). A survey of household incomes (Table 8.5) showed massive regional differences in wealth at county and local levels between the South-East and the rest of the UK.

Regional disparities in wealth in the UK largely result from differences in employment opportunities. The decline of industry in specialised industrial regions such as the North-west and Wales created high levels of unemployment. Meanwhile, the growth of the service sector and the more diverse manufacturing base in the South has kept unemployment relatively low there. However, it is not just industrial regions which have lagged

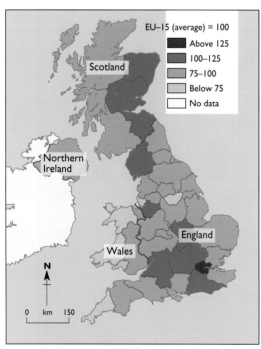

EU–15 (average) = 100

■	Above 125
■	100–125
■	75–100
■	Below 75
□	No data

Scotland

Northern Ireland

England

Wales

N

0   km   150

▲ **Figure 8.22** *Regional GDP per capita in the UK, 2000*

▼ **Table 8.5** *Average annual household incomes by county, 1999*

Top 5 counties	
Surrey	£29 700
Berkshire	£28 100
Buckinghamshire	£27 900
Hertfordshire	£26 900
Outer London	£25 600
Bottom 5 counties	
Tyne and Wear	£17 400
Cornwall	£17 400
South Yorkshire	£17 500
Isle of Wight	£17 700
Mid-Glamorgan	£17 700

behind. Regions dependent on seasonal activities like tourism, or low-wage industries such as agriculture and fishing (e.g. Cornwall, north-west Scotland) are some of the poorest in the UK.

Regional policies in the UK

The primary aim of regional policy is to reduce unemployment in areas where it is persistently high. Regional policy in the UK has a long history. It dates back to the 1930s and the Great Depression. Since then, all governments have adopted regional policies, though with varying commitment. Since the early 1980s, spending on regional policy has fallen steeply. By the 1990s, the cost of regional policy was less than 0.1 per cent of GDP. Details of the current regional policy are given in Box 4.

In spite of nearly 70 years of regional policy, regional disparities in the UK remain as high as ever. Even so, regional policy has had a positive impact in a number of ways:

- It has helped to diversify regional economies and improve infrastructure (e.g. by building new roads, factories, etc.).
- It has secured inward investment, including high levels of FDI in Scotland, Wales, Northern Ireland and northern England.
- It has created thousands of jobs.

Box 4 UK regional policy: types of assistance available

- In January 2000, the government overhauled its regional policy measures and introduced a new assisted map (Figure 8.23). Three types of assisted area were defined. Tier 1 areas had the most acute economic problems and qualified for the highest level of assistance. These areas – Cornwall, South Yorkshire, Merseyside, West Wales and the valleys – correspond exactly to the EU's Objective 1 areas. Tier 2 areas, however, are not identical to Objective 2 areas. This is because the UK's regional policies address different needs, i.e. investment in businesses either to create and/or safeguard jobs and increase competitiveness. Tier 3 areas are eligible for enterprise grant assistance for small businesses, as are Tier 1 and Tier 2 areas.

- The main policy instruments is regional selective aid (RSA). RSA grants, which are discretionary, are available to manufacturing and service industries in assisted areas. Tier 1 areas get the highest levels of assistance; up to 15 per cent of the cost of projects of £500 000 and above. Grants are for land purchase, site preparation, building and plant machinery. In Tier 2 areas, RSA grants are worth up to 10 per cent of project costs. Enterprise grants, available in all three areas, provide financial assistance for small business enterprises. They are available only for projects costing less than £500 000, and their maximum value is £7 000. Only businesses that employ up to 250 people are eligible for enterprise grants.

- Regional enterprise grants (REGs) are available to very small firms. They provide money for capital investment and financial assistance of up to 50 per cent of the cost of projects.

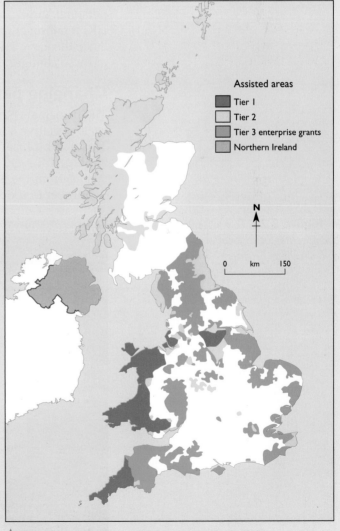

▲ **Figure 8.23** *UK regional policy: assisted areas, 2000–2006*

Key Terms

Retailing: The business of selling to customers for their personal use.

Comparison goods/services: Goods (e.g. footwear and clothing) and services (e.g. lawyer) which are compared for value by customers.

Regional development agencies

In 1999, the UK government set up Regional Development Agencies (RDAs) for the English regions. Previously, only Scotland, Wales and Northern Ireland had their own development agencies, and, as a result, they received more generous funding than the English regions. The RDAs will have funding of £1 billion a year with responsibility for narrowing the wealth gap between the regions, as well as for building factories and clearing derelict land.

EU structural funds

A number of UK industrial regions qualify for financial assistance from the EU's two structural funds – the European Regional Development Fund (ERDF) and the European Social Fund (ESF). Five types of areas are eligible for funding, but only two – Objective 1 and Objective 2 areas – cover urban and industrial problem regions (see Box 2 page 258). Assisted areas in the UK currently receive twice as much financial aid from the EU's structural funds as from the UK government's own regional policy.

8.5 Tertiary economic activities: retailing

Retailing is the business of selling goods and services to customers for their personal, family and household use. As one of the largest industries in the UK, retailing employs 2.3 million people (10 per cent of the workforce). Since the early 1970s, retailing has undergone profound changes. The nature of these changes, their causes and their effects are the subject of this section.

Changing patterns of retailing in the UK

The principal change in the geography of retailing in the last 30 years has been a relative shift from central urban areas to edge-of-town locations. This decentralisation of retailing occurred in three stages:

■ **Stage 1**: Large, free-standing food superstores relocated at edge-of-town sites from the early 1970s (Figure 8.24). Between 1971 and 1992, the number of food superstores increased from 21 to 769.

► **Figure 8.24** *This large Asda foodstore has been located on an out-of-town site near Brighton.*

DESIGNER BARGAIN BOOM

In the past five years there has been a quiet revolution in shopping habits. Once considered a chore, shopping has become a leisure activity – sparking a boom in factory outlet centres which offer low prices on designer goods.

The trend looks likely to continue. There are 28 outlets in Britain and another 19 will open over the next few years.

Developers claim that they have a different ethos from conventional shops in that they only sell last season's stock, seconds, discontinued lines and returned goods which would be otherwise be surplus to retailers' requirements. Most outlets offer discounts ranging between 30 per cent and 70 per cent off high street prices.

Ian Watters, director of MEPC, which has developed six factory outlets in Britain, said that the concept has been a success as it has focused on shopping as a leisure activity.

'People are prepared to be in their car for between an hour and a half to two hours to visit these places – a lot longer than the journey to conventional shopping centres. Shopping is increasingly becoming a leisure activity and the demand for factory outlet shopping reflects this.'

The international property adviser, DTZ Debenham Thorpe, believes that the amount of space devoted to factory retail outlets in Europe will double in the next year. Similar ventures are being planned in Germany and France.

One of the most successful British factory outlets is Cheshire Oaks, near Ellesmere Port, which opened with 32 shops in March 1995. Today, it has more than 100 shops and space for 3,200 cars. It contains high street shops ranging from Timberland to Tie Rack, Gap, Next, Viyella, Helly Hansen and Principles. Every shop guarantees a discount.

Bill Russell, the manager of Cheshire Oaks, said that there were 6 million visitors last year and that numbers were set to increase. 'We feel that Cheshire Oaks is so successful because there are a lot of fashion-conscious people living within a two-hour drive who also like a good bargain.

'The ethos of the shopper tends to be that they have money in their pocket and they don't know what they are going to buy with it but they are going to find a bargain.'

But retailers in Chester – eight miles away – are feeling the pinch.

Bob Clough–Parker, secretary of the city's chamber of trade, said that traders had lost up to 6 per cent of sales – an estimated £15 million – since Cheshire Oaks had opened. This figure was expected to double over the next three years.

'We were always anxious when we learned there would be an out of town development. It was obviously going to present something of a threat to city centre retailing, although there is still a vibrant retail economy in Chester. Clearly it has to be regarded as a threat to the economy but at the same time, it presents a challenge for the business community.'

HOW PRICES COMPARE:

Baby Gap cashmere cardigan
- High Street: £72 ■ Factory outlet: £29.99

Helly Hansen jackets
- High Street: £119 ■ Factory outlet: £70

Liz Clairborne designer jackets
- High Street: £170 ■ Factory outlet: £85

Nike trainers
- High Street: £100 ■ Factory outlet: £85

Benetton sweatshirts
- High Street: £45 ■ Factory outlet: £25

▲ **Figure 8.25** *Factory outlets have an impact on traditional centres of retailing (Source: The Guardian, 30 October, 1999)*

- **Stage 2**: Retail warehouses selling bulky durable goods such as DIY, furniture, carpets, and so on, moved to edge-of-town locations in the 1980s. Initially, these units clustered together in unplanned retail parks, but more recent retail parks have been carefully planned and include food superstores and fast-food outlets to attract the maximum number of shoppers.
- **Stage 3**: Regional shopping centres (e.g. Meadowhall, Lakeside) were built. These sell a full range of high-street durables such as clothing, footwear and other **comparison goods**. These planned, very extensive shopping centres compete directly with retailers in the CBD (see Box 5).

Today, around one-fifth of all shopping expenditure takes place in out-of-town stores. Even so, shopping in town centres still accounts for nearly half of total retail spending. Chains (multiples) such as WH Smith, Marks & Spencer, and Boots, with branches nationwide, dominate retail spending. A strong trend in the last 25 years has been the growth of large multiple retailers at the expense of independent shops.

During the 1990s, a number of new types of retailing appeared, including warehouse clubs, factory outlets, **teleshopping** and shopping on the internet. Warehouse clubs are businesses which specialise in bulk sales of reduced-price, quality goods. Access is limited to subscribing customers. Factory outlets (Figure 8.25) are groups of shops selling seconds or end-of-line goods. Both warehouse clubs and factory outlets have out-of-centre locations, extensive parking, fast-food outlets and children's play areas. These new types of retailing will further modify shopping behaviour and the geography of urban retailing in future.

The causes of retail change in the UK

There are several explanations for retail change and specifically for the decentralisation of retail services in the last 30 years:

---

10 Read the newspaper article (Figure 8.24) on the growth of factory outlets in the UK.

a What is the main attraction of factory outlets for shoppers?

b How do factory outlets promote shopping as a leisure activity?

c What impact has the Cheshire Oaks factory outlet centre had on retailing in Chester?

Key Terms

Anchor tenant: A major retailer whose presence in a planned shopping centre attracts large numbers of customers and helps to ensure its success.

- The rising prosperity of most British people. One result has been increased levels of retail spending and demand for more retail services.
- Economies of scale have driven the demand for larger retail units. In large stores the fixed cost of wages, rents, heating, local taxes, etc. can be spread over a much greater volume of sales, thus raising profits. Meanwhile, large stores require extensive sites which are often unavailable in city centres.
- Retailing has followed the suburbanisation of populations. As the better-off have moved from inner city areas to the suburbs, retailing has decentralised to be nearer consumers and the greatest market potential.
- Congestion in city centres has made central shopping areas less accessible and less attractive to consumers.
- With the growth in private car ownership, car-borne shopping (which enables one-stop shopping and bulk purchasing) has become the norm. Suburban sites are relatively congestion-free and provide large areas of free parking. Ease of access and parking are major attractions for shoppers.
- The new malls, retail parks, etc., which have on-site leisure facilities and landscaping are more attractive than many older CBDs, and are also perceived as being safer. They have helped to make shopping the second most popular leisure activity (after watching television).

▶ **Figure 8.26** *The urban shopping hierarchy of Sunderland*

11 Describe and explain three possible factors that influenced the rise and shape of the catchment of shopping centres in Figure 8.26.

12 From Figure 8.26
a Describe the hierarchy of shopping centres in Sunderland, Tyne and Wear.
b Explain why a shopping hierarchy is an effective way of supplying people with the goods and services they need.
c Study data on types of housing (i.e. economic status) and retail floorspace in the housing areas in Table 8.6. Comment on the inequality of access to retail services in Sunderland (Figure 8.26). Use the Chi-squared test to assist your analysis.

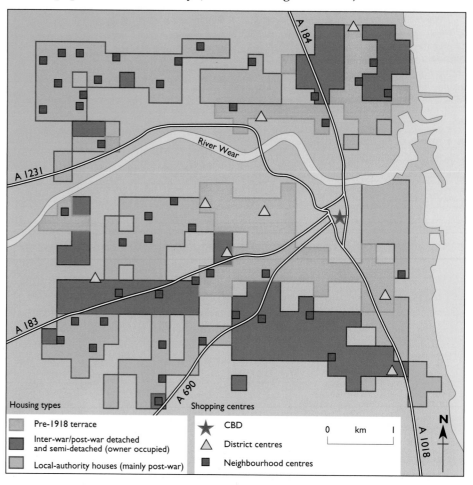

Housing types
- Pre-1918 terrace
- Inter-war/post-war detached and semi-detached (owner occupied)
- Local-authority houses (mainly post-war)

Shopping centres
- ★ CBD
- △ District centres
- ■ Neighbourhood centres

▶ **Table 8.6** *Types of housing and retail floorspace in Sunderland*

	Percentage of housing types in the city	Retail floorspace (ft^2) in each type of housing area
Terrace	25	260 222
Semi-detached/detached (owner occupied)	29	107 446
Local authority (post-1945)	46	92 553

325

282

Box 5 Retail parks and regional shopping centres

Retail park: Lower Swansea valley

■ One of the largest unplanned retail parks is the Lower Swansea valley in South Wales. Easily accessible and located close to the M4, the park developed in the early 1970s from a cluster of superstores. It expanded rapidly during the 1980s when the site became part of the Swansea Enterprise Zone (EZ). Retail units locating in the EZ were exempt from local taxes for ten years and had fewer planning restrictions. Most retailers locating in the park in the 1980s occupied large retail warehouses and specialised in DIY, furniture and furnishings. By the late 1980s, the total retail floorspace in the park was equal to one-fifth of that in Swansea's CBD. Its continued growth threatened the viability of retailing in the city centre.

Regional shopping centres

A number of regional shopping centres have been established in the UK since the late 1970s (Figure 8.25). These centres are:

■ Purpose-built with enclosed shopping malls (often on two levels).
■ Large, with around 100 000m^2 of retail floorspace.
■ Based on car-borne shopping, with up to 10 000 free parking spaces.
■ Dominated by comparison-goods retailing and high-street retailers, with one or two department stores or variety stores as **anchor tenants**.
■ In direct competition with shops in the central areas of towns and cities.
■ People go there (event places) for a day out because there are also cafes, cinemas and restaurants.

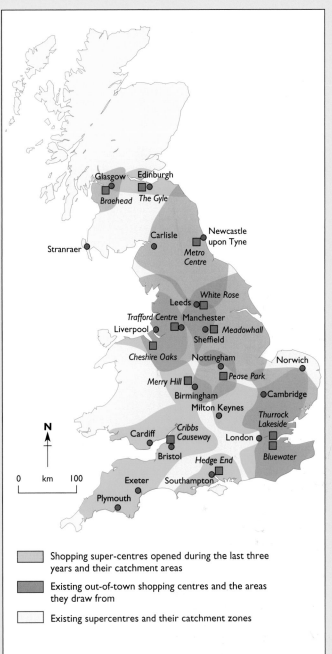

▲ **Figure 8.27** *Regional shopping centres and their catchment areas in the UK*

Meadowhall

■ Meadowhall (Figure 8.28) on the eastern outskirts of Sheffield, is a typical example of a regional shopping centre. Opened in 1990, it was built on derelict land in the Don Valley. It has 280 shops (including all the leading high-street multiples). It is conveniently located alongside the M1, serves a catchment of 9 million people living within an hour's drive time, and receives 22 million visitors a year.

◀ **Figure 8.28** *Meadowhall shopping centre*

The economic impacts of retail change

Fifty years ago the hierarchy of urban shopping centres comprised the CBD, district centres, neighbourhood centres and local/corner shops. The CBD was the main source of comparison goods and services, and provided for the occasional shopping needs of most urban residents. The district centres met the demand for weekly **convenience shopping**, while neighbourhood centres and corner shops provided for daily needs. This hierarchy, based on public transport and walking, meant that the goods and services needed most were often available locally.

The traditional hierarchy, however, is no longer an accurate description of retail provision in UK cities. In the last 30 years, several new types of shopping centre, including free-standing food superstores, retail parks and regional shopping centres, have modified the traditional intra-urban hierarchy. These new types of shopping centres reflect changes in shopping behaviour made possible by personal mobility (private cars), and the almost universal ownership of domestic appliances such as fridges and freezers.

The growth of food superstores operated by multiples has been at the expense of thousands of small independent retailers and corner/local shops. Independent food retailers such as butchers, greengrocers and bakers have been hardest hit (Figure 8.28).

Following the suburbanisation of superstores and retail warehouses, a third wave of retail decentralisation brought about the establishment of regional shopping centres at edge-of-town locations. These centres, dominated by comparison-goods retailing, were in direct competition with high-street retailers. In some parts of the UK, regional shopping centres and other large edge-of-town centres threatened the viability of high-street shopping in nearby town centres. For example, the Metro Centre on Tyneside had a severe impact on Gateshead; Merry Hill in the West Midlands devastated the centre of Dudley and even Sheffield's city centre was badly hit by Meadowhall. The decline of retailing in town centres is evident by the loss of major high-street multiples, and increases in the number of vacant units, charity shops, discount stores and amusement arcades.

▼ **Figure 8.29** *Out-of-town shopping has had a huge impact on smaller towns and villages, causing closures of independent retail outlets such as 'corner shops'.*

The social impacts of retail change

New shopping centres in the suburbs are sometimes poorly served by public transport. They rely on the car-owning population. People too old or too young to drive, or too poor to own a car, are doubly penalised. On the one hand, competition may force the closure of local shops within walking distance of people's homes. On the other they cannot benefit from the wider range of goods and lower prices in food superstores, retail parks and regional shopping centres. Increasingly, new edge-of-town centres are beginning to recognise these problems. Some provide a special bus service targeted at less mobile shoppers.

The environmental impacts of retail change

Suburban shopping centres have increased the number of car journeys, causing more congestion and rising levels of air pollution. Most journeys are over short distances and are inefficient. The development of suburban shopping centres in the UK has also contributed to **urban sprawl**. Edge-of-town centres are single-storey buildings requiring extensive parking, so the demand for large sites is often at the expense of the green belt.

Planners' responses to retail change

Concern about the impact of the suburbanisation of retailing led to the government issuing guidelines to planners in the mid-1990s. As a result, local authority planners adopted two strategies to tackle the suburbanisation problem: first they developed policies which aimed to make town centres more attractive to shoppers (Figure 8.30); and second, they placed restrictions on new retail developments in the suburbs and on greenfield sites. As part of a policy of regenerating town centres, local authority structure plans promote:

- A greater diversity of land use, including a wide range of employment and more housing in apartments over shops.
- A greater use of town centres in the evening, with more cafes, restaurants, and other service retailers as well as good public transport.
- Better parking facilities (including Park-and-Ride schemes) to make it easier for motorists to leave their cars outside town centres.
- Improved public transport facilities, making town centres more accessible to the non-car-owning public, and encouraging more people to leave their cars at home.
- Town-centre management, with managers responsible for improving services, the appearance (e.g. new street furniture), design, safety (e.g. pedestrianisation) and security of town centres.

New edge-of-town proposals will get approval only if they do not affect the vitality and viability of existing town centres and suburban centres. Planners will also look closely at the scope for access by different transport media, and at the extra car usage they are expected to generate. Current policies mean that it is extremely unlikely that any new regional shopping centres will be built in the near future. Indeed, all large edge-of-town proposals of over 20000m^2 must be approved by the Secretary of State.

▼ **Figure 8.30** *A refurbished town centre shopping precinct in Bradford*

Figure 8.31 *The City of London: the world's leading financial centre*

8.6 Quaternary and quinary economic activities

Quarternary activities are also known as producer services. They provide services such as accounting and banking for other economic activities (e.g. manufacturing), commerce and business. In contrast, quinary activities provide services directly to individual consumers. Examples of quinary services include education, health care and tourism. In reality, the distinction between quatenary and quinary services is not always clear cut. Banking and legal services can be both producer and consumer services.

Producer services

Producer services are a leading and rapidly growing sector of the post-industrial economy. In the UK, this sector includes sections J and K of the UK's **Standard Industrial Classification** (Figure 8.1). In 2000, producer services employed 4.9 million people (19 per cent of the workforce) which is more than manufacturing. Producer services comprise a diverse range of activities including banking, advertising, accountancy and legal services.

Financial services

Financial services make a huge contribution to the UK economy. In 2000, they employed over 1 million people; contributed 5 per cent to British GDP; and generated net export earnings worth £31 billion. Financial services are one of the fastest growing areas of the UK economy. They offer highly skilled, well-paid jobs and attract well-qualified graduates. London dominates the financial services sector in the UK. In fact, London is one of the world's three biggest international financial centres (Figure 8.31), accounting for more than one-third of all foreign exchange dealing. London is also a major centre for trade in shares and bonds and for insurance and fund management.

Despite London's dominance, financial services have become an important economic sector in several other UK cities. Leeds, Manchester, Birmingham and Glasgow all benefited from the expansion of financial services in the 1990s. Leeds was the only major UK city to record an increase in total employment in the 1990s. Leeds serves mainly local and regional markets. However, in areas of legal services and accountancy, it competes at the national level. Several leading international accountancy firms (Price Waterhouse, Coopers & Lybrand, Arthur Andersen) have established a significant presence in the city. As a result, Leeds shows signs of developing as a glocal centre for accountancy. Apart from London, the only other UK cities with recognisable global service functions are Manchester, Birmingham, Glasgow and Edinburgh.

Globalisation and producer services

Economic activity has become increasingly globalised in the past 30 years. This restructuring of the world economy, known as globalisation, has been powered by TNCs. Today the world's largest 600 companies generate 20 per cent of global production. In addition, the shift to services and finance on a global scale has given major impetus to the growth of world cities.

Globalisation is only possible through the emergence of world cities (Figure 8.32). **World cities** are the command and control centres of the global economy (Figure 8.33). New York, London, Tokyo and Paris are at the apex of the global urban hierarchy. These four world cities, together with the lower-order cities such as Chicago, Los Angeles, Hong Kong, and Singapore, dominate global producer services. In this way they effectively control the global economy.

Figure 8.32 *Shinjuku, Tokyo. Tokyo, together with New York and London dominates the global hierachy. It is a prime centre for global accountancy, advertising, banking and legal services.*

▼ Table 8.7 *World cities and the number of corporate headquarters located there, 1995*

City	Headquarters of the world's top companies
Tokyo	86
London	28
Paris	28
New York	25
Osaka	20
Frankfurt	8
Chicago	7
Washington DC	6
Atlanta	5
Amsterdam	4

13 Describe and explain the advantages to TNCs of a location in a world city.

▼ Figure 8.33 *World cities (After: Baverstock, Smith and Taylor, 1999)*

Defining world cities

World cities have two outstanding characteristics: they are the headquarters of many TNCs (Table 8.7), and they support large concentrations of producer services. Lower down the global urban hierarchy, cities serve smaller areas such as countries, regions and sub-regions. These cities have weaker connections with the global economy and less power to influence it. For example, in the USA, New York is a truy global city but San Francisco and Seattle are secondary global cities, their influence extending only as far as the Pacific Rim of Asia.

The headquartersof large TNCs are disproportionately concentrated in the world's largest cities. TNC headquarters support a wide range of jobs, including high-level decision-making and management. TNCs are also major customers for producer services such as accounting, banking and advertising. Of course, not all TNCs locate in major world cities. Two of the world's biggest TNCs – Microsoft and Boeing – have their headquarters in Seattle rather than New York.

Financial and other producer services have increasingly become concentrated in world cities. The agglomeration of many producer services reflects the importance of inter-linkages with other producer service firms and the corporate headquarters of manufacturing and other service firms. Despite the key role of information technology in promoting the globalisation of service activities, face-to-face contact between managers and decision-makers remains crucial. The classification of world cities in Figure 8.33 is based on their importance as producer services and financial centres

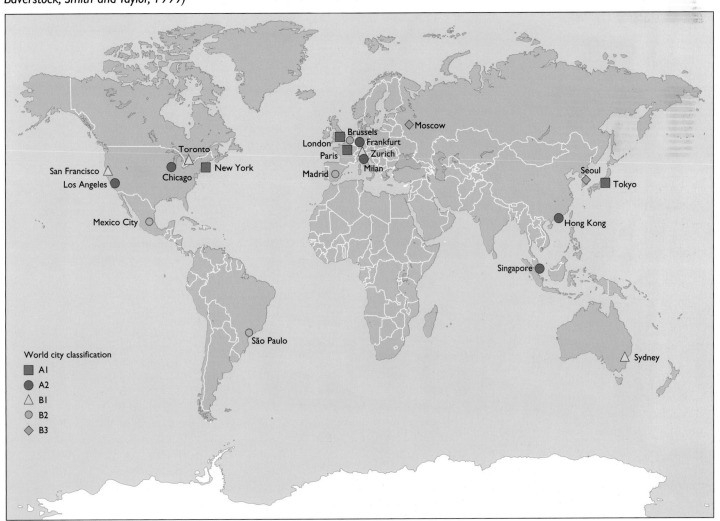

Research and development

Research and development (**R & D**) and design functions are knowledge-based industries. They employ highly-qualified engineers, scientists and technicians. The manufacturing side of production normally operates in a different place.

Many knowledge-based firms and the R & D departments of large firms are located in science parks. Science parks are often purpose-built, with attractive buildings, ample parking and landscaped surroundings (Figure 8.34). Most science parks locate in proximity to universities and research institutions with which they have strong links.

The Cambridge Science Park, established in 1972, was the first in the UK.

▼ **Figure 8.34** *York University Science Park*

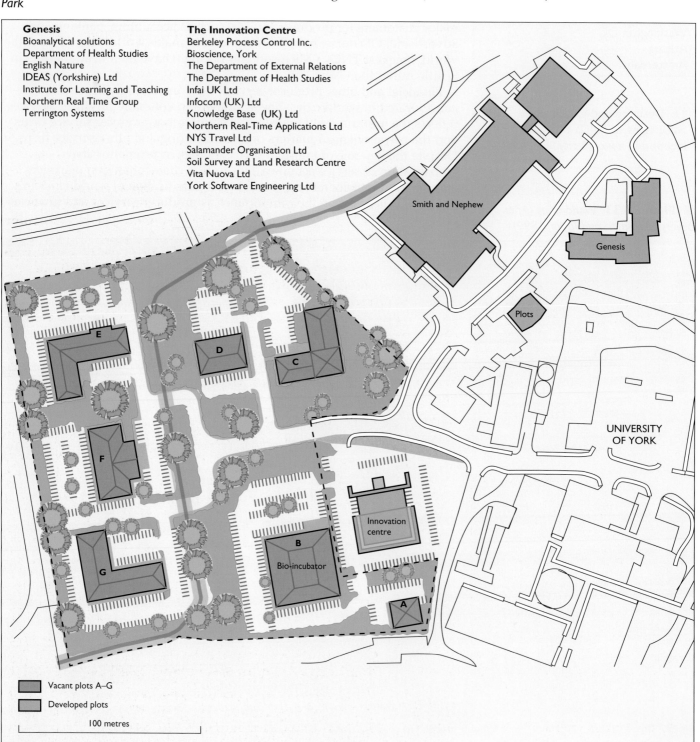

Genesis
Bioanalytical solutions
Department of Health Studies
English Nature
IDEAS (Yorkshire) Ltd
Institute for Learning and Teaching
Northern Real Time Group
Terrington Systems

The Innovation Centre
Berkeley Process Control Inc.
Bioscience, York
The Department of External Relations
The Department of Health Studies
Infai UK Ltd
Infocom (UK) Ltd
Knowledge Base (UK) Ltd
Northern Real-Time Applications Ltd
NYS Travel Ltd
Salamander Organisation Ltd
Soil Survey and Land Research Centre
Vita Nuova Ltd
York Software Engineering Ltd

Vacant plots A–G
Developed plots
100 metres

Today, there are 50 science parks in the UK. They house nearly 1400 knowledge-based industries and employ 25000 highly-skilled people. Some parks, like that in Cambridge, occupy **greenfield** sites on the edge of town. Others, such as Bradford's science park, are located on inner city sites close to universities.

The York Science Park (Figure 8.34) is one of the newest. Opened in 1995, it occupies nearly 10 hectares on the York University campus. It aims to attract knowledge-based enterprises, offering purpose-built facilities and strong links with the university's scientists and engineers. One major multinational firm, Smith & Nephew, which recently relocated its Group Research Centre for worldwide healthcare to York.

8.7 Economic activities in the twenty-first century

Dynamism and change, so characteristic of the world economy between 1970 and 2000, are likely to continue unabated during the twenty-first century. In MEDCs, the relative importance of manufacturing industry will decline further (Figure 8.35). Industries which rely on low technology and cheap labour will relocate in LEDCs, leaving MEDCs to concentrate on innovative technology, research, development and design. But the continued decline of employment in manufacturing will be more than offset by further growth in the service sector. Today services, rather than manufacturing, drive the global economy. Already, services account for two-thirds of output and 80 per cent of employment in rich countries like the UK.

In MEDCs, far-reaching changes are currently affecting service industries. Economies are increasingly dominated by financial services, media and communications industries. Microsoft is now one of the world's most valued company. Mergers between media and communications giants such as AOL and Time Warner dominate the news headlines. The value of the mobile phone and internet companies far exceeds those of many well known manufacturing firms.

14 Why do the production, headquarter and R & D functions of a company often have different locations?

▼ **Figure 8.35** *Cambridge Science Park houses many knowledge-based companies involved in Research and Development.*

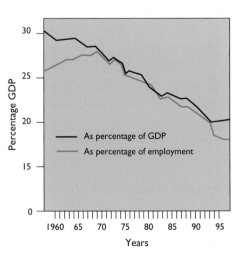

▲ **Figure 8.36** *The decline of manufacturing industry in MEDCs*

At the start of the twenty-first century, the potential of the internet for buying and selling is beginning to be realised. **On-line shopping** for items as varied as cars, books and clothes poses a serious threat to high street retailing and the jobs it supports. Retailers may find it more profitable to close high street stores and deal with customers on-line. As a result of information technology, banks have already closed hundreds of high street branches. With on-line banking, more closures will follow. However, one lesson that economic history teaches us is that technological change usually creates more jobs than it destroys. Despite massive deindustrialisation in the UK in the 1980s, there are more jobs in the UK today than at any time in the last one hundred years.

In LEDCs, agriculture's relative contribution to employment and GDP will fall as rural populations migrate in increasing numbers to towns and cities. Meanwhile, globalisation will cause TNCs to seek out new centres of production and new markets in LEDCs, providing much needed jobs. Hopefully, the expansion of trade through the relaxation of trade barriers will encourage economic expansion in the world's poorer countries. Overall, we can look forward to greater interdependence between countries and, in the course of the century, to the emergence of a true global economy.

THE TOP TEN FASTEST GROWING COMPANIES IN EUROPE

Company Name	Estd.	Sector	Turnover last year (millions US$)	Percentage turnover change since 1993	Real employment growth	Percentage employment Change since 1993/94
Realtech AG	1994	Services (IT and Internet)	17.4	336900	102	4550
BUW Unternehmensgruppe	1993	Services (IT and Internet)	4.6	37671	93	4650
Lernout and Hauspie Speech Products	1987	Services (IT and Internet)	203.2	38918	1705	1795
Key Tech Products Ltd	1993	Manufacturing (electronic and other electrical)	14.1	18866	275	1100
Brokat Infosystems AG	1994	Services (IT and Internet)	31.6	7895	506	8433
Progressive Computer Recruitment Ltd.	1990	Services (personal and business)	91.1	13142	188	1567
Genion Fahrtzeugtechnik GmbH	1994	Services (engineering and management)	6.2	11115	51	1275
Iona Technologies PLC	1991	Services (IT and Internet)	110.2	6515	588	4900
Adam Associates	1994	Services (personal and business)	9.7	8084	48	2400
Dyson Appliances	1991	Manufacturing (electronic and other electrical)	271.7	7199	886	2953

▲ **Table 8.8** *The ten fastest growing companies in Europe*

15 Study Table 8.8. which gives a profile of Europe's fastest growing companies in 2000.

a Write a paragraph summarising the main characteristics of Europe's fastest growing companies.

b On the evidence of the information in Table 8.8, comment on the possible future direction of economic activity in Europe and in MEDCs in general.

Summary

There are five broad sectors of economic activity. The primary sector covers activities which procure raw materials, energy and food. Manufacturing comprises the secondary sector, producing semi-finished and finished goods. The tertiary sector provides services such as transport, power supply, etc. to industry. The quaternary sector provides producer-oriented services such as banking, insurance, finance, etc. The quinary sector includes those services provided directly to the individual consumers, e.g. health care, education.

As a country's economy grows, a shift in employment takes place between the five economic sectors. Primary activities (especially agriculture) dominate pre-industrial economies. Industrial economies experience a shift in employment from the primary sector to the secondary sector. Service activities are the leading employer in the post-industrial economies of MEDCs.

Major changes occurred in the UK coal industry between 1970 and 2000. Output and employment declined steeply between 1970 and 2000. This decline was due to competition from other fuels (e.g. gas) and cheap foreign imports of coal. The decline of mining had a devastating effect on the economies and social fabric of many mining communities in the coalfields.

Coal production often has an adverse impact on the environment.

The environmental effects of coal mining are felt at global and local scales. Coal-burning power stations and the release of CO_2 contribute to global warming. So_2 emissions from power stations cause acid rain and pollute the atmosphere at a local scale. The local effects of deep mining and open-casting (spoil heaps, subsidence, dust, etc.) can be severe in long established coalfield regions.

Globalisation of the world economy has been the major influence on industrial change in both MEDCs and LEDCs since 1970. Manufacturing is increasingly dominated by the activities of TNCs which operate at a global scale. The growth of manufacturing in many LEDCs has resulted from inward investment by TNCs, exploiting new markets and lower production costs. Industrial growth in many regions in MEDCs has been sustained by FDI in industrial sectors such as motor vehicles and high-tech.

Globalisation has both advantages and disadvantages for countries and regions. FDI provides jobs and skills, and benefits a country's balance of trade. It also creates branch-plant economies, increases dependence and leaves countries and regions vulnerable to disinvestment by TNCs. Globalisation has exposed many traditional industries in MEDCs to competition and contributed to deindustrialisation during the 1970s and 1980s.

Governments play a key role in the location of industry. At the national scale, governments offer grants and loans to attract FDI from TNCs. Governments define assisted areas and operate policies which aim to reduce regional disparities in income and wealth. The EU operates similar policies through its Structural Funds.

The location and structure of retailing has undergone significant change since 1970. There has been a decline of small independent retailers and the growth of large multiples. In addition, this period has seen the creation of new urban shopping hierarchies with the trend to out-of-centre developments such as retail parks, suburban malls, free-standing superstores, factory outlets, etc.

Producer services are a rapidly growing economic sector in post-industrial economies. Producer services make a huge contribution to employment and economic activity in many leading cities in MEDCs. At the global scale, world cities such as London and New York, provide financial services and headquarters for TNCs. This has led to the emergence of a global hierarchy of cities. The largest world cities provide services which drive the global economy.

Chapter 9

GEOGRAPHICAL SKILLS AND TECHNIQUES

This book has introduced you to the main areas of study in Physical and Human Geography. As well as text, there are maps, photographs and diagrams, data and graphs which all help us to understand the topics more clearly. This final section shows you how gain the skills and techniques needed to use this range of resources effectively. These skills are not learned in isolation from the rest of Geography. They are the basic tools which geographers use to understand people and their environment, as well as many of the links between one aspect of geography and another.

There are many opportunities throughout this book to use the skills and techniques described in this final section and they are indicated by the **258** sign.

Many geographers will use these skills and techniques in their independent research and fieldwork investigations. This individual work needs to be well planned and carefully structured and there are specific guidelines in this section to help you write up your work well.

The aim of this section is to show you how to use the geographical tools effectively and make full use of all the resources presented throughout this book. Geographers should be competent in drawing a variety of maps and graphs and appreciate how the limitations of each technique affect the interpretation. Similarly, using data in statistical analysis will help us to interpret evidence in greater depth as well as reach appropriate conclusions. Of course, those conclusions are also based on our understanding of the themes discussed in the earlier chapters and the links we make between the different topics. Remember that everything presented in this book is included for a purpose, to enhance our understanding of geography, so learn to use the resources with confidence!

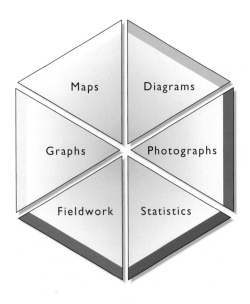

Skills and techniques for geographical investigations

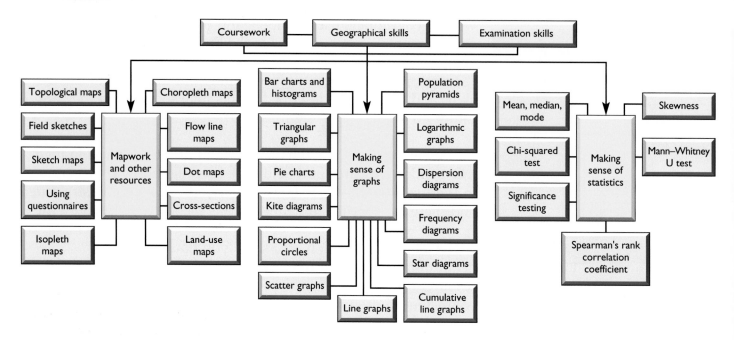

9.1 Introduction

Box I Geographical skills

Geographical skills include:
- understanding maps and photographs
- presenting and understanding diagrams
- sketching
- annotating
- using data
- drawing and understanding graphs
- using and interpreting statistics.

Geographers use maps, photographs, diagrams, sketches, data, graphs and statistics to illustrate, analyse and support a range of geographical ideas. Many examples have been used in the earlier sections of this book. The resources have a purpose and show different ways of presenting information.

Geographical skills are not confined to writing up fieldwork. A variety of skills is needed in different situations and with different data. You may be tested on many of these skills in your examination papers, especially in structured questions or data-response questions. Consequently, many of the examples and exercises in this chapter link to different AS topics. They are not necessarily linked to fieldwork exercises that you might carry out. A connecting symbol **258** is used to link up the skill to another section in this book.

Fieldwork is an important aspect of geography and is required for all AS specifications. The first part of this chapter shows you how to write up your work effectively so that you have a clear record of what you did. Your fieldwork experience is often an excellent source of case study material for other examination answers.

Some specifications examine your skills and fieldwork experience using a geographical skills paper. Others examine fieldwork and skills by coursework or project. It is important that you conduct your fieldwork in a structured and organised way and write it up carefully so that you can gain maximum marks. It is essential, too, that you give enough time to the early stages of planning. Table 9.1 shows how your fieldwork experience will be examined at AS level.

▶ **Table 9.1** *Fieldwork assessment at AS level*

Specification	AS: Mode of examination	% of total AS marks
Edexcel A	Coursework: Personal Enquiry 2500 words OR Applied Geographical Skills Paper	40%
Edexcel B	Coursework: Environmental investigation 2500 words	33.3%
AQA A	Geographical Skills Paper	30%
AQA B	Fieldwork and skills tested in exam paper in Unit 1: The Dynamics of Change	40%
OCR A	Geographical Investigation: written examination AND coursework 1000 words	30%
OCR B	Geographical Investigation: written examination	40%

9.2 Planning your coursework

There are six key elements for successful fieldwork:
- a topic that interests you
- a clear and focused title that allows for some analysis
- a thorough appreciation of the mark scheme
- plenty of varied, relevant and good quality data
- rigorous data collection methods
- a clear structure to the written work.

Choosing a topic

The topic usually focuses on one or two key questions. Some geographical principles or theory must be applied to a specific location in order to answer these questions. The topic should have the potential for plenty of data collection and analysis. Even if you have a limited choice of topic for your class fieldwork, you will be expected to make your work different from the work of other students. Having an interest in the subject matter can make all the difference.

Developing a title

All fieldwork must have an aim, which is reflected in the title. The aim provides a focus for the written report and its conclusion. Your response to the aim should be within the word limit of your study.

Phrasing the title as a question could help to direct the focus of the study, but it could also restrict the kind of response. A 'closed' question, such as *Should Sainsbury's build a new store at Castle Marina?* requires a 'yes' or 'no' answer. It leaves little room to develop any ideas. On the other hand, an 'open' question such as *What is the likely impact of building a new Sainsbury's store at Castle Marina?* allows room for evaluation. There is an opportunity to discuss the issues that arise as a result of building the store.

Setting the hypothesis

Hypotheses help you to structure the data collection and the written report. More successful fieldwork enquiries usually have a series of short hypotheses which, taken together, fulfil the aim. A hypothesis is a simple statement which is investigated to test its accuracy. For example, if the aim is *An analysis of downstream changes in hydraulic variables along the Aber Anafon* then your investigation might focus on the following hypotheses:
- the width:depth ratio increases downstream
- streams become more efficient downstream
- as velocity increases, discharge increases.

Box 2 Phrasing the title

Compare these titles and their effects:

1 A study of sand dunes
This title could be about any aspect of a dunes system.

2 Investigating change in abiotic factors across the Aberffraw sand dunes.
This title is focused on one aspect of a specific dune system. It guides you towards considering 'change' in your analysis and conclusions, and helps you to determine what data to collect.

64

Taken together, an investigation of these hypotheses will fulfil your overall aim and it will be much easier to produce an organised and logical investigation.

The aim and hypotheses must show geographical understanding. Sensible geographical reasoning, theoretical ideas or models should be used to support the analysis and a brief explanatory section should be included. You may start a paragraph with 'Theory suggests', and then identify and discuss the relationship between two sets of data. Remember that just because two sets of data can be graphed, does not necessarily mean they are linked together in some way.

The mark scheme

Read the mark scheme before you begin your fieldwork. The mark scheme explains what the examiners are looking for and how marks are allocated. It will show you how to reach maximum marks (Table 9.2).

Make sure you cover all the aspects described in the mark scheme. The more detail given in the report, the better the grade, but there may be a penalty for overlength work. Pay particular attention to the word limit. Use appropriate geographical terms and be concise with your text.

▼ Table 9.2 *Extract from a typical mark scheme for data collection and how to attain the band*

Marks	Description	To attain the band
7–9 marks (max)	The student follows a systematic research programme and collects sufficient data to meet the aims of the fieldwork, making a range of accurate observations and measurements. The data collection methods, including sampling, are justified and significant factors affecting them are taken into account.	■ make sure that you have different types of data and enough data to test for its significance ■ show how your data was accurately collected ■ explain, or justify, why you selected a particular sampling procedure ■ consider and discuss all the possible factors that may have affected the data you collected.
5–6 marks	The student makes accurate observations and measurements. The data collection methods, including sampling, are explained, and some factors affecting them are commented upon.	■ explain your data collection methods without justifying the choice ■ comment on only some of the factors affecting the data collection ■ your data is not clearly linked to your fieldwork aims.

Collecting data

You should be able to explain each of the decisions you make regarding data collection.

Box 3 The proposal form

Identify the following elements on the proposal form:
■ Location of study area
■ Hypotheses / key questions to be investigated
■ Data to be collected which is relevant for each hypothesis
■ Sampling techniques and size of your sample
■ Analytical techniques you intend to use with your data.
■ Include a risk assessment of your fieldwork. Do not take risks when collecting your data.

The proposal form

It may be necessary to submit a proposal form to the examination board for approval if your fieldwork is to be written up and submitted as part of your examination. The proposal form ensures that you are proceeding with the correct hypotheses, data collection and analyses. It encourages careful planning before you begin the data collection and helps you to avoid any future pitfalls. The more care taken over planning and writing a proposal, the more effective the fieldwork will be.

Use an appropriate scale for your study location. The area should be large enough to find differences but you should not be overwhelmed by too many sites. Allow enough time to carry out a thorough investigation and plan how to manage the time effectively. A pilot study is a useful means of assessing the time needed.

Your analysis is only as good as the data you collect. Ensure that there are minimal limitations to your data collection and make sure that the data is collected as carefully and accurately as possible.

Primary data

Pimary data is data which you collect yourself. There are two types of primary data: **quantitative** data can be specifically measured, whereas **qualitative** data is descriptive. Both have their uses but many enquiries focus more on quantitative information.

Qualitative data is subjective but it may well lead to data that can be measured. Some information needs to be seen for you to form an opinion or evaluation (Figure 9.1). You should conduct such work yourself at each site in order to be consistent.

▼ **Figure 9.1** *Bi-polar analysis to evaluate the environmental quality of an urban area*

	+2	+1	–1	–2	
High density houses					Low density houses
Congested					Little traffic
Area poorly looked after					Area well looked after
Low property values					High property values
Lack of open space/gardens					Plenty of open space/ spacious gardens
Mainly on-street parking					Cars off street when parked
Noisy					Quiet

While interviews with experts may be relevant, they can be difficult to incorporate effectively into coursework. Transcripts of interviews can use up the word limit, but quotations or extracts from your interviews can be very effective. This type of information is best used as an aid to your analysis and conclusions rather than as an end in itself. The expert's views should help to explain your own observations and data. Plan your questions carefully so that you find about what you really need to know.

There are safety hazards in towns, in villages and in the countryside. The sites of your data collection must be safe and manageable. For example, it is unrealistic to survey an entire large city area such as Manchester. Conversely, you are not likely to find measurable changes downstream if the sites along the river are only 50m apart. You should also be sure to collect all the data you need if you are working away from home or school. There may be very few opportunities to return to collect additional information.

Sampling

On most fieldwork investigations you will not be able to measure or record everything in your study area. The data population should be sampled. Give good reasons for the choice of sampling methods in your coursework. Several types of sampling are possible. Your choice depends on the type of data you are investigating and the purpose of your data collection.

Random sampling involves using random number tables to select data or sites for collecting information. It eliminates bias in your sample. Random sampling can be used when the area or population being surveyed is the same and has no different sub-sets. Selecting pebbles to measure is often done using this method.

Systematic sampling uses data from equally distributed collection sites in the study area. Use this method if you expect a variable to change over distance (e.g. vegetation over sand dunes) or time (e.g. number of pedestrians though the day).

Stratified sampling is used when the study area or population has different sub-sets and data needs to be collected from each sub-set, for example, gender or age groups using a leisure centre, or different types of land use in a town. The larger the sub-set, the more samples should be taken from it.

The size of the sample is also important. The larger the sample size the more reliable your analysis and conclusions. Sample size affects the statistical tests that can be applied. Plan your statistical analysis and make sure that the data you collect is appropriate. There is little point in spending a day on a river measuring at only three sites if you intend to correlate variables using Spearman's Rank, which requires at least ten sets of data to be meaningful. If you wish to apply a Chi Squared test you will probably need at least 50 pieces of data to divide into 4–6 groups (Table 9.3).

▼ **Table 9.3** *Type of data needed for the most common statistical techniques*

Statistical technique	Purpose	Type of data required
Spearman's Rank correlation coefficient (page 323)	To see if there is a relationship between two sets of data	10–30 pairs of data; the data should have very few numbers of equal rank in the list
Chi Squared test (page 325)	To see if there is a difference between similar sets of data	Data must be place into groups of five or more elements
Mann-Whitney U test (page 328)	To see if there is a difference between two separate populations	sample sizes of 5–20 pieces of data; both sets of data must have similarly shaped distributions, i.e. they should show roughly the same pattern of values
Quartiles and inter-quartile range (page 322)	To indicate the spread of data around the median value. This helps you to analyse the range or spread of a set of results you have collected	Any measured data

The size of a questionnaire sample is often problematic. At AS level it will be impossible for you to get a representative sample of shoppers in your local large supermarket, and you may have to wait a long time to match this number with people using a local village shop. You need to be practical, as well as objective. Allocate an appropriate amount of time, perhaps one hour, within which you could collect 30 questionnaire responses as your primary data. Use the results from other students as secondary data and acknowledge them in your written report. Conduct a pilot study to make sure that you get appropriate information.

Secondary data

Secondary data is data which you do not collect yourself. This includes:
■ census data
■ data from local councils, such as traffic flows or environmental surveys
■ data from other pieces of fieldwork
■ data from other students in your group
■ data from websites.

Using secondary data can often give you another perspective to analyse. If similar data is available for a different location, or different period, then you may explore the similarities and differences. Your own primary data collection, timing and location must be comparable with the secondary data you acquire, so it is better to begin with secondary data and match the primary data collection to it. The amount of secondary and primary data should be balanced.

The pilot study

A pilot study allows you to test a small amount of data to ensure that the appropriate procedures are followed and responses are being collected. Visit the fieldwork sites and collect a small amount of every type of data that has been planned in the enquiry. Test the data collection methods to sort out any problems that may have arisen. For example, the pilot study might highlight issues with the choice of sites you have selected for the data collection, or the appropriateness of the data you need for your hypothesis.

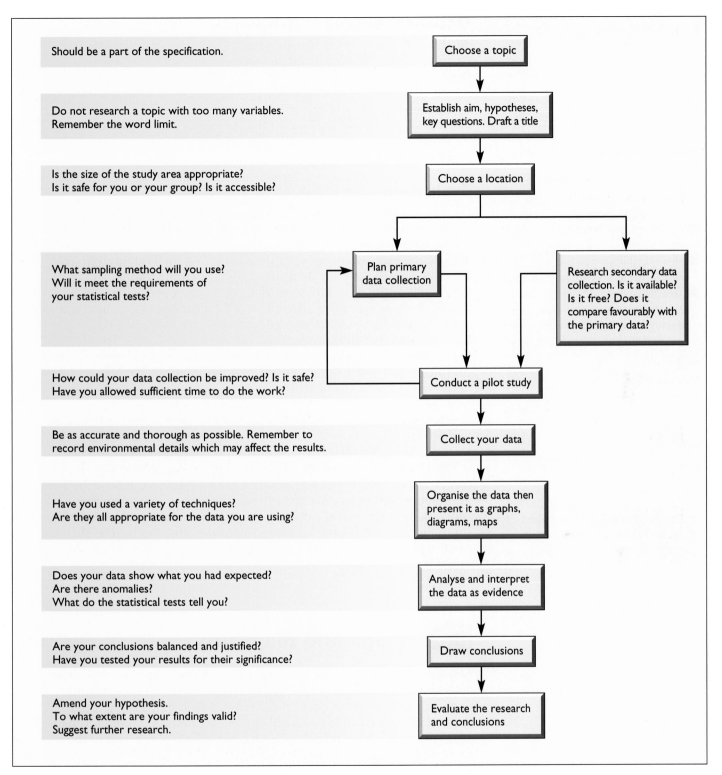

Should be a part of the specification.

Choose a topic

Do not research a topic with too many variables. Remember the word limit.

Establish aim, hypotheses, key questions. Draft a title

Is the size of the study area appropriate? Is it safe for you or your group? Is it accessible?

Choose a location

What sampling method will you use? Will it meet the requirements of your statistical tests?

Plan primary data collection

Research secondary data collection. Is it available? Is it free? Does it compare favourably with the primary data?

How could your data collection be improved? Is it safe? Have you allowed sufficient time to do the work?

Conduct a pilot study

Be as accurate and thorough as possible. Remember to record environmental details which may affect the results.

Collect your data

Have you used a variety of techniques? Are they all appropriate for the data you are using?

Organise the data then present it as graphs, diagrams, maps

Does your data show what you had expected? Are there anomalies? What do the statistical tests tell you?

Analyse and interpret the data as evidence

Are your conclusions balanced and justified? Have you tested your results for their significance?

Draw conclusions

Amend your hypothesis. To what extent are your findings valid? Suggest further research.

Evaluate the research and conclusions

▲ **Figure 9.2** *The Fieldwork Campaign Trail*

Data presentation

Data may be presented in a variety of ways. The rest of this chapter will help you to select different and use appropriate techniques of setting out the information.

Generally, all illustrations should use appropriate annotations. Text may be replaced by detailed and effectively annotated diagrams, particularly if the word count is in danger of being exceeded. Diagrams, graphs, maps and photographs are included for a purpose, which should be clear from the annotations and supporting captions. Each item should be numbered in sequence and carry a reference to it in the main text.

Data analysis and conclusions

Analysis is an important element of any fieldwork enquiry. Intrepret data in terms of the key aim, hypotheses and questions that were formulated at the beginning of the process.

The quality of the analysis may be restricted by the data. Data from a limited number of locations will not allow much choice in selecting an appropriate statistical technique. This will also limit the extent to which the information can be can analysed. The data must be suitable for the statistical test you intend to apply.

Evaluation

Data is evaluated or judged according to the validity of its geographical results. When doing the assessment, consider the validity in terms of the following questions:

■ How representative was your data/sampling method/sample size?
■ Did the timing of your data collection affect your results?
■ Was there sufficient data to reach a significant conclusion?
■ What factors may have affected your data collection, e.g. heavy rainfall, Christmas shopping, local circumstances such as market day?

Each evaluation draws conclusions, which may be partial, tentative or incomplete. Partial conclusions are biased in some way; they are not completely impartial. Tentative conclusions are drawn if there is still some doubt about the assessment. They are usually experimental, showing that more research needs to be done. It may not possible to reach a full or complete conclusion at all. If only a small aspect of the topic was investigated, a limited database was used, or all the variables which affected the topic were not used, then the conclusion will probably remain incomplete.

Fieldwork conclusions should be statistically significant in order to reach comprehensive conclusions. Data may give unexpected results. If the information had been collected carefully and accurately, then the conclusions it yields will not be wrong. The research may, however, contradict an established theory. In such cases, the conclusions should discuss possible reasons for the contradiction, suggest how the theory might be modified, and recommend further investigations.

9.2 Maps and diagrams

Maps show the location of geographical features relative to each other. They are supported by other reference features, which must include a title, scale, key and north, or directional, arrow. Many maps carry copyright restrictions and it is not advisable to simply photocopy them for the report. Maps should be customised to show details that are relevant to the research topic, thus unnecessary details should be left out. Maps may be created on computer using a suitable ICT package. Make sure you have the appropriate skills to use such software effectively. There are various categories of maps, each with a different purpose.

Choropleth maps

Choropleth maps use degrees of shading of one colour to represent density. Data is divided into categories with the densest or darkest shade representing the highest value category. Lower values have progressively lighter shading. There are 4–6 categories on one map. These categories should be of fairly equal size with no value appearing in more than one category (Figure 9.3).

► **Figure 9.3** *Choropleth map showing proposed distribution of new houses in Greater Manchester, 1996–2011*

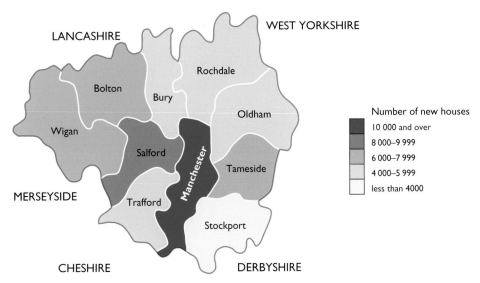

LANCASHIRE

WEST YORKSHIRE

Rochdale
Bolton
Bury
Oldham
Wigan
Salford
Manchester
Tameside
Trafford
Stockport

MERSEYSIDE

CHESHIRE

DERBYSHIRE

Number of new houses

10 000 and over
8 000–9 999
6 000–7 999
4 000–5 999
less than 4000

Hint

When drawing a choropleth map, place a piece of graph paper under your sheet of paper and use it as a guide to draw neat, equally spaced lines.

▼ **Figure 9.4** *Types of graded shading for a choropleth map*

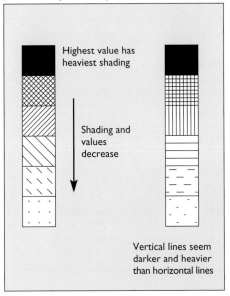

Highest value has heaviest shading

Shading and values decrease

Vertical lines seem darker and heavier than horizontal lines

Cross-hatching may be used instead of colour for shading (Figure 9.4). Keep to vertical and horizontal lines, or to diagonal lines but do not mix the two sets. Usually one colour is used so that the intensity of the shading is evident. Some atlases use tones of two related colours for shading. Do not use white as a category of shading. White implies that there is no data available, that the cartographer has forgotten to fill in an area, or that ideas for colours and types of shading have been used up.

You will discover when you complete activities 1 and 2 below, that choropleth maps give a good visual impression of pattern, but:
- they do not show precise data
- the visual image depends on how the categories are created; a very different picture is created if the values are changed
- the shaded areas appear to have equal values throughout with a sudden change at their boundaries
- many small changes within large areas are hidden.

► **Figure 9.5** *Outline map of regions in the UK*

▼ **Table 9.4** *Balance of internal migration in the UK, 1999*

Region	Balance of internal migration (thousands)
Wales	+2.1
Scotland	−1.2
Northern Ireland	−0.8
North East	−4.7
North West and Merseyside	−12.0
Yorkshire and the Humber	−5.0
East Midlands	+10.9
West Midlands	−7.3
East Anglia	+18.5
London	−46.7
South East	+18.6
South West	+27.7

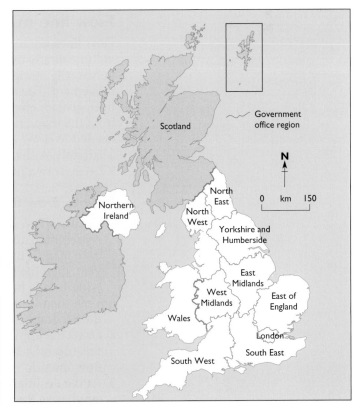

Scotland

Government office region

N

Northern Ireland

North East

North West

Yorkshire and Humberside

East Midlands

West Midlands

East of England

Wales

London

South East

South West

0 km 150

1 Use the data in Table 9.3 and Figure 9.5 to map the balance of internal migration in regions of the United Kingdom.

2 Use the data in Table 9.4 and Figure 9.6 to represent international immigration to the UK 1997.

181

► **Figure 9.6** *Outline map of countries in Europe*

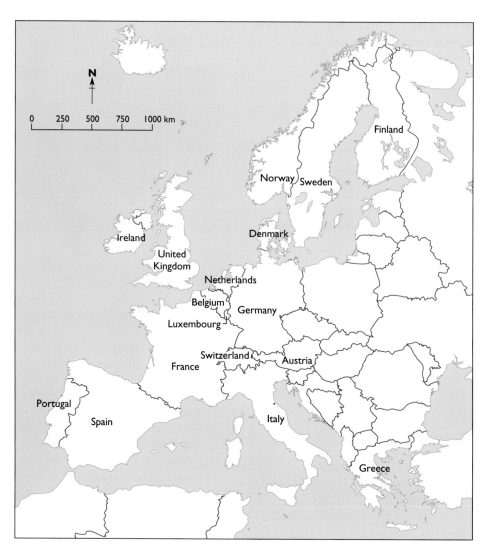

▼ **Table 9.5** *Immigration to the UK from European countries, 1997*

Country of citizenship	Number of Immigrants (thousands)
Belgium	0.5
Denmark	1.2
Germany	7.7
Greece	8.9
Spain	3.2
France	20.6
Ireland	1.4
Italy	2.6
Luxembourg	0.0
Netherlands	4.7
Austria	0.7
Portugal	2.1
Finland	2.8
Sweden	4.3
Norway	1.2
Switzerland	0.2

Flow line maps

Flow lines are bands of different widths drawn on a map to show the total volume of any movement between locations (Figure 9.7). The lines are not necessarily straight and they show as accurately as possible the direction of movement. The thicker the line, the greater the amount of movement. Flow lines give an effective visual image but it is difficult to read precise data from the map and they can be awkward to draw. A specific drawing tool is needed to draw a clear line which can be made thicker, easily and accurately. A sharp pencil creates too thin a line. The information in Figure 9.7 was drawn in using data from Table 9.5.

How to draw the flow lines

1 Decide on a value for the thinnest line. Look at the range of data to be represented and calculate the maximum width of line that will fit on the base map. If the narrowest line, say 1 mm, represents 1000 persons per year, then the widest line will be 85 mm. This would probably be too wide for the base map. If the narrowest line represents 5000 persons per year, then the widest line will be more manageable at 17 mm. However, the accuracy in the lowest values will be lost.

2 Draw the appropriate width of lines from London to each region. Plan this carefully because the map could become congested. Make sure the flow lines do not cross each other; they may be curved if necessary.

3 At the destination region, finish each line with an arrow head to emphasise the direction of movement from London.

► Figure 9.7 *Flow line map to show migration from London to other UK regions*

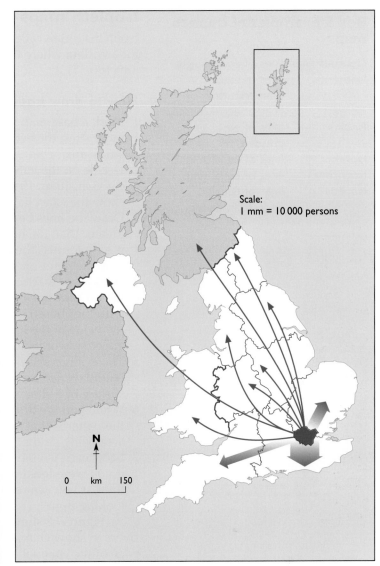

Scale:
1 mm = 10 000 persons

N

0   km   150

▼ Table 9.5 *Outmigration from London, 1998*

Destination region	Number of persons leaving London (thousands)
Wales	5
Scotland	7
Northern Ireland	2
North East	4
North West and Merseyside	10
Yorkshire and the Humber	8
East Midlands	10
West Midlands	8
East Anglia	55
South East	85
South West	20

Dot maps

The distribution of data can be plotted accurately on a map using dots, where one dot has a specific value. Typically, dot maps are used for population or livestock distribution. In chapter 5, Figure 5.10 shows the distribution of population in Nepal and the Ganges plain. The dots have been placed where there are the greatest numbers of people, i.e. along rivers, avoiding areas of low rainfall and high altitude, and where irrigation agriculture is possible. The dots have been placed precisely using other geographical maps and data as reference. The dot map, therefore, shows a much more informative pattern of population distribution than other maps, such as a choropleth map.

Dot maps have two main disadvantages. It is difficult to choose a value for a dot which gives a fair representation of the information. The value of each dot is important. In Figure 5.10a, one dot represents 5000 people and low density areas are over represented. In large settlements there would be too many dots and the map becomes crowded as the dots merge. In this example the urban areas have been represented by proportional circles to enhance the clarity.

The second disadvantage is the impracticality of drawing identical dots. It takes great skill to draw identical dots neatly and accurately. Using a fibre-tipped pen may help.

180

159

Box 5 Examples of isopleth maps

Contour lines	lines of equal height
Isotherms	lines of equal temperature
Isohyets	lines of equal rainfall
Isobars	lines of equal air pressure
Isovels	lines of equal velocity
Isochrones	lines of equal time
Isopeds	lines of equal pedestrian density

Isopleth maps

Isopleths (often called isolines) are lines of equal value. They can be drawn for any data where there is a spread of values at precise points across an area (Box 5).

How to draw isopleths

1 Mark a base map with the observed values for a selection of sites. The more sites, the better. They sites should be fairly evenly spread over the study area.
2 Decide on the values for the isolines. They are always at equal intervals, e.g. 5 m, 10 m, 15 m, and so on. It helps if there are some points that show the exact value of your isolines but this might not be the case.
3 Draw the lines beginning with the highest value so that all numbers greater than the highest-value-isoline lie within this line. Be observant: it is easy to miss the odd points that do not fit the expected pattern. The map may end up showing small 'cells' for the highest value.
4 Work your way to the outside of the study area, drawing in the other lines. Draw lines directly through points that have values which coincide with the chosen interval. Where there are no exact points representing the value of the line, interpolate and locate the line between points on either side of its value (Figure 9.8).
5 The isolines should join up to form a shape and must not end before the boundary. Isolines may end at the coast, or a vertical cliff, or at the surface of a stream (Figure 9.8).
6 While you are drawing the isolines, use geographical reasoning to justify the patterns that are created. Does the pattern match the data and the area being studied? Is the pattern what was expected?
7 Label each isoline with its value. Break the line and write the number as if you were looking inwards towards the higher values. Thus on contour lines, if you can read the height on a map the right way up, you are looking uphill.

As Figure 9.8 shows, isolines may be drawn through cross-sections of rivers to show changes in velocity. The lines must be drawn as far as the edge of the channel, since velocity does not stop in mid-river after all.

As activity 3 indicates, it is the interpretation of an isoline map which gains the most marks. The analysis should include comments on the factors of the study area that shaped the resultant pattern. Factors that influence isolines may be physical features, road network, accessibility, building density, land values, or environmental quality. The scale of the study area is also important. Isolines may reveal changes over a relatively small area.

► **Figure 9.8** *Velocity readings across the River Dulais, near Swansea*

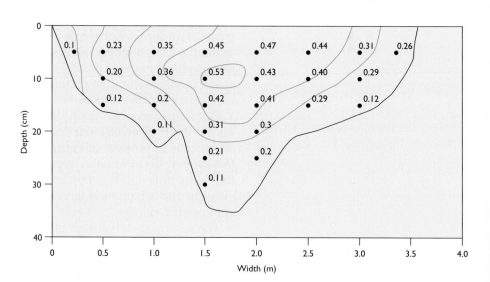

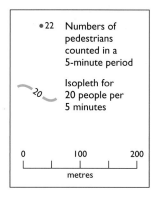

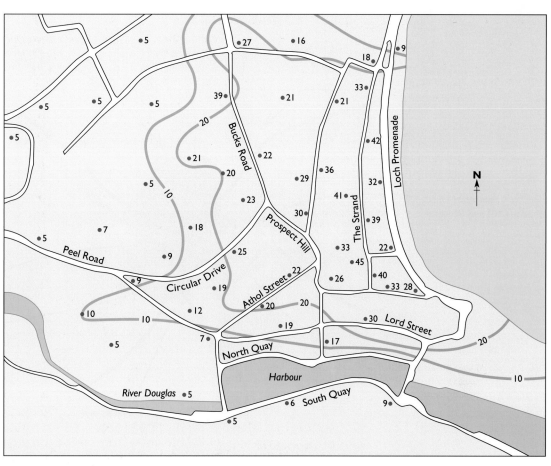

▲ **Figure 9.9** *Pedestrian counts at various locations in Douglas*

3 These questions are taken from a geographical skills examination paper.

a Complete Figure 9.9 by plotting the missing isopleths for 30 and 40 persons per 5 minutes. (4 marks)

b Describe and suggest reasons for the pedestrian flows you have mapped in Figure 9.9. (6 marks)

Topological maps

Topological maps are arranged to show clearly the links between places rather than their precise geographical position. The most famous topological map is that of the London Underground system (Figure 9.10).

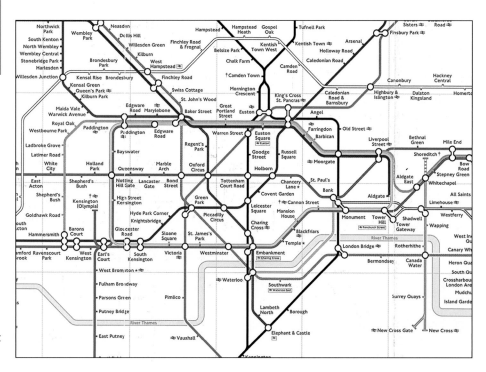

► **Figure 9.10** *Topological map: extract of the London underground system*

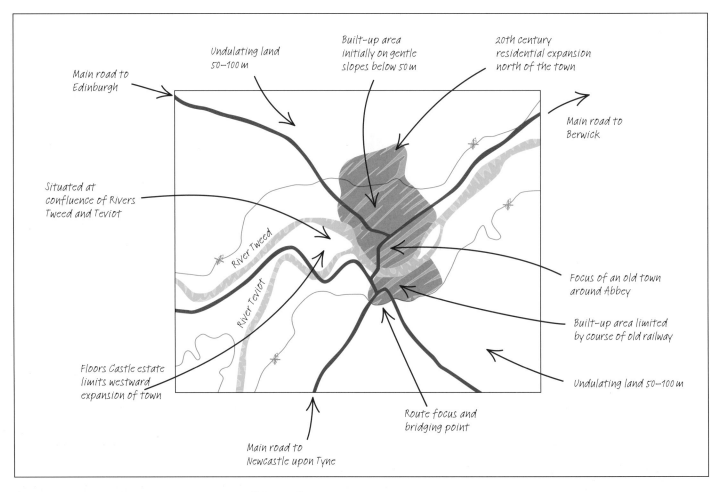

Main road to Edinburgh

Undulating land 50–100 m

Built-up area initially on gentle slopes below 50 m

20th century residential expansion north of the town

Main road to Berwick

Situated at confluence of Rivers Tweed and Teviot

River Tweed

River Teviot

Focus of an old town around Abbey

Built-up area limited by course of old railway

Floors Castle estate limits westward expansion of town

Undulating land 50–100 m

Route focus and bridging point

Main road to Newcastle upon Tyne

▲ Figure 9.11 *Sketch map to show site and situation of Kelso*

250

Sketch maps

Sketch maps identify and highlight the main geographical features of a location for a specific purpose. They exclude much of the unnecessary detail found on published maps. In an examination context, they show examiners how much students understand about a landscape.

The title of a sketch map should show its purpose, for example, the map might show the location of study sites in a fieldwork exercise, the distribution of land use in an urban area, the location of a business park, or the site of a village. A sketch map is not a replica of a published map. It is a selection of relevant information from a map usually drawn by hand.

How to draw a sketch map

1 Draw a frame the same shape as the area to be sketched.
2 Using feint lines, divide the frame into sections to help you transfer information from the grid on the published map.
3 Add annotations. Many sketch maps have some extended description or explanation added which relates to the title and draws the reader's attention to the main features. Do not cross the leader lines that link the annotations to points on the map. Now try activity 4.

Cross-sections

A cross-section shows what the land would look like if you could take a slice through it and view it from the side. Geographers usually draw cross-sections through valleys, hills or sand dunes. A sketch cross-section involves some guesswork using the contours as a guide (Figure 9.12). An accurate cross-section requires detailed manipulation of contours or slope angles. Cross-sections may also be digrammatic.

12

19

62

74

4 Draw an annotated sketch map to show the site and situation of Ebbw Vale. Page 308 shows cross-sections from the same area.

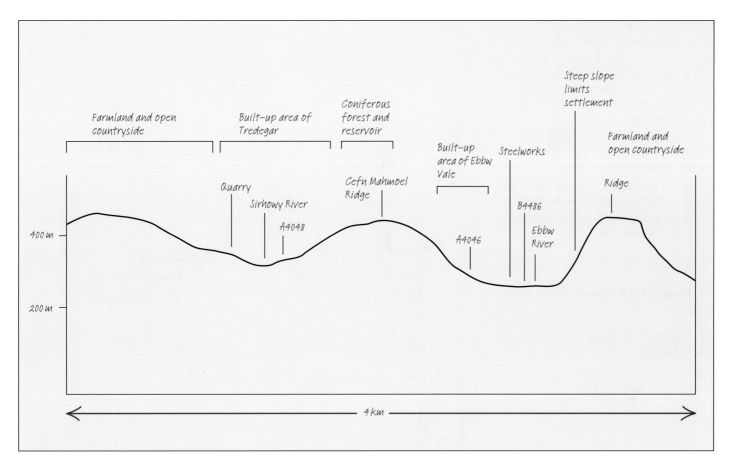

▲ Figure 9.12 *Sketch cross-section through Tredegar and Ebbw Vale*

How to draw an accurate cross-section

1 Place the straight side of a piece of paper between the appropriate points of the map you are using.
2 Identify the height of the land at the start and finish points.
3 Working from left to right, mark in the contour points between the chosen points of your section. Each time the paper crosses a contour line, note its precise position and value in metres. Where the contour lines are very close together, you may choose to use only the bold contours, i.e. every 50 m on a 1:50 000 map and every 25 m on a 1:25 000 map.
  If your cross-section includes the top of a hill or at the bottom of a valley, then you must have two entries on the same contour line.
4 Transfer the contour details into a frame on graph paper. Label the vertical scale (Figure 9.13). The horizontal distance is already determined by the distance you have marked across the map. The vertical scale is more problematic and should not be the same as the horizontal scale. The same scale would result in a cross-section that is almost flat.
  Do not exaggerate the scale either, else rolling countryside might appear as steep mountain slopes. However if the actual landscape is fairly flat you may have to exaggerate the topography a little to show some of its features.
5 Place your straight piece of paper with its marking below the frame you have drawn. Transfer the position of each contour vertically up from the base line and mark it with a dot. Join the dots to reveal an accurate cross-section. Note that valleys and hilltops are never absolutely flat; draw a rounded profile for these between the dots but without crossing the next contour level on the vertical scale.
6 Label and annotate the cross-section to show relevant geographical features, such as rivers, hilltops, vegetation, settlement, or transport routes.

Long sections are slices of a landscape taken from high land to low land, usually down a valley. They are constructed in the same way as accurate cross-sections but tend to be rather longer than valley cross-sections.

River cross-sections are often drawn for fieldwork purposes. The most common problem is that students make the vertical scale too large. Stream cross-sections should be in proportion and should be as realistic as possible.

► **Figure 9.13** *How to draw a cross-section*

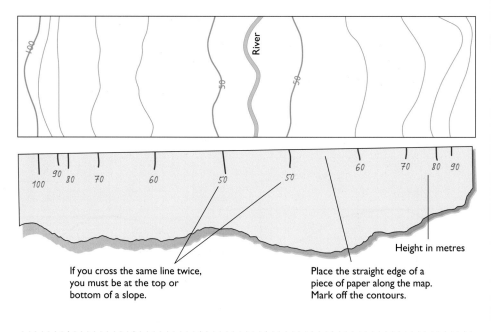

If you cross the same line twice, you must be at the top or bottom of a slope.

Place the straight edge of a piece of paper along the map. Mark off the contours.

Height in metres

► **Figure 9.14** *Long section down the Afon Crawnon*

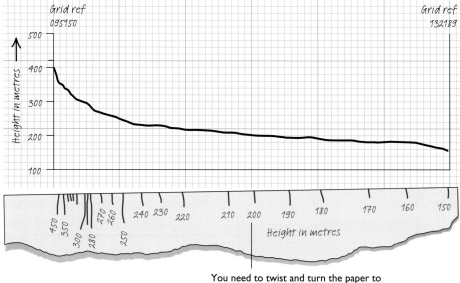

You need to twist and turn the paper to match the meanders in the river.

► **Figure 9.15** *Cross-section across the Nant Trefil*

Remember you need to round up the line. Land is rarely absolutely flat

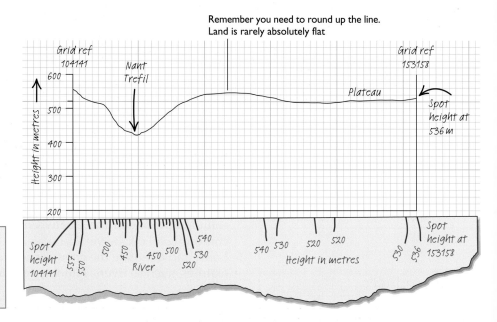

5 Draw a long section down the Afon Crawnon from 095148 to 133188 using the Ordnance Survey extract on page 6.

6

a Describe and suggest reasons for the changes in land use with distance from the Palace Pier in Brighton.

b To what extent has topography influenced land use in Brighton?

c Suggest why some land in the east has become designated as 'land in transition'. What might this land be used for in the future?

Land-use maps

As their name suggests, these maps show the various types of uses of land in an area. There have been three major published surveys of land use across the United Kingdom: in 1930, 1960, and 1996, at a scale of 1:25 000. The most recent land-use surveys have been much smaller in scale and produced for a specific purpose, such as the Land-use – UK project 2000 (Figure 9.16). The main classification of urban, industrial and agricultural land is sub-divided to give a substantial amount of detail field by field and street by street. These maps may also be produced on a smaller scale (Figure 9.17). Land-use maps become out of date very quickly but they provide an excellent base from which to investigate land-use change. It can be difficult to follow the contour lines on some land-use maps so an Ordnance Survey map is also needed to help.

Goad maps are more detailed land-use maps of most urban areas. Some local authorities have land-use maps of agricultural areas. Like all specialist maps, they are an aid to the interpretation of the landscape rather than an end in themselves.

▼ **Figure 9.16** *Land-use map of the Brighton and Hove area*

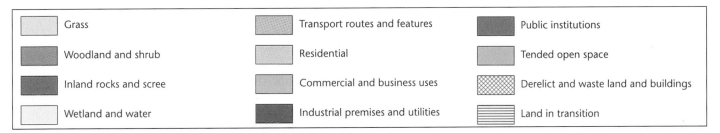

Grass	Transport routes and features	Public institutions
Woodland and shrub	Residential	Tended open space
Inland rocks and scree	Commercial and business uses	Derelict and waste land and buildings
Wetland and water	Industrial premises and utilities	Land in transition

55

(Source: Land-Use UK Project, based on data supplied by Ordnance Survey)

► **Figure 9.17** *Agricultural land-use in Oxfordshire, 1998*

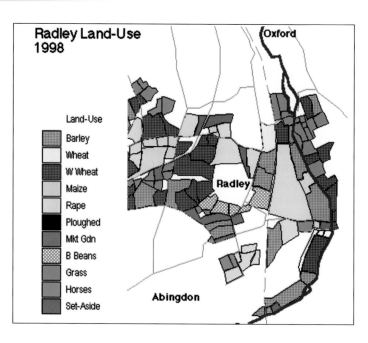

Designing and using questionnaires

Most geographers are familiar with questionnaires. They are typically used to collect data on socio-economic characteristics of a population, spatial patterns, or patterns of behaviour. The usefulness of a questionnaire survey depends on the quality of the questions asked and the size of the sample. It is absolutely essential to conduct a pilot survey. Check that the questions will deliver the kind of information that is needed to answer the research.

Types of information and questions

The information received from questionnaires may range from quantitative to qualitative. Quantitative data is easy to manipulate and interpret, but qualitative data may give more insight into a particular topic.

Closed questions are most frequently used in questionnaires because they give respondents a limited choice of answer. Each question should have an 'other' category to provide for the unusual or unexpected answer. However, if the closed questions are inappropriate, many respondents will select the 'other' category, making the research meaningless.

Open questions allow the respondent to give a personal answer, such as an opinion. While these may give some insight, the answers are often very difficult to classify and manipulate qualitatively.

Scaled questions are used where respondents choose a position on a given scale, e.g. from –3 to +3, or from 1 to 4. Avoid an uneven number of categories because most respondents will choose the middle value for safety. Responses to scaled questions are used in a bi-polar analysis (see Figure 9.1).

The layout of the questionnaire should make it easy to collate the information. Number each question. Make sure there is room for responses on the right-hand side of the page so that results can be transferred quickly to a spreadsheet.

Questionnaire sampling

You must be able to justify the sample size of your questionnaires. Time available to collect the data is important. Fifty questionnaires will take some time to complete, but in a busy street will only cover a small proportion of the total population. On the other hand, 50 questionnaires at a village shop could take hours. A small survey can, at best, yield very tentative conclusions.

Box 6 The main problems with questionnaires

- Students try to reach definitive conclusions from tiny sample sizes.
- Responses to questions depend on a range of social, economic and cultural factors that are not directly measured in the questionnaire.
- The sample population is so varied in terms of the range of socio-economic characteristics that often any summary data is inconclusive.
- Effective sampling of respondents is very difficult. It is difficult to select random respondents.
- Often, too many questions are asked and busy people may not be co-operative.

26
85

Consider the sampling strategy carefully. Will you select people randomly, using random number tables? Will you really be able to select every tenth person who passes by? Or will you select those people who look as if they will be obliging with their time and responses?

Delivering and collecting questionnaires to houses or offices also needs careful planning. Choose the sampling procedure and justify your choice. Remember your personal safety; it is better to collect data with a friend or in small groups.

Field sketches

A field sketch is a drawing of a landscape which highlights its most important geographical features. The sketches should keep a sense of proportion and show the necessary information. As with all diagrams, field sketches have a purpose, and that influences what you draw.

28
255

▼ **Figure 9.18** *Field sketch of area shown in Figure 7.35b*

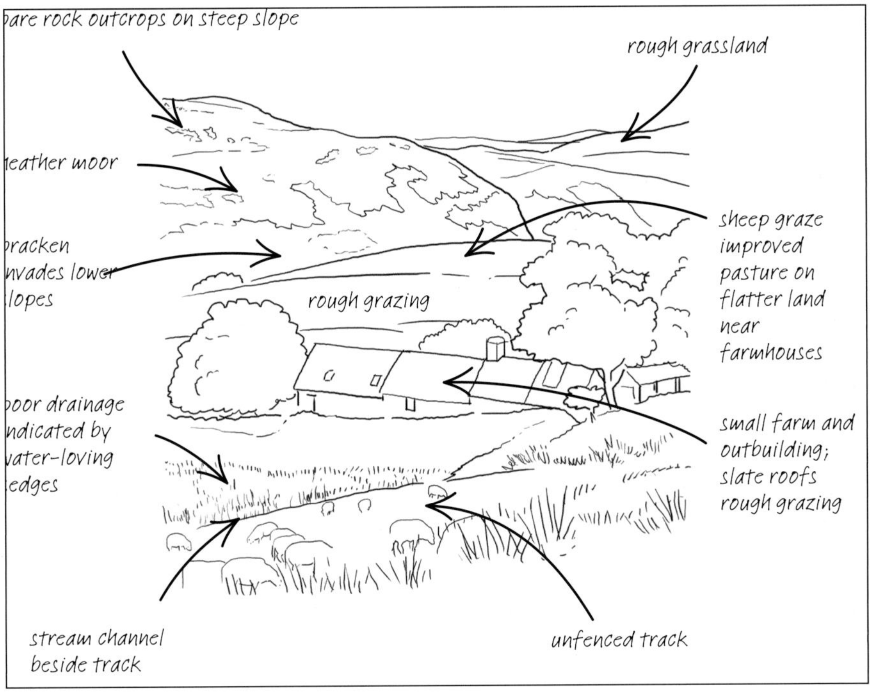

- bare rock outcrops on steep slope
- rough grassland
- heather moor
- bracken invades lower slopes
- rough grazing
- sheep graze improved pasture on flatter land near farmhouses
- poor drainage indicated by water-loving sedges
- small farm and outbuilding; slate roofs rough grazing
- stream channel beside track
- unfenced track

1 Draw an appropriate square or rectangle.
2 Divide your frame into three sections: sky or background, middle ground, and foreground.
3 Choose some features in the landscape that act as markers for each section of the sketch. As you draw in the information, think about what you see and how it has developed, i.e. the processes that operate on the landscape. For example, in human landscapes processes of infill, sprawl, degradation or renewal may be important in understanding its character.
4 Annotate your sketch. This shows others, notably examiners, what you have observed and understood about the actual environment. Leader lines should not cross each other and all annotations should be relevant to the aim of the sketch.

9.4 Making sense of graphs

Experienced geographers are familiar with a range of graphical techniques. The graphical method you use should be appropriate and the analysis and interpretation of your graphs should show depth. Graphs illustrate data in a visual way as an aid to interpretation. Similar types of information may be plotted on the same set of axes. All graphic labels should be easy to read and consistent within the text in which they are published. Two or more graphs may be placed on the same page where information is to be compared and interrelationships are being investigated. ICT packages are available that will help to collate data and present it in graphic form. A clean, neat presentation of graphs is more likely to be easily interpreted.

35
50
188

Line graphs

Line graphs are one of the most common ways of representing data which shows change over time or space, or where there are continuous measurements, e.g. temperature patterns, traffic flows, crop yields, changes in stream discharge. They can be an effective way of comparing trends in similar data on one graph.

114
133
138

► **Figure 9.19** *Changes in life expectancy in selected African countries, 1955–2000*

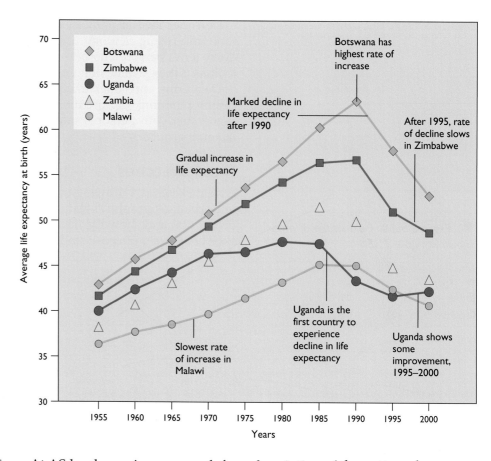

7

a Graph the data in Table 9.6 which shows change in the relative value of farm production in the UK, 1950–1999.

b Annotate the graph to identify and suggest reasons for the changes in farm production between 1950 and 1999.

At AS level examiners may ask for a description of the pattern shown on the graph. Describe the general trend shown by the data and identify any anomalies, i.e. any odd features which do not fit the general pattern (Figure 9.19). Use the data from the graph to illustrate what you mean in your answer. Look out for positions on the graph where the line changes gradient, i.e. the rate of change varies, and use this for your description.

Where students are asked to 'suggest reasons for the pattern you have identified', the response should draw on the theory and general knowledge of the subject.

▼ **Table 9.6** *Value of farm output in the UK, 1950–1999*

Farm type	1950	1975	1989	1999
Farm crops	20.6	25.8	26.3	29.6
Horticulture	10.9	10.2	12.0	13.2
Livestock	68.5	64.0	61.7	57.2

(Source: National Statistics 2001)

Cumulative line graphs

These are graphs in which successive plots are made by adding more data. The final value is the total of all the data. A typical representation of this is in the Lorenz curve which represents the distribution of population in a region or country.

Scatter graphs

Scatter graphs are a quick and easy way to show a correlation (relationship) between two variables. Geographers frequently look at patterns between pairs of variables, e.g. land values and pedestrian density, or soil acidity and distance from the sea.

Figure 9.20 identifies key questions in the design and use of scatter graphs. Two elements are of particular importance.

■ Decide which variable causes change to the other. The one that causes the change is the independent variable and should be placed on the horizontal or x-axis. The other variable is the dependent variable and is placed on the vertical or y-axis, e.g. distance from the sea causes change in soil acidity, so distance is placed on the x-axis and soil pH on the y-axis.

■ Avoid the temptation to 'join the dots'. Connect the points only if there is some logical connection between them, or if their values are linked in some way. There is not necessarily a correlation between two variables simply because they have been placed on the same graph.

If you are using an ICT package to draw a scatter graph and best-fit line then you must make sure that the computer draws something that is geographically sensible. You might be caught out if there are anomalies in your data which the best fit line tries to include.

▶ **Figure 9.20** *Scatter graph showing the correlation between soil acidity and distance from the sea, in metres*

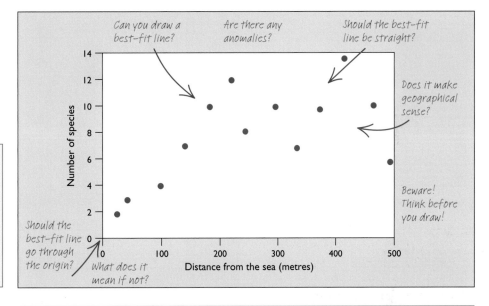

8 Using the information in Table 9.7,
a Draw and annotate a scatter graph to show the relationship between the quality of housing and distance from the centre of Lima, Peru.
b Suggest reasons for the relationship you have graphed.

▶ **Table 9.7** *The quality of housing and distance from the centre of Lima, Peru*

District	Distance from the centre of Lima (km)	% of homes with no services (water, electricity or sewerage)	Brief description of district
Lima	0	0.6	High density city centre slum
Rimac	1	1.6	Slum housing along river liable to flood
La Victoria	2	0.3	Old central slum, many migrants
El Agustino	3	3.1	Rural migrants' housing on very steep hillside
San Isidro	7	0.0	High-class apartments for wealthy professional elite and middle class. Prestige offices, banks and embassies
Miraflores	10	0.1	
Surco	12	0.7	
Barranco	13	0.3	
Ate	13	10.7	High density shanty town of migrants on steep slopes
Carabayllo	15	8.2	Squatters in shacks on scrub vegetation
Lurigancho	20	9.7	Shanty town gradually converted to permanent housing
Villa El Salvador	25	8.6	Edge of city shacks made from rush matting and cardboard

(Source: Instituto Nacional de Estadìstica, Peru)

313

Dispersion diagrams

Dispersion diagrams are like a scatter graph with only a vertical axis. They show how a single set of data is spread out. Analyse the spread of the data using the median and inter-quartile range (see page 322).

Choose a convenient vertical scale to be drawn on graph paper. Plot the data in a vertical line a little way to the right of the y-axis. Any equal values should be plotted to the right of the first figure forming a short line (Figure 9.21).

> **9** Draw a dispersion graph for the south-facing slope using data from Table 9.8.

► **Figure 9.21** *A dispersion diagram of slope angles on a north-facing slope*

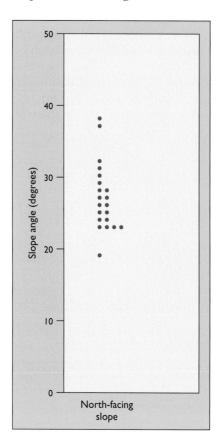

▼ **Table 9.8** *Slope angles at points on the north- and south-facing slopes of a valley*

North-facing slopes angles (°)	South-facing slopes angles (°)
19	11
23	13
23	14
23	15
23	15
24	15
24	16
25	16
25	16
26	16
26	17
27	17
27	18
28	19
28	19
29	19
30	20
31	20
32	21
37	21
38	24

Bar charts and histograms

Bar charts are easy to construct and use. The horizontal x-axis represents single categories for data such as months of the year, or places where data was collected. The bars should not be drawn joined to each other. The vertical y-axis usually has the quantitative scale. If using percentage values, make sure that each column adds up to 100 per cent in total.

Bar charts may be simple, i.e. represent one set of data such as monthly rainfall, or they may be divided (compound), i.e. to represent data which is subdivided, such as proportion of people in different occupations in wards of a town. For each location the subdivisions in a divided (compound) bar chart must be drawn in the same vertical order (Figure 9.22).

Histograms are similar to bar charts with a quantitative vertical y-axis scale. However the horizontal x-axis also has quantitative categories, such as pebbles sizes or percentage vegetation cover. Technically, it is the area of the block on the histogram which represents the value of the data, but usually for geographers the width of each column is the same across the chart so the area loses its importance. The data must be classified into groups along the x-axis. When you draw a histogram, as in exercise 10, select the groups so that they represent the same size range. The visual impact of the histogram will depend on the group size you choose.

114

239

> **10**
> **a** Use the information in Table 9.9 to produce a bar chart showing the distribution of world steel production 2001.
> **b** What conclusions can you draw from the graph, that the table does not show?

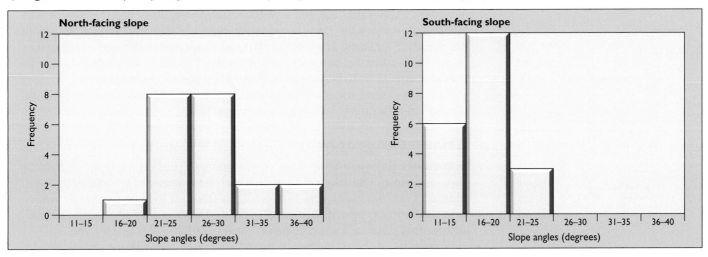

▼ **Figure 9.22** *Frequency diagrams to show slope angles on a north- and south-facing slope*

11

a Complete a divided (compound) histogram to show the percentage of different types of farms in regions of the United Kingdom. use the information in Table 9.10.

b Use an atlas and your geographical knowledge to suggest how the pattern of farming is influenced by the climate of the United Kingdom. Chapter 4 and Chapter 7 should give you some ideas.

12 Look at Figure 9.22.

a Describe the differences shown on each slope and suggest reasons for the differences you have identified.

b Compare the methods of representing data using a histogram and a dispersion diagram (Figure 9.20). What are the advantages and disadvantages of each method?

c List the statements you can make about the two slopes using each method of representation.

▼ **Table 9.9** *Distribution of world steel production, 2001*

Country	Million metric tonnes crude steel production	World rank
China	148.9	1
Japan	102.9	2
United States	90.1	3
Russia	59.0	4
Germany	44.8	5
South Korea	43.9	6
Ukraine	33.1	7
India	27.3	8
Brazil	26.7	9
Italy	26.7	10
France	19.3	11
Taiwan	17.2	12
Spain	16.5	13
Turkey	15.3	14
Canada	15.0	15
UK	13.7	16
Mexico	13.3	17
Belgium	10.8	18
South Africa	8.8	19
Poland	8.8	20

(Source: www.worldsteel.com)

▼ **Table 9.10** *Percentage of different types of farms in regions of the United Kingdom*

% of total land use	Cereals	Horticulture & Poultry	Pigs	Dairy	Cattle & Sheep	Mixed	Other
North East	20	2	2	5	40	10	21
North West	6	4	3	20	33	4	30
Yorks & Humberside	27	3	5	8	26	9	22
East Midlands	36	5	3	7	20	8	21
West Midlands	15	5	3	12	30	9	26
East Anglia	51	8	5	1	8	6	21
London	17	18	4	3	11	5	42
South East	19	10	3	4	23	7	34
South West	9	5	3	15	32	7	29

(Source: DEFRA 2002)

125
134
261
216

Frequency diagrams

Frequency diagrams are histograms which show how often values of a given size occur. One typical use is for showing the different categories of pebble sizes or slope angles. Each column represents the same range of sizes. Make sure the ranges do not overlap. Columns should be drawn without spaces between them because the data is continuous.

You can extend the analysis of data by looking at skewness of the data shown on a frequency diagram (see page 323).

Triangular graphs

Sometimes there are three data components to plot on a graph, instead of two. Examples include soil texture which has elements of sand, silt and clay, and employment structure which is divided into primary, secondary and tertiary components. Instead of using the two axes of a conventional graph, this data calls for three axes on an equilateral triangle. The components of the data should be in percentages and therefore must add up to 100 per cent.

Figure 9.22 shows how to construct the triangle, using data for three components (primary, secondary and tertiary). Look at each component in turn. Each has a base line from 0 to 100 per cent, which form one side of the triangle. Primary sector percentages are shown in blue, secondary in red and tertiary in green. Follow these lines to read each component in turn. Turn the page around if it makes it easier to read the graph with their respective base lines at the bottom. The towns are plotted where the relevant graph lines intersect.

▼ **Figure 9.23** *Constructing a triangular graph*

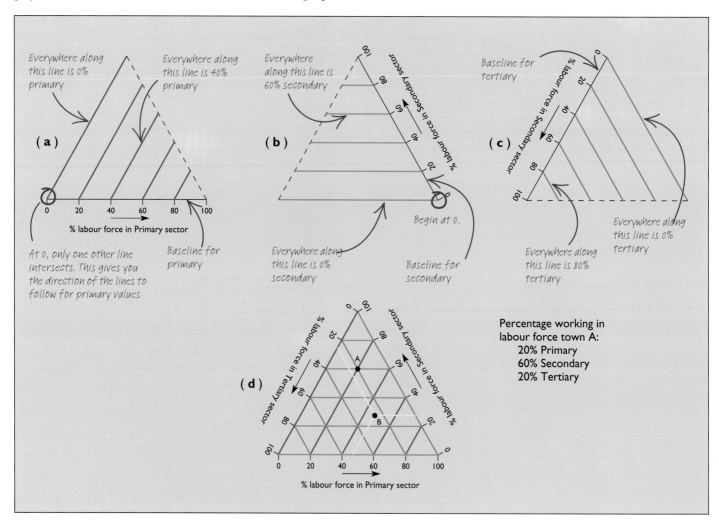

▼ **Table 9.11** *The purpose of building dams in Spain*

Purpose of dam building	% of dams built
Multi-purpose	25
Irrigation	30
Hydropower	21
Water supply	20
Flood control	2
Other	2

12 Using the information in Table 9.12, draw a pie chart to show the purpose of building dams in Spain.

13 Look at the examples in Figure 9.24 and discuss why (b) has not used the appropriate data.

Pie charts

A pie chart is a circle divided into component parts like slices of a pie. Pie charts are useful for data divided into component parts, especially if there are four or more components. However, with too many components it becomes difficult to compare each segment. To draw a pie chart you must know the total value of all the component parts. Use the formula in Box 7 to calculate the size of each slice of the pie.

When drawing in the slices on the circle, start with the largest component at '12 o' clock' and work clockwise around the circle. Shade or colour each component and use a key if necessary. Use the same key to compare information in several pie charts.

Box 7 Calculation for pie chart

$$\text{Angle for each component} = \frac{\text{value of one component}}{\text{total value of all components}} \times 360°$$

Each component of the pie is a proportion of 360°.

Position pie charts on the same page if they are to be compared. Pie charts may also be used on top of a base map to show location. As such, they are useful for comparing information between locations, e.g. the proportion of agricultural land uses in different regions of the UK.

The advantage of located pie graphs is that a lot of information can be shown on one diagram. However, the information may not be easily, nor accurately, read from these charts. If the circles are too large, they may obscure much of the base map.

155

212

239

317

▼ **Figure 9.24a** *Activities undertaken by visitors to Hartfield*

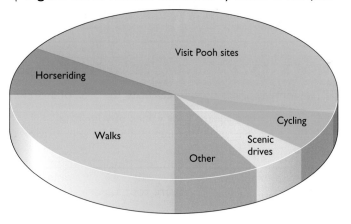

b *Reasons for people visiting Hartfield*

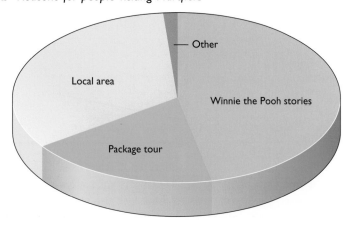

Proportional circles

Proportional circles are a more complicated type of pie chart that show more data. The circles are of different sizes to represent a range of data values. The real value of each data set is proportional to the area of the circle. Each proportional circle may be divided as a pie chart to show the extra information. They may be positioned on a map to show data for the area in which they are located.

Proportional circles can be complicated to draw. It is not always easy to read specific data values from them, and the circles can obscure large areas of a map or even overlap each other.

▼ **Figure 9.25** *World employment by sector for selected countries*

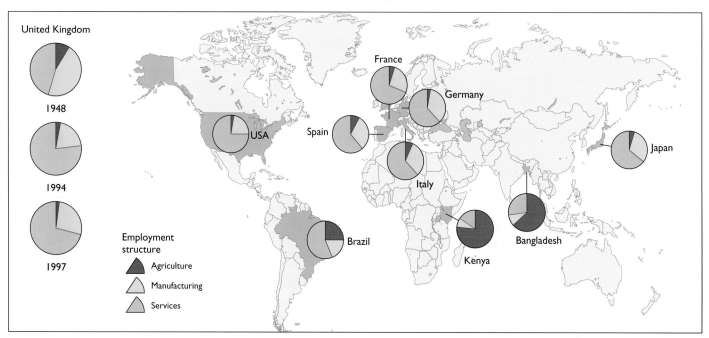

Box 8 How to calculate the areas of proportional circles

The formula πr^2 is used to calculate the area of a circle.

1 Calculate the square root values of your data.
2 Draw a linear scale that includes all the square root values in your range of data.

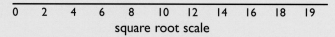

3 At the right-hand end of your scale, draw the largest circle you wish to use on your diagram or map. (Use the formula πr^2 to calculate its area.) Join the centre of that circle to the left-hand end of the square root scale line. This will give you a range of radii to use for drawing circles.

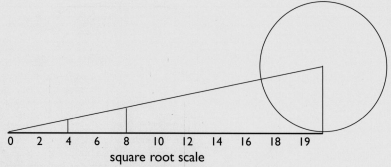

4 For each piece of data, locate the square root along your base line and measure the size of each circle you need to draw. Use the square root values you calculated in step 1.

318

15 Using the information in Table 9.12

a Draw proportional circles to represent the size of domestic tourism in the United Kingdom in 1990, 1995 and 1999.

b Describe and suggest reasons for the changes you have represented.

▼ **Figure 9.26a** *How to draw a star diagram*

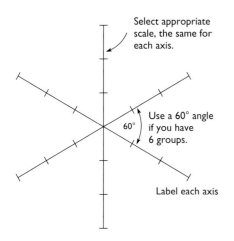

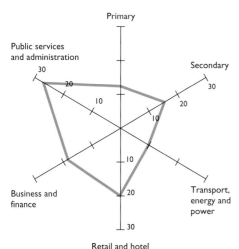

▼ **Table 9.12** *Domestic tourism in the United Kingdom (million trips)*

Year	Holiday trips	Visiting friends and relatives	Business and work trips	Total
1990	58.4	20.2	11.5	90.1
1995	66.2	34.6	34.6	115.6
1999	75.3	47.5	17.3	140.1

Star diagrams

Star diagrams can represent a range of 4–6 components in one data set. Specific values can be read from the graph. Each axis represents one component and as the scale on each axis is the same, there is some comparison between components. Star diagrams are used to compare data. When the points on each axis are joined, the shape provides a visual impression of the balance between the different components in different places.

The axes on a star diagram may typically represent wind direction, or the orientation of glacial deposits, or categories of the standard industrial classification for different locations.

Population pyramids

Population pyramids are used to illustrate the age structure of the population of a country, region, town, or a particular ethnic group. The population pyramid is an important predictive tool for identifying the population structure of the future. This enables governments to plan population policies more effectively.

The central axis of the population pyramid usually consists of five-year age groups: 0–4, 5–9, 10–14, etc. The progress of a particular population group can be monitored as it ages and moves up the pyramid. The units of the horizontal scale on the population pyramid may show the percentage of male and female populations in each age group, or the actual numbers of males and females. Some pyramids provide only a summary of the population structure and have broader age groups, in which case it is the comparison between locations which is important. Be careful when you interpret such diagrams because the range of the age groups varies considerably. The size of the 15–59 age group, for instance, will almost inevitably be larger than the 0–14 group.

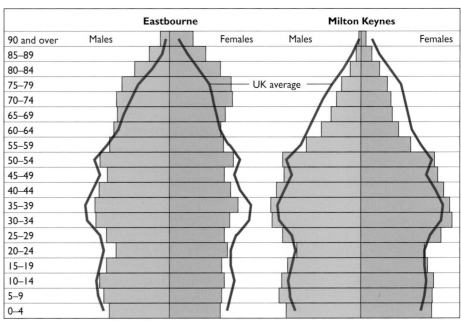

► **Figure 9.27** *Population pyramids for Eastbourne, Milton Keynes and London, UK, 2001 (Source: UK census, 30 September 2002)*

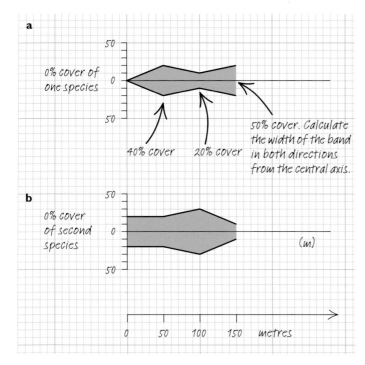

Kite diagrams

Kite diagrams are usually used to represent changes in the type and amount of different vegetation species over a transect, e.g. across sand dunes or down a slope. They give a good visual impression of the relationship between different species as well as the degree of change and can be easily linked to line graphs or bar charts of other indicators that also change with distance.

The thickness of the line on the kite diagram represents the number of species or the percentage cover of each species. Using graph paper, allocate a row to each species. The centre of each row represents 0 and the full thickness represents 100 per cent cover, or the largest number of occurrences in the data set.

◀ **Figure 9.28a, b** *How to draw a kite diagram*

▼ **Figure 9.29** *Kite diagram to show vegetation change at Newborough, Anglesey*

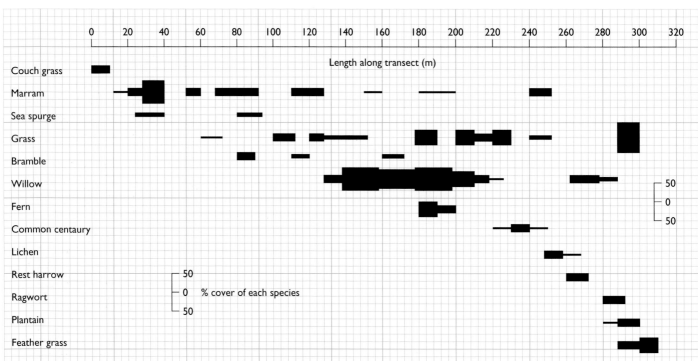

Logarithmic graphs

Logarithmic graphs (log graphs) look more complicated than they really are. These graphs can show a wide range of data on one graph without losing the clarity of the smallest values. Log graphs are also used where a comparison of rates of change are required.

Log graph paper is divided into cycles divided by a base line. An important characteristic of log graphs is that they always begin with the number 1 to a power of 10, i.e. 0.01, 0.1, 1, 10, 100, 1000, 10 000 (Figure 9.30). Each cycle represents a ten-fold increase in value on the scale beginning at each base line. Log graph paper may show a log scale on both axes using log–log paper (Figure 9.30a), or a linear scale on the x-axis using semi-log or log–normal paper (Figure 9.30b).

57

▼ **Figure 9.30** *Types of log paper*

Figure 9.30a *Log–log graph paper*

1 cycle
1 cycle
1 cycle
1 cycle
2 examples of values on a log scale. After one cycle, the values increase by a factor of 10.
The graph must begin with 1 to a power of 10.

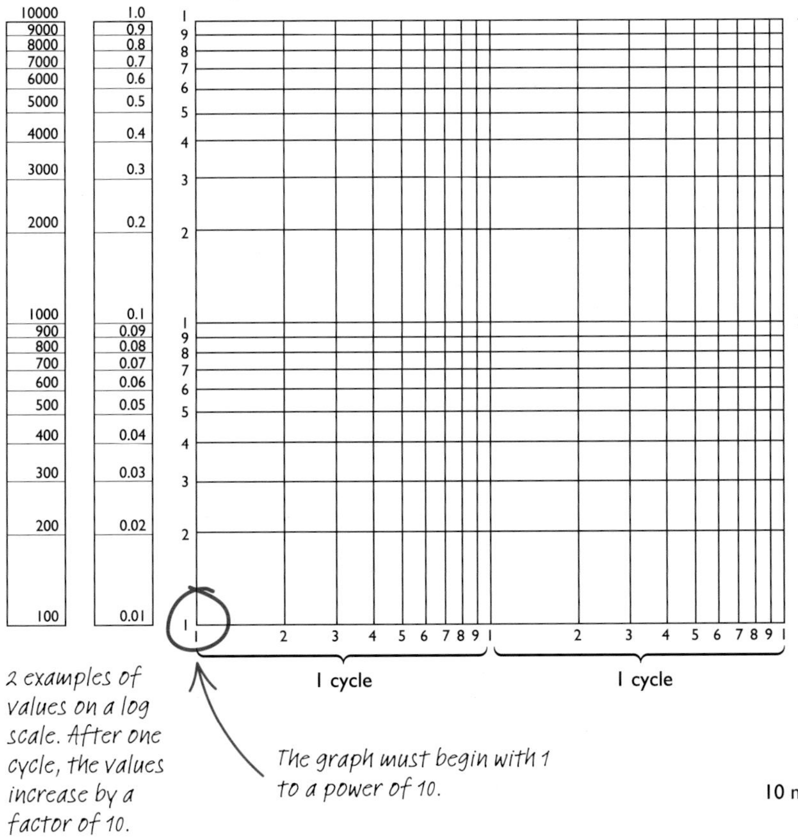

Figure 9.30b *Semi-log or log–normal paper*

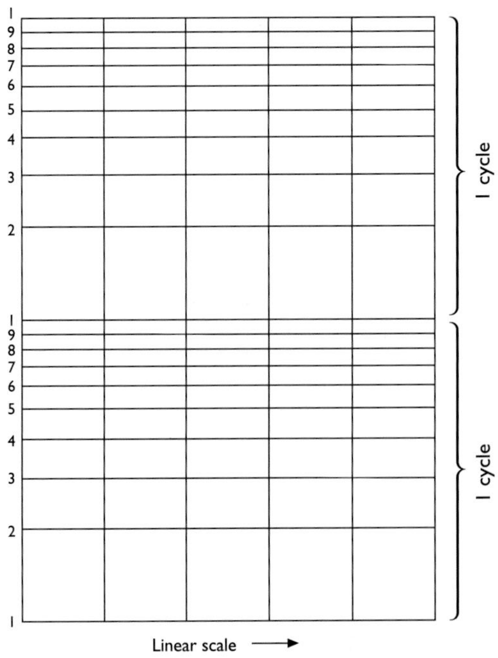

1 cycle
1 cycle
Linear scale

10 million
1 million
100 000
Population size
Rank of city

►▼ **Figure 9.31** *Log graph distribution of city sizes in Argentina, 1991 (Source: INDEC, Censo Nacional de Poblacion y Vivienda, 1991)*

Population of cities in Argentina, 1991	
City	Population (thousands)
Buenos Aires	1 1295
Cordoba	1 208
Rosario	1 118
Mendoza	773
La Plata	642
Tucuman	622
Mar del Plata	512
Santa Fe	406
Salta	370
San Juan	352

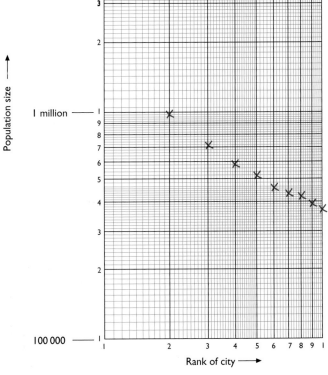

▼ **Table 9.13** *Top ten cities in the United Kingdom*

Rank	City	Population 2001
1	London	7 172 036
2	Birmingham	977 091
3	Leeds	715 404
4	Glasgow City	577 869
5	Sheffield	513 234
6	Bradford	467 668
7	Edinburgh	448 624
8	Liverpool	439 476
9	Manchester	392 819
10	Kirklees	388 576

16
a Use the information in Table 9.13 to plot the city size distribution for the United Kingdom on a sheet of log–log graph paper.
b Describe the distribution shown on your graph.

9.5 Making sense of statistics

Box 9 How to calculate the mean, median and mode

Mean the arithmetic average value calculated from a data set of any size.

mean = total of all values in the data set / number of pieces of data

Median the middle value of a range of data
median = the middle value when all the data is placed in numerical order. If there is an even number of values the median is half-way between the two middle values.

Mode the value in the data set that occurs most often. Together, these identical values are known as the modal class.

Box 10 How to calculate the inter-quartile range

■ Place the data in descending order
The quartile = number of pieces of data ÷ 4

■ Use the quartile number count from the highest value for the upper quartile.

■ Count from the bottom upwards to give the lower quartile.

■ The range between upper and lower quartiles forms the inter-quartile range, i.e. the middle part of the sample.

■ When the data set is divided by 4, a whole number can be easily identified. If the division does not result in a whole number, make an informed guess at where the quartiles should be.

This section is a guide to the basic statistical techniques listed in the AS Geography specifications. Statistics can be challenging for some students, but the ability to use them marks an important transition from GCSE to AS level.

Statistical analysis increases the depth at which data can be interpreted. It enables you to consider data in more detail and evaluate it more effectively. Statistical techniques require data to be recorded in a particular form. An analysis of the information leads to further comment and interpretation. This interpretation of the results is an important part of the written fieldwork report.

Descriptive statistics

Descriptive statistics summarise data. The mean, median and mode are the easiest statistics to compute and are a useful start to comparing sets of data.

The range of data values and the uniformity of a sample is important. The range gives a measure of the difference between the largest and smallest figures in a data set. The smaller the range, the more uniform the data set, and vice versa. Geographers can compare the extent of the range as a start to analysing data in more detail. Look at Table 9.8 on page 314: the range of data is 19°–38° on the north facing-slope and 11°–24° on the south-facing slope.

At times the extremes of a wide range are not typical of the data set as a whole. It would be better then to analyse the range of the middle values, or the inter-quartile range.

Skewness

Data can be further analysed by looking at the skewness of the information on a frequency diagram (Figure 9.32). The shape of the graphic information will show whether the distribution is normal, positively skewed, or negatively skewed. A normal distribution has a symmetrical shape, and the mean, median and mode occur at the same point. Skewed distributions are not symmetrical in shape. When the modal class, i.e. the largest category, occurs towards the left of the distribution it is said to have a positive skew. If the modal class lies towards the right side of the distribution, it has a negative skew.

▼ **Figure 9.32** *Frequency distributions*

a Normal distribution

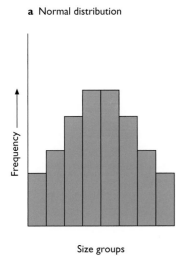

Size groups

b Positively skewed

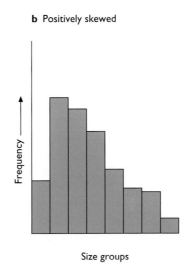

Size groups

c Negatively skewed

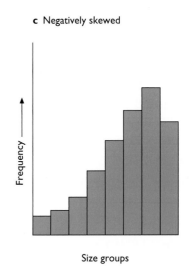

Size groups

Inferential statistics

Inferential statistics investigate relationships between sets of data or observations.

Spearman's Rank Correlation coefficient

Spearman's Rank coefficient is a number that summarises the relationship between two sets of data. The calculation for this statistical test is based on the rank of data rather than actual values. The test itself is used to investigate whether or not there is a linear relationship between two sets of data, i.e whether a change in one variable is matched by a change in the other. The test is not appropriate for a relationship that is not linear, or where the best-fit line is not straight (Figure 9.33).

To carry out a Spearman's Rank Correlation coefficient test, there should be between 10 and 30 pairs of data in the set for the results to be reliable. If there are more than 30 pairs, then the data should be sampled. Also, no more than two pairs of data may be ranked equally. The main inaccuracy of this test is in the ranking of data rather than the use of measured values.

At the start of the investigation, state the null hypothesis (H_0) that there is no significant relationship between the sets of data. You may have to accept the alternative hypothesis (H_1) that there is a relationship between the sets of data, based on evidence from your calculation.

Draw a scatter graph of the sets of data being studied to get a visual impression of the relationship before you begin the calculation. Determine from this whether there appears to be a linear relationship.

The coefficient may be positive or negative (Figure 9.33a, b) and the calculated figure must be between +1 and −1. (Figure 9.33c, d). If the final reading is not between +1 and −1, then a mistake has been made. Check the calculation again.

Box 11 Formula for Spearman's Rank Correlation coefficient

$$r_s = 1 - \left[\frac{6 \times \Sigma d^2}{n^3 - n} \right]$$

where r_s = Spearman's Rank Correlation coefficient
n = number of pairs of data

---

a

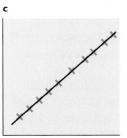

The best fit line here is positive, i.e. an increase in one variable is matched by an increase in the other.

b

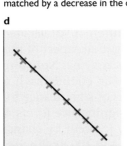

Here the best fit line is negative. There is an inverse relationship. An increase in one variable is matched by a decrease in the other.

c **d**

This is a perfect relationship. All the points lie on the line, Spearman Rank Coefficient r_s = +1

This is a perfect relationship, but the Spearman Rank Coefficient r_s = −1

◀ **Figure 9.33** *Line of best fit*

52

323

Box 12 Spearman's Rank correlation: a worked example

1 Look at the information in Table 9.15.

2 Organise the data into a table as shown in Table 9.16:

3 Rank each set of data from the highest to the lowest value.

4 Calculate the difference between the ranks. This must be done in the same direction i.e. 1 − 11 = −10, 7 − 5 = +2

5 Square the difference between the ranks to get rid of the minus sign.

6 Find the sum of all the ranks squared (Σd^2)

7 Apply the formula $r_s = 1 - \left[\dfrac{6 \times \Sigma d^2}{n^3 - n} \right]$

$$r_s = 1 - \left[\frac{6 \times 997}{3360} \right]$$
$$= 1 - 1.78$$
$$r_s = -0.78$$

Remember: the r_s value must be between +1 and −1.

8 Use Figure 9.34 to read off the significance of the positive and negative correlation. The closer the r_s value is to 1 the stronger the relationship appears to be. However, it cannot be said with any certainty whether a similar relationship would exist with different data. To do this, you must go further and test this r_s value for significance.

▼ **Table 9.14** *Influence of female literacy on fertility rates*
(Source: Human Development Report 1994 (To be updated)

Country	Female literacy rate (%) 1994	Total Fertility rate
United Kingdom	99.0	1.8
United States	99.0	2.1
Japan	99.0	1.4
Rep of Korea	96.8	1.3
Spain	97.1	1.2
Mexico	86.7	2.7
Malaysia	77.5	3.4
Saudi Arabia	47.6	6.4
South Africa	81.2	4.0
Indonesia	77.1	2.5
China	70.9	1.0
Kenya	52.3	5.5
Bangladesh	24.3	2.9
Uganda	48.7	7.1
Burkina Faso	8.6	7.2

▼ **Table 9.15** *Table of organised information*

Country	Female literacy rate (%)	Rank	Total fertility Rate	Rank	Difference between the ranks	Difference squared
United Kingdom	99.0	1	1.8	11	−10	100
United States	99.0	1	2.1	10	−9	81
Japan	99.0	1	1.4	12	−11	121
Rep of Korea	96.8	5	1.3	13	−8	64
Spain	97.1	4	1.2	14	−10	100
Mexico	86.7	6	2.7	8	−2	4
Malaysia	77.5	8	3.4	6	+2	4
Saudi Arabia	47.6	13	6.4	3	+10	100
South Africa	81.2	7	4.0	5	+2	4
Indonesia	77.1	9	2.5	9	0	0
China	70.9	10	1.0	15	−5	25
Kenya	52.3	11	5.5	4	+7	49
Bangladesh	24.3	14	2.9	7	+7	49
Uganda	48.7	12	7.1	2	+10	100
Burkina Faso	8.6	15	7.2	1	+14	196
					Sum of d² (Σd^2) =	997

Significance testing

Geographers cannot reach reliable conclusions about relationships without testing the statistic for significance. The statistic, in this case the r_s value, should show whether the relationship has occurred by chance, or whether it would be replicated if different values are used. The significance test ignores the positive or negative aspect of the correlation. It focuses entirely on how reliable the results may be. Note too that 'significance levels' are also sometimes referred to as 'confidence limits'.

Figure 9.34 shows levels of significance for the Spearman rank statistic. The bold lines represent significance levels, i.e. the number of times out of 100 a relationship is likely to occur by chance. A significance level of 1 per cent means that there is a 1 in 100 likelihood of a relationship occurring by chance. This means that the relationship will occur 99 times out of 100, if the calculation was done with different data.

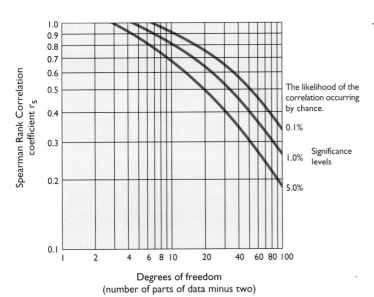

◀ **Figure 9.34** *The significance of Spearman's rank r_s values*

The likelihood of the correlation occurring by chance.

0.1%

1.0% } Significance levels

5.0%

Box 13
Formula for chi-squared test

$$\chi^2 = \Sigma \frac{(O - E)^2}{E}$$

where χ^2 = chi-squared value
        O = observed values
        E = expected values

The value of 1 per cent, or 1 time in 100, is within acceptable limits for definitive statements to be made about about the relationship. The readng shows a high level of confidence and is very significant. Thus for the data in Table 9.15, we can say with a high degree of certainty that there is a relationship between rates of female literacy and the total fertility rate.

Similarly, if the r_s value lies above the 0.1 per cent significance level, only 0.1 times in 100 is there the likelihood of the relationship occurring by chance. This represents a very high degree of reliability.

If the r_s value lies between the 5 per cent significance level and the 1 per cent level, there is a 5 per cent likelihood of the relationship occurring by chance. This is the limit of risk. Below the 5 per cent significance level there is too much risk of the relationship occurring by chance to be able to make reliable statements about the relationship. If the fieldwork or test were conducted again, the same result would be seen only 95 times out of 100. This is not enough to be scientifically sure of the conclusions. The 5 per cent level is therefore called the **rejection level**.

A table of r_s values (Table 9.16) may be used instead of a graph. This shows the critical values for r_s at different confidence limits. You should learn to use the table, but if you picture in your mind the significance lines from Figure 9.34, that will help you.

The results of all investigations must be explained in the written report. This is easier if the investigation shows up predictable results. If the results are unexpected, then the entire investigation should be re-examined: perhaps the theory was wrong; perhaps there were flaws in the data collection. Such re-examination forms a challenging part of any fieldwork, which is why so many marks are allocated to it.

Chi-squared test

The chi-squared test is used to examine whether there is any significant difference between two sets of data. Data for the chi-squared test must be presented in groups, with at least five occurrences in each group. There is no limit to the number of groups, but using only two or three makes the calculations more manageable.

This statistical method tests observed values against expected values as if there was no difference between the groups. There are two types of chi-squared test:
- one-tailed test – one set of data is compared against a theoretical pattern to see if there is a significant difference within the groups
- two-tailed test - one set of data is compared with another set to check if there is a significant difference between the two sets of data.

▼ **Table 9.16** *Critical values for the significance of r_s*

Number of pairs of data	r_s values	
	95% significance level*	99% significance level**
8	0.64	0.83
9	0.60	0.78
10	0.56	0.75
11	0.52	0.74
12	0.50	0.70
13	0.48	0.67
14	0.46	0.64
15	0.44	0.62
16	0.43	0.60
17	0.41	0.58
18	0.40	0.56
19	0.39	0.55
20	0.38	0.53
21	0.37	0.52
22	0.36	0.51
23	0.35	0.50
24	0.34	0.49
25	0.34	0.48
26	0.33	0.47
27	0.32	0.46
28	0.32	0.45
29	0.31	0.44
30	0.31	0.43

*(may also written as 0.05 i.e. 5/100)
**(may also written as 0.01 i.e. 1/100)

Box 14 The one-tailed test

Sample question: Look at the data in Table 9.17. Would you expect there to be a difference between the number of streams per km^2 in areas with different geology?

▼ **Table 9.17** *Number of streams in areas of different of geology*

Geology	Number of streams per km^2 (the observed values)
Sandstone	24
Limestone	9
Granite	36
Clay	51

The total number of streams in the sample is 120.

State the null hypothesis.

H_0 states that there is no significant difference between the number of streams on different types of geology.

Stage 1: Work out the necessary details so that the chi-squared formula can be applied.
Calculate the expected values (E).
If there were no differences in number of streams between the geological groups, how many streams would occur in each one?

$$E = \frac{\text{total number of streams}}{\text{number of groups}}$$
$$= \frac{(24 + 9 + 36 + 51)}{4}$$
$$= 30$$

This expected value (E) applies to each of the geology categories.

Stage 2: For each geological category, apply the formula $\frac{(O - E)^2}{E}$

Add up the results for all four categories of geology.

$$\chi^2 = \Sigma\frac{(O - E)^2}{E}$$
$$= \frac{(24 - 30)^2}{30} + \frac{(9 - 30)^2}{30} + \frac{(36 - 30)^2}{30} + \frac{(51 - 30)^2}{30}$$
$$= 1.2 + 28.0 + 1.2 + 4.0$$
$$= 34.4$$

■ Test the significance of the result by calculating the degrees of freedom for a one-tailed test. Use the formula:
Degrees of freedom df = (number of rows − 1).
In this example, df = 3.

■ Explain the results, using the information in Table 9.17. This result is significant at the 99 per cent level; 99 times out of 100 there will be a difference in the number of streams per km^2 on different geology. The amended hypothesis (H_1) is therefore true.

▼ **Table 9.18** *Critical values for the significance of the chi-squared statistic*

Degrees of freedom	Significance level	
	95% significance level*	99% significance level**
1	3.84	6.64
2	5.99	9.21
3	7.82	11.34
4	9.49	13.28
5	11.08	15.09
6	12.59	16.81
7	14.07	18.48
8	15.51	20.09
9	16.92	21.67
10	18.31	23.21
11	19.68	24.72
12	21.03	26.22
13	22.36	27.69
14	23.68	29.14
15	25.00	30.58
16	26.30	32.00
17	27.59	33.41
18	28.87	34.80
19	30.14	36.19
20	37.57	37.57
21	32.67	38.93
22	33.92	40.29
23	35.18	41.64
24	36.43	42.98
25	37.65	44.31
26	35.88	45.64
27	40.11	46.96
28	41.34	48.28
29	42.56	45.59
30	43.77	50.89
40	55.76	63.69
50	67.51	76.15
60	79.08	88.38
70	90.53	100.43
80	101.88	112.33
90	113.13	124.12
100	124.34	135.81

*(may also be written as 0.05 i.e. $^5/_{100}$)

**(may also be written as 0.01 i.e. $^1/_{100}$)

Box 15 The two-tailed test

Sample question: Look at the information in Table 9.19 carefully. Is there a significant difference between environmental quality in areas with different income in Merida?

■ State the null hypothesis.
H_0 states that there is no significant difference between the environmental quality and income level in areas of Merida.

■ Arrange the information as shown in Table 9.20.

▼ **Table 9.19** *Environment and income in Merida, Mexico*
The data in the table is based on a random sample of 75 inhabitants in the city of Merida, in Mexico. It shows the income levels and index of environmental quality for their neighbourhoods.

Index of environmental quality (max = 100)	Number of pepople in each income level		
	Very low	Low	Medium/ High
81–100	6	11	13
56–80	32	5	8

(Source: Burke, Case Studies of the Third World)

▼ **Table 9.20** *Arranged information*

Index of environmental quality (max = 100)	Income level						
	Very low		Low		Medium/High		
	Observed values O	Expected values E	Observed values O	Expected values E	Observed values O	Expected values E	Total in row
81–100	6	A	11	6.4	13	8.4	30
56–80	32	B	5	9.6	8	12.6	45
Total in column	38		16		21		75

■ **Stage 1:** Calculate the expected values for each cell of observed values. Use the formula:

$$\text{Expected value} = \frac{\text{total of the column} \times \text{total of the row}}{\text{total of the whole sample}}$$

for cell A, the expected value is $\frac{38 \times 30}{75} = 15.2$

for cell B, the expected value is $\frac{38 \times 45}{75} = 22.8$

■ **Stage 2:** Apply the chi-squared formula for each of the observed values in turn.

$$\chi^2 = \Sigma \frac{(O - E)^2}{E}$$

$$= \frac{(6 - 15.2)^2}{15.2} + \frac{(32 - 22.8)^2}{22.8} + \frac{(11 - 6.4)^2}{6.4} + \frac{(5 - 9.6)^2}{9.6} + \frac{(13 - 8.4)^2}{8.4} + \frac{(8 - 12.6)^2}{12.6}$$

$$= 19.1$$

■ Interpret the chi-squared result and test for significance. Use Table 9.19.

■ The horizontal axis of Figure 9.34 refers to degrees of freedom. Calculate the degrees of freedom (df) for a two-tailed test.
Degrees of freedom, df = (number of rows −1) × (number of columns − 1).

■ Read the chi-squared value on the vertical axis of a graph of χ values. The 5 per cent, 1 per cent and 0.1 per cent confidence limits (significance levels) are shown on the graph. The chi-squared value must be above the confidence limit line for it to apply.
For Merida, the chi-squared value is significant at the 99 per cent level therefore the amended hypothesis (H_1) is true. There is a significant difference between environmental quality in areas of different income in Merida.

Mann-Whitney U test

This test is used to investigate whether there is a significant difference between two separate populations or sets of data where both have a similar distribution. The Mann-Whitney U test compares ranked data along two transects, e.g. inside a woodland and outside a woodland, or bedload size at a meander bend and at a riffle. The sample size of each population should be between 5 and 20 but they do not have to be the same. The test focuses on the difference between the two medians of the samples.

Box 16 How to apply the Mann-Whitney U test

Sample question: Table 9.21 shows pebble sizes for two sites across a river, at a meander bend and at a riffle. Is there a significant difference in the sizes of the pebbles?

▼ **Table 9.21** *Pebble sizes at a meander and at a riffle of the same river*

Meander		Riffle	
Size of bedload at the meander (mm)	Rank	Size of bedload at the riffle (mm)	Rank
x	r_x	y	r_y
15.9	19	10.6	15
11.3	16	8.6	13
16.2	20	5.2	6
13.2	17	6.9	9
15.3	18	4.3	2
5.1	5	9.7	14
5.9	7	8.2	12
6.3	8	7.4	10
4.7	4	7.9	11
4.2	1	4.6	3
Number in sample $N_x = 10$		Number in sample $N_y = 10$	
Total of rank scores for $r_x = 115$		Total of rank scores for $r_y = 95$	

1 State the null hyphothesis.
   H_0 states that there is no significant difference between pebbles sizes at a meander and at a riffle.

2 Rank the data from highest to the lowest values using the whole sample.

3 Add the rank scores for each sample.

4 Apply the two formulas for the two sets of data:
   In meander sample rx, x refers to sample X
   $$U_x = N_x \times N_y + \frac{N_x (N_x - 1)}{2} - r_x$$

   In riffle sample ry, y refers to sample Y
   $$U_y = N_y \times N_y + \frac{N_y (N_y - 1)}{2} - r_y$$

   $$U_x = 10 \times 10 + \frac{10(10 - 1)}{2} - 115$$

   $$= 510$$

   $$U_y = 10 \times 10 + \frac{10(10 - 1)}{2} - 95$$

   $$= 490$$

5 Use the smaller of the two U values to determine the significance. Read off the critical value on Table 9.22.

6 Write up the conclusions, remembering that this test examines the differences between the medians, i.e. the middle values not the average values. In this case, we can accept H_1, that there is a significant difference between pebble sizes at a meander and a riffle at the 0.05 level.

▼ **Table 9.22** *Table of critical values for Mann-Whitney U test*

Critical values of U at p = 0.05

Reject H_0 if the critical value of U is equal to or less than the appropriate critical value.

n_1	1	2	3	4	5	6	7	8	9	10	11	12	13	14	15	16	17	18	19	20
1	-	-	-	-	-	-	-	-	-	-	-	-	-	-	-	-	-	-	-	-
2	-	-	-	-	-	-	-	0	0	0	0	1	1	1	1	1	2	2	2	2
3	-	-	-	-	0	1	1	2	2	3	3	4	4	5	5	6	6	7	7	8
4	-	-	-	0	1	2	3	4	4	5	6	7	8	9	10	11	11	12	13	13
5	-	-	0	1	2	3	5	6	7	8	9	11	12	13	14	15	17	18	19	20
6	-	-	1	2	3	5	6	8	10	11	13	14	16	17	19	21	22	24	25	27
7	-	-	1	3	5	6	8	10	12	14	16	18	20	22	24	26	28	30	32	34
8	-	0	2	4	6	8	10	13	15	17	19	22	24	26	29	31	34	36	38	41
9	-	0	2	4	7	10	12	15	17	20	23	26	28	31	34	37	39	42	45	48
10	-	0	3	5	8	11	14	17	20	23	26	29	33	36	39	42	45	48	52	55
11	-	0	3	6	9	13	16	19	23	26	30	33	37	40	44	47	51	55	58	62
12	-	1	4	7	11	14	18	22	26	29	33	37	41	45	49	53	57	61	65	69
13	-	1	4	8	12	16	20	24	28	33	37	41	45	50	54	59	63	67	72	76
14	-	1	5	9	13	17	22	26	31	36	40	45	50	55	59	64	67	74	78	83
15	-	1	5	10	14	19	24	29	34	39	44	49	54	59	64	70	75	80	85	90
16	-	1	6	11	15	21	26	31	37	42	47	53	59	64	70	75	81	86	92	98
17	-	2	6	11	17	22	28	34	39	45	51	57	63	67	75	81	87	93	99	105
18	-	2	7	12	18	24	30	36	42	48	55	61	67	74	80	86	93	99	106	112
19	-	2	7	13	19	25	32	38	45	52	58	65	72	78	85	92	99	106	113	119
20	-	2	8	13	20	27	34	41	48	55	62	69	76	83	90	98	105	112	119	127

(Source: Methods of Statistical Analysis of Fieldwork Data, by John & Richardson)

Summary

Geographical skills are an important part of AS Geography.

Geographical skills are not confined to coursework or fieldwork; good skills are necessary to gain a good grade on the written examinations papers.

Coursework needs careful planning and a clear focus on a manageable topic.

Hypothesis testing is an effective way of organising ideas in coursework.

Coursework needs good quality data that is carefully sampled and meets the needs of the analytical techniques which will be used.

All maps, graphs and diagrams have a purpose and a geographical message.

Students should learn to use, understand and interpret a variety of geographical resources.

Students should appreciate the advantages and limitations of a range of geographical techniques.

It is important to be able to apply learned geographical skills to new situations, or new data.

Statistical tests help to analyse data in greater depth.

Different statistical tests are used for different purposes and require different types of data.

Significance testing is an important final stage in statistical analysis.

Glossary

Chapter 1 The earth's crust at work

Asthenosphere: A zone in the upper mantle of the planet earth where rocks become plastic and are easily deformed. Lies at a depth of 100-350 km.

Atoll: A circular coral reef or string of coral islands surrounding a lagoon.

Chemical weathering *see Box 2, page 26*

Composite volcano: A volcano which is progressively built up by layers of ash and lava.

Convergent plate boundary: The zone of contact between two tectonic plates that are moving towards each other.

Craton: A large unit of the earth's crust that has experienced long-term tectonic stability.

Divergent plate boundary: The zone of contact between two tectonic plates that are moving away from each other.

Earth's crust: The surface skin of the lithosphere.

Earthquake: A shock, or series of shocks, caused by energy released by sudden fracturing along a section of a fault. The shock waves move through the earth's interior and are recorded by seismometers.

Epeirogenic: Processes associated with tectonic activity in the earth's crust.

Epicentre *see page 10*

Erosion *see page 25*

Exogenic: Processes associated with weathering and erosion at or near the earth's surface.

Fault *see page 10*

Fissure: A linear crack in the earth's crust, which at the largest scale, may allow magma to seep out.

Focus *see page 10*

Graben: A trough formed by the sinking of a block of crustal material between two parallel faults.

Horst: An uplifted block of crustal material bounded by faults.

Hot spot: Localities in the earth's crust where plumes of magma rise from the upper mantle to give persistent volcanic activity.

Igneous *see page 21*

Landform: An individual element or feature of the natural landscape. Landforms vary in size from small meanders to large volcanoes.

Landscape (natural): The assemblage of landforms that make up a particular environment. *Note: Landscape can also include elements resulting from human activity.*

Lithosphere: The rigid outer layer of the earth, comprising the earth's crust and the solid upper part of the mantle.

Magma: Molten material, generally containing gases, that forms when temperatures rise and melting occurs in the upper mantle or the crust of the earth.

Mass movement: The downslope movement of material with a set of processes ranging from very slow creep to sudden rockfalls. *See also mass wasting.*

Mass wasting: The movement of regolith downslope by gravity without the aid of a transporting medium.

Mechanical weathering *see Box 2, page 26*

Ocean trench: A long, narrow, very deep basin in the ocean floor. Often arcuate in shape and associated with subduction zones.

Plate tectonics model *see page 10*

Plateau: A mainly level area of elevated land, often made from a layer of magma.

Pluton: A body of intrusive igneous rock embedded within the earth's crust.

Pyroclastic rocks: Rocks made up of solid fragments of igneous material ejected during an explosive volcanic eruption, deposited and subsequently consolidated.

Regolith: The veneer of loose, unconsolidated, often weathered material, that overlies bedrock.

Richter scale: A scale, based on the recorded amplitude of seismic waves, used for measuring the amount of energy released by an earthquake.

Sea-floor spreading: The lateral movement of the oceanic crust away from mid-ocean ridges, i.e. at divergent plate boundaries.

Seamount: A submerged volcanic mountain standing more than 1000 m above the seabed.

Shear strength: The maximum resistance of a material when stress is applied to it.

Shear stress: The amount of stress applied to a material.

Shield volcano: A broad volcano built up from the repeated eruption of basic lava to form a low dome (shield).

Stable slope: A slope where slope evolution is slow because shear strength exceeds shear stress, and resisting forces exceed driving forces.

Subduction zone: The zone along a convergent plate boundary where a section of an oceanic plate is being dragged beneath a continental plate.

Tephra: The solid pyroclastic material ejected during an eruption.

Transform plate boundary: The zone of contact between two tectonic plates that are moving past each other. Also known as a *conservative* boundary.

Unstable slope: A slope where slope evolution is relatively rapid because shear stresses exceed shear strength, and driving forces exceed resisting forces.

Volcano: A mountain formed from volcanic material ejected from a vent in a central crater.

Weathering *see page 25*

Chapter 2 Hydrology and rivers

Antecedent soil moisture: The amount of water that exists in the soil before a rainfall event.

Aquicludes: An impermeable rock which cannot hold water or allow it to pass through.

Aquifers: A rock that contains water in its pore spaces.

Bankfull discharge *see page 49*

Basal removal: The removal of debris or slope material from the foot of a slope, e.g. a river cliff or sea cliff.

Base flow *see page 48*

Base level: The lowest level to which rivers can erode. This is sea level for most rivers.

Capacity: The total sediment load of a river.

Channelisation *see page 70*

Clearwater erosion: Erosion due to a reduced sediment load. This gives the river extra energy to erode.

Closed system *see page 40*

Competence: The size of the largest sediment particle which can be carried by a river at a given discharge. Like capacity, this will vary in time and space.

Cross-profile *see page 60*

Deposition *see page 58*

Discharge *see page 42*

Drainage basin *see page 42*

Drainage density: The length of a stream channel (km) per unit area (km_2).

Equilibrium *see page 58*

Evaporation *see page 42*

Evapotranspiration *see page 45*

Flood discharge *see page 49*

Flooding *see page 53*

Floodplain *see page 65*

Flows *see page 40*

Gorge: A steep-sided river valley usually resulting from resistant rock types, a concentration of river erosion vertically, or slow weathering processes on the valley slopes.

Graded long profile *see page 58*

Groundwater *see page 45*

Hard-engineering *see page 70*

Heliocoidal flow *see page 64*

Hydrograph *see page 49*

Hydrology *see page 40*

Infiltration *see page 45*

Inputs *see page 40*

Interception *see page 45*

Interception loss *see page 45*

Interflow: The hydrological processes that move water to the river channel more slowly than overland flow and more quickly than base flow (groundwater flow).

Isostasy *see page 67*

Knickpoint: A change in gradient on the long profile of a river due to a change in base level.

Lag time: The time (hours or days) between the peak rainfall and the peak discharge recorded.

Lateral erosion: Erosion of the river banks sideways.

Levee: A ridge of sediment parallel to a river channel formed by deposition of coarser material during flooding.

Long profile: The graph of a river from its source to its mouth.

Meanders *see page 64*

Open system *see page 40*

Outputs *see page 40*

Overland flow *see page 45*

Percolation *see page 45*

Permeable *see page 46*

Pervious *see page 46*

Planform *see page 63*

Plunge pool: A depression in the river bed at the base of a waterfall formed by high amounts of river erosion.

Point bar *see page 65*

Pools *see page 65*

Porosity *see page 46*

Potential evapotranspiration: The amount of water which would evaporate and transpire if it was available.

Precipitation: The deposition of water from the atmosphere in liquid (rain) or solid (snow) form.

Reaches: A section of river channel downstream.

Realignment *see page 70*

Recession limb: The part of the storm hydrograph where discharge is decreasing. The recession limb is usually less steep than the rising limb.

Recurrence interval *see page 53*

Regime *see page 42*

Resectioning *see page 70*

Riffles *see page 65*

Rising limb: The part of the storm hydrograph where discharge is increasing.

River cliff *see page 65*

River terrace: A remnant of an earlier floodplain level.

Runoff: The proportion of precipitation which becomes channel flow.

Sediment load *see page 56*

Sinuosity *see page 63*

Sinuosity index: The channel length divided by the valley floor distance *see Figure 2.34, page 61*

Soft-engineering *see page 70*

Stemflow *see page 45*

Stores *see page 40*

Storm hydrograph: A hydrograph for a single rainfall event.

Thalweg: The line of maximum depth along a river channel.

Throughfall *see page 45*

Throughflow *see page 45*

Transpiration *see page 42*

Vertical erosion: Erosion of the river bed downwards.

Watershed *see page 42*

Chapter 3 Coasts

Accordant (Pacific) coasts *see page 88*

Accretion *see page 91*

Arch: A natural rock bridge formed by the marine erosion of cliffs on hardrock, upland coastlines.

Backwash: Water from a breaking wave which flows under gravity down a beach and returns to the sea.

Barrier beach island: An offshore beach of sand and shingle which is built by longshore drift and is exposed even at high tide.

Barrier beach: Long narrow swash-aligned beach across a bay or coastal inlet. On the landward side of the beach there may be one or more lagoons.

Beach face: The steeply sloping section towards the back of a beach which faces seaward.

Berm: A prominent ridge which marks the uppermost part of a beach face.

Blow hole: A vertical shaft on a cliff top caused by roof collapse of a cave. In stormy conditions, sea spray may be ejected through the blow hole.

Blow-out: Huge localised erosion of a coastal dune by the wind. Blow-outs often occur where fragile dune vegetation has been damaged, exposing the sand to wind erosion.

Breakpoint bar: An offshore bar of sand and shingle which runs parallel to the coastline.

Cave: A large cavity formed in cliffs between the high water mark and the cliff foot. Caves are erosional features which occur where wave action exploits lines of weakness (joints, bedding planes, etc.).

Centrifugal force: The outward force that acts at the Earth's surface as a result of the Earth's rotation on its axis.

Cliff profile *see page 82*

Coherent rocks *see page 82*

Discordant (Atlantic) coasts *see page 88*

Dominant waves *see page 79*

Drift-aligned beach *see page 91*

Equilibrium *see page 77*

Fetch *see page 79*

Geo: A long narrow inlet on a cliffed coastline formed by roof collapse in a cave running at right angles to the shore.

Glacial outwash *see page 94*

Glacio-eustacy: Absolute changes in sea level in response to glacial periods. During a glacial, water accumulates in ice sheets and the sea level falls. During warmer inter-glacials, shrinking ice sheets produce a rise in sea level.

Glacio-isostacy: The localised and relative change in sea-level which occurs during glacials and inter-glacials by the crust's response to loading and unloading of ice.

Halosere: The vegetation succession that occurs on coastal mudflats and salt marshes.

Hard-engineering *see Box 4, page 101*

Incoherent rocks *see page 82*

Intertidal areas *see page 102*

Isostatic recovery: The rising of the crust when ice sheets melt in inter-glacials (see also glacio-isostacy).

Lithology *see page 82*

Longshore drift *see page 92*

Mudflats: Accretions of silt and clay in creaks and estuaries. Mudflats are generally under water.

Negative feedback *see page 77*

Percolation rate *see page 91*

Pioneer communities *see page 94*

Plant succession *see page 96*

Positive feedback *see page 77*

Raised beach: A former beach which now lies several metres above modern sea level.

Recurved laterals: Shingle ridges which form the hook-like distal ends of spits.

Ridges: Broad and gently elevated sections of sand beaches aligned parallel to the shoreline.

Runnels: Broad and gentle depressions which separate ridges on sand beaches.

Salt marshes: Former coastal mudflats which have been colonised by vegetation and are inundated at spring tides.

Saltation: The transport of sand grains by wind, producing a skipping motion.

Sediment budget *see page 105*

Sediment cells: Stretches of coastline with self-contained cycles of erosion and sediment deposition.

Shore platforms: Extensive rocky platforms which slope gently seawards at the base of cliffs.

Slacks: Depressions between dune ridges where the water table reaches the surface.

Slope-over-wall cliffs: Cliffs with a long convex upper slope, and a short, steep lower slope (wall). The upper slope has been formed by sub-aerial processes while the lower slope has been formed by marine erosion.

Soft-engineering *see Box 4, page 101*

Spits: Beaches attached to the coastline only at one end.

Stacks: Stacks are the remains of cliffs which have retreated inland and often the result of collapsed arches.

Structure *see page 88*

Sub-aerial processes: Weathering and mass movement processes such as mudslides and rotational slides.

Surfing breakers *see page 79*

Surging breakers *see page 79*

Sustainable management *see page 102*

Swash: The forward movement of water up a beach which follows a breaking wave.

Swash-aligned beach *see page 92*

Tidal range *see page 80*

Tombolo: A beach which joins an island to the mainland.

Transgression: A period of eustatic sea-level rise which causes widespread submergence of coasts.

Zonation *see page 96*

Chapter 4 Atmosphere and ecosystems

Absolute humidity *see page 119*

Atmosphere: The envelope of gases (mainly nitrogen and oxygen) surrounding the earth.

Biodiversity: The range and number of species in an ecosystem.

Biomass: The total mass of living organisms in an ecosystem.

Biomes *see page 109*

Biosphere: The living component – plants and animals – of planet Earth.

Climatic climax: The ecosystem that is the end product of succession in a specific climate.

Condensation: The process that involves water changing from a gas to a liquid.

Conduction: The movement of heat energy through solids.

Convection: The movement of liquids or gases as a result of heating.

DOM *see page 134*

El Niño event *see page 144*

Ecosystems: The living and non-living components of a specific environment.

Ecozones *see page 107*

Equilibrium: A state of dynamic balance in an environmental system.

Germination: The process of initial growth in a plant from seed.

Global warming *see page 110*

Greenhouse effect *see page 112*

Habitat: The environment within which a species lives.

Hygroscopic nuclei *see page 119*

Keystone species: A species that is crucial in structure and function to an ecosystem.

Lithosphere: The rigid outer layer of the earth, comprising the earth's crust and the solid upper part of the mantle.

Litter *see page 134*

Ozone hole *see page 112*

Peturbation: A significant disturbance within an ecosystem

Photosynthesis: The process by which green plants take in sunlight, carbon dioxide and water to produce oxygen, tissue and water.

Plagioclimax: A climax community modified and controlled by human activities.

Precipitation *see Box 2, page 120*

Primary succession *see page 131*

Radiation: The transfer of heat by the emission of rays of energy.

Regolith: The surface mantle of weathered debris and soil overlying bedrock.

Relative humidity *see page 119*

Resilience *see page 126*

Resistance *see page 126*

Secondary succession *see page 131*

Seral: A stage in ecosystem succession.

Succession *see page 126*

Transpiration: The loss of water from the leaves of plants.

Trophic levels: Specific levels within a food chain or food

Chapter 5 Population patterns and processes

Age-sex (population) pyramids: A popular method of displaying population structure in terms of age and gender by the use of a set of horizontal bar graphs.

Age-specific death rate: The proportion of an age cohort who die while in a particular age group.

Apartheid: The controversial so-called 'equal but separate' policy in South Africa in force until 1994. It segregated the white and black populations, allocating the majority black populations their own townships (homelands), restricting their rights and maintaining their economic dependency.

Birth rate *see page 161*

Census: A survey of population taken by a government at intervals of a number of years, The UK government takes a census every 10 years.

Counter-urbanisation: The process of the movement of people away from large cities to smaller towns and rural environments.

Death rate *see page 161*

Demographic Transition *see page 161*

Demography *see page 152*

Development: Economic and social progress that leads to an improvement in the quality of life for an increasing proportion of the population.

Development indicator: A measure that is used to assess the process of development.

Economic migration: Movement whose primary motivation is economic benefit.

Emigration (out-migration): The movement of people away from an area.

GDP: Gross domestic product.

Immigration (in-migration): The movement of people into an area.

Infant mortality rate (IMR) *see page 165*

LEDCs: Less Economically Developed Countries.

Life expectancy *see page 165*

MEDCs: More Economically Developed Countries.

Migration *see page 161*

Natural change *see page 161*

Net migration: The balance between immigration and emigration.

NICs: Newly Industrialised Countries.

Population change *see page 160*

Population density *see page 158*

Population distribution *see page 158*

Population dynamics: The changes and trends in population over time and space.

Population Geography *see page 152*

Population structure (age-sex structure) *see page 161*

'Pull' factors: Factors which motivate people to move into a place, i.e. perceived attractions.

'Push' factors: Factors which motivate people to move out, i.e. perceived disbenefits.

Replacement level fertility (RLF) *see page 165*

Step-wise migration: The process by which migration involves two or more moves to the long-term destination.

Total fertility rate (TFR) *see page 165*

Chapter 6 The urban world

Bid-rent model: A model (graph) that plots the decline of land values with increasing distance from the CBD and the relationship to land-use distribution.

Brownfield sites *see page 199*

CBD: Central Business District

Compact city: A city developed at relatively high densities, with minimal low density sprawl.

Concentration *see page 187*

Conurbation: A cluster of urban settlements whose built-up areas have merged to form a multinuclear urban agglomeration.

Decentralisation: The outward movement of economic activities and service functions away from the urban core. Often associated with suburbanisation.

Dispersion *see page 187*

Edge city: A recent form of urban settlement that has appeared on the outer edges of large metropolitan areas, and had a considerable degree of social and economic autonomy.

Enveloping: The modernising and improvement of the outer 'skin' of a house: roof, walls, windows, doors, etc.

Extension: The outward expansion of a town or city.

Gentrification: A process by which run-down houses in a declining neighbourhood are bought and improved by more affluent people. The neighbourhood becomes fashionable and property prices rise.

Green wedges *see page 200*

Greenbelt: A zone surrounding conurbations and major cities designated by legislation to reduce urban sprawl and protect urban fringe countryside.

Greenfield sites *see page 199*

Industrial city: A generally monocentric compact urban form that evolved in parallel with the process of industrialisation.

Inner city: A zone of mixed land uses, and relatively high density development surrounding the business core of a city.

Intensification: The increase in development density within a city.

LEDCs: Less Economically Developed Countries.

MEDCs: More Economically Developed Countries.

Megacity: The name given to huge urban agglomerations with populations of at least 10 million.

Monocentric: Having a single central core.

Morphology: The pattern formed by the distribution of lane uses within a town or city.

Multicentric: Having more than one core area.

New town: A planned urban centre, designed to be free-standing, self-contained economically, and socially balanced.

NICs: Newly Industrialised Countries

Overpopulation: When a population is so large, in relation to available resources and levels of technology, that a majority of people do not enjoy acceptable living standards.

Post-industrial city: A multicentric, sprawling urban form that is evolving with the growth of decentralised industries, especially tertiary and quaternary industries, and with the continued process of suburbanisation.

Primate city: An urban centre that dominates a country's urban system, with a population much greater than that of the next-largest city.

Redevelopment: The rebuilding of decaying. The redevelopment of older areas in a city usually involves extensive demolition, and replacement by modern layouts and buildings.

Redistribution *see page 187*

Regeneration *see page 208*

Regional city: The pattern of functional zones linking urban and rural environments into a sub-regional systems.

Renovation: The physical upgrading and modernisation of older areas of a city.

Replacement: The modernisation of an area by replacing obsolescent buildings and land uses.

Right-to-buy: A policy introduced by the Conservative government in 1980 that gives existing tenants of council properties the right to buy their homes.

Rural buffers *see page 200*

Sites-and-services scheme: An approach popular in some LEDCs where the government provides the land, infrastructure and utilities while families then build their own homes

Social area: A geographical [spatial] unit that has distinctive socio-economic characteristics.

Social segregation: The geographical [spatial] separation of different socio-economic and cultural groups.

Spread city: Multicentric (multinuclear) urban form, extending over a wide area and developed at relatively low densities.

Strategic gaps *see page 200*

Suburbanisation: The process of the outward spread of largely residential districts, developed at relatively low densities.

Urban agglomeration: See 'conurbation'.

Urbanisation *see page 187*

Chapter 7 Rural environments

Access *see page 258*

Agribusiness *see page 242*

Agriculture *see page 228*

Area of Outstanding Natural Beauty (AONB): An area given special designation to protect its environment.

Biodiversity: The existence of a wide variety of plants and animals in their natural environments.

Bottom-up approach *see page 224*

Capital-intensive: An economic activity that requires high inputs of capital in relation to output.

CDR: Complex, diverse and risk-prone smallholder farming systems.

Commercial agriculture *see page 228*

Countryside Agency (CA): The UK government body created in 1999 to replace the Countryside Commission (CoCo), with responsibility for encouraging and enabling policies and initiatives for the sustainable management of rural environments in England.

Dispersed settlement *see page 243*

Diversification: The broadening of the economic base of a rural region.

Ecotourism *see page 236*

Environmentally Sensitive Area (ESA) *see page 249*

Fallow: The resting of arable land for one or more seasons.

Gatekeeper *see page 253*

Green Revolution *see Box 2, page 224*

Intercropping: The planting of several crops between each other on the same plot.

Labour-intensive: An economic activity that employs large numbers of people in relation to output.

MAFF: Ministry of Agriculture, Fisheries and Food

Nomadic: A way of life without permanent settlements.

Nucleated settlement *see page 243*

Penetration road: A road built through a previously undeveloped region which improves access and the potential for settlement.

Productivity: The yield from a given unit of land over a certain time period.

Quota: A form of production control or rationing where individual businesses are given a set limit to their production of certain products.

Right-to-Roam: A campaign to increase rights of access to open and unenclosed areas of the countryside. In 2000, the UK government introduced legislation to increase such access.

Rural *see page 227*

Rural Development Agency (RDA): A UK government funded agency that replaced the Rural Development Commission in 1999. Its primary aim is to encourage and enable economic development in rural area .

Set-aside: A UK government policy where arable farmers are required to leave a proportion of their arable land fallow for a year or more, and for which they receive a compensatory grant. The aim is to control overproduction.

Settlement hierarchy *see page 246*

Shifting cultivation *see page 230*

Sphere of influence: The area served by a central place.

Stakeholder *see page 253*

Subsistence agriculture *see page 228*

Sustainability *see Box 1, page 226*

The Common Agricultural Policy (CAP): The agricultural policy of the European Union (EU) which economically supports farming within the EU by protecting farmers from cheap imports, by guaranteeing prices for farm products and by giving grants to farmers in poorer areas.

The Council for the Preservation of Rural England (CPRE): A voluntary organisation which campaigns for conservation and access in the UK countryside.

Top-down approach *see page 224*

Chapter 8 Economic activities

Assisted area: Areas suffering a range of economic problems (unemployment, declining industrial structure, etc.) and targeted by the government for financial assistance to assist economic regeneration

Anchor tenant *see page 282*

Automated production: Manufacturing by machine rather than by human labour.

Biodiversity: The existence of a wide variety of plants and animals in their natural environments.

Comparison goods/services *see page 280*

Convenience shopping: The frequent purchase of low order goods and services, usually at the nearest retail outlet.

Deep-mining *see page 266*

Deindustrialisation *see page 273*

Desertification: The transformation of fertile land into desert as a result of intensive farming, soil erosion, etc.

EU Regional aid *see page 271*

Foreign direct investment (FDI) *see page 274*

Fossil fuels: Any naturally occurring fuel, such as coal, natural gas and oil, formed by the decomposition of prehistoric plants and animals.

Greenfield sites: Sites developed on mainly agricultural land – i.e. sites which have not been previously built on.

Globalisation: The location of production by TNCs in several different countries or continents. The purpose of globalisation is for firms to source materials, make products and supply markets worldwide.

Industrial stage: An economic stage where a country's employment and GNP are dominated by manufacturing and mining activities. In the last 30 years, NICs such as South Korea and Taiwan have passed through this stage.

Infrastructure: The basic network of transport links, and services such as electricity, gas and water.

LEDCs: Less Economically Developed Countries

Maquiladoras: Branch plants located in the US-Mexican border region which are the result of inward investment primarily by US and East Asian TNCs.

MEDCs: More Economically Developed Countries

Multiple retailer: A large retail company with outlets in many shopping centres, e.g Boots the chemist. Multiples account for the bulk of retail turnover in the UK and and their importance is increasing.

Non-renewable resources: Natural resources such as fossil fuels and mineral ores which once used cannot be replaced.

On-line shopping: Shopping over the internet.

Open-cast mining *see page 266*

Post-industrial stage: An economic stage where a country's employment and GNP are dominated by service activities. Most MEDCs are in the post-industrial stage, having passed through the early industrial stage by the mid-twentieth century.

Pre-industrial stage: An economic stage where a country's employment and GNP are dominated by primary activities, especially agriculture. Many of the poorest LEDCs are still in this stage.

Producer services: Quaternary activities such as financial services, advertising, consultancy, etc. which provide services to other firms and organisations rather than to individual consumers.

R&D: Research and Development

Regeneration: The economic, social and environmental improvement of a declining, ageing district.

Regional shopping centres: Very large enclosed shopping centres selling mainly comparison goods (e.g. the Trafford Centre), located on out-of-centre sites. In terms of size and function, regional shopping centres are comparable to the central shopping areas of cities.

Reindustrialisation: The regeneration of manufacturing industry in older industrial regions between 1980 and 2000, largely brought about by foreign direct investment.

Renewable resources: A natural resource which is either inexhaustible (e.g. solar energy) or follows a biological cycle (e.g. timber) or a physical cycle (e.g. water) of continuous renewal.

Retailing *see page 280*

Science parks: Purpose-built areas, often on the edge-of-town, dedicated to research and development and knowledge-based industries.

Sectoral shift *see page 263*

Sustainable production: Production which meets current demand without compromising the ability of future generations to meet their own needs.

Teleshopping: Ordering retail goods by telephone.

Transnational Corporations (TNCs) *see page 275*

World cities: Cities such as New York, London and Tokyo which are the control and command centres of the global economy.

Chapter 9 Skills and techniques for geographical investigations

Bar chart: A series of 'bars' that rise from an axis. The bars are of equal width but their height/length is proportional to the information they represent.

Bi-polar analysis *see page 297*

Chi-squared test: Test to see if there is any significant difference between two sets of data. *see Box 13 page 326*

Choropleth map: Map that is shaded or coloured to show average information for specific areas. The information is grouped.

Cross-section: A carefully drawn 'side-view' through a specific section of land area.

Cumulative line graph: Line on a graph plotted along the increasing (cumulative) value of the data being represented.

Dispersion diagram: Graph that shows the spread of one set of information; a type of scatter graph.

Dot map: Dots plotted on a map to show the distribution of data. Each dot has a specific value.

Flow line map: Map using bands of varying widths to represent movement between locations.

Frequency diagram *see page 316*

Histogram: Bars on a graph based on two sets of quantitative information. Data is plotted into classes before drawn on the graph.

Hypothesis: A statement formed at the start as the focus of an investigation and tested for its accuracy.

Inter-quartile range *see Box 10, page 323*

Isopleth map: Map using lines that link areas of equal-value characteristics.

Kite diagram *see page 321*

Land-use map: Map that shows different types of human activity in a particular area.

Line graph: Line on a graph that shows continuous information, e.g. change with time.

Logarithmic graph: Graph that compares rates of change on axes that maintain the clarity of the smallest the smallest values. Uses logarithmic (log) paper.

Mann–Whitney U test: Tests whether there is a significant difference between two separate sets of data of similar distribution. *see Box 16 page 328.*

Mean *see Box 9, page 323*

Median *see Box 9, page 323*

Mode *see Box 9, page 323*

Pie chart: The area of a circle divided into segments, so that each segment represents particular information.

Population pyramid: The population structure of a region shown as two sets of histograms, one for males and one for females. The shape is usually in the form of a pyramid. Also called *age–sex pyramid.*

Primary data: Data which you collect yourself.

Proportional circles: Data represented as a circle where the value of the area is proportional to the information it shows; usually plotted as a series of circles to show a range of information.

Qualitative data: Data that can be clearly measured.

Quantitative data: Data that is descriptive and usually subjective.

Sampling: The process of testing a small representative section of the fieldwork area being studied.

Scatter graph: Values plotted individually on a graph according to the relationship between two variables.

Secondary data: Data assembled from research done by others, e.g. census data, other fieldwork reports, book and website information.

Significance test: A test of reliability, to see whether relationship between variables exists by chance.

Sketch map: A freehand drawing of a map area that highlights its main geographical features.

Skewness: The degree to which information shown on a frequency diagram veers from being a 'normal' (symmetrical) distribution.

Spearman's Rank correlation: Tests whether there is a linear relationship between two sets of ranked data. *see Box 11, page 324*

Star diagram *see page 320*

Topological map: Representative map that shows spatial patterns or links between places.

Triangular graph: Graph that shows three sets of variables; uses three axes in the shape of a triangle.

Index